PC Audio Editing with Adob

PC Audio Editing with Adobe Audition 2.0

Broadcast, desktop and CD audio production

Roger Derry M.I.B.S.

AMSTERDAM • BOSTON • HEIDELBERG • LONDON • NEW YORK • OXFORD
PARIS • SAN DIEGO • SAN FRANCISCO • SINGAPORE • SYDNEY • TOKYO

Focal Press is an imprint of Elsevier

Focal Press is an imprint of Elsevier
Linacre House, Jordan Hill, Oxford OX2 8DP
30 Corporate Drive, Suite 400, Burlington, MA 01803

First edition 2006

Copyright © 2006, Roger Derry. Published by Elsevier Ltd. All rights reserved.

The rights of Roger Derry to be identified as the author of this work has been asserted
in accordance with the Copyright, Designs and Patents Act 1988

No part of this publication may be reproduced, stored in a retrieval system or transmitted in
any form or by any means electronic, mechanical, photocopying, recording or otherwise
without the prior written permission of the publisher

Permissions may be sought directly from Elsevier's Science & Technology Rights
Department in Oxford, UK: phone (+44) (0) 1865 843830; fax (+44) (0) 1865 853333;
email: permissions@elsevier.com. Alternatively you can submit your request online
by visiting the Elsevier web site at http://elsevier.com/local/permissions, and selecting
Obtaining permission to use Elsevier material

Notice
No responsibility is assumed by the publisher for any injury and/or damage to persons or
property as a matter of products liability, negligence or otherwise, or from any use or operation
of any methods, products, instructions or ideas contained in the material herein. Because of
rapid advances in the medical sciences, in particular, independent verification of diagnoses
and drug dosages should be made

British Library Cataloguing in Publication Data
A catalogue record for this book is available from the British Library

Library of Congress Cataloguing in Publication Data
A catalogue record for this book is available from the Library of Congress

ISBN-13: 978-0-240-51969-2
ISBN-10: 0-2405-1969-8

For information on all Focal Press publications visit our
website at www.focalpress.com

Typeset by Charon Tec Ltd, Chennai, India
www.charontec.com
Printed and bound in Great Britain

06 07 08 09 10   10 9 8 7 6 5 4 3 2 1

Working together to grow
libraries in developing countries

www.elsevier.com | www.bookaid.org | www.sabre.org

ELSEVIER   BOOK AID International   Sabre Foundation

To
Rosemarie
With many thanks for her help and tolerance

Contents

Foreword		xiii
Preface		xv
1	**Visual editing**	1
2	**Some technical bits**	5
	2.1 Loudness, decibels and frequencies	5
	2.2 Hearing safety	7
	2.3 Analog and digital audio	8
	2.4 Time code	12
3	**Transfer**	14
	3.1 Reviewing material	14
	3.2 Head alignment	16
	3.3 Digital	16
	3.4 Analog	22
	3.5 Recording	24
4	**Editing**	29
	4.1 Introduction	29
	4.2 Loading a file	30
	4.3 Transport	35
	4.4 Making an edit	38
	4.5 Visual editing	39
5	**Quarrying material**	42
	5.1 Introduction	42
	5.2 Blue bar blues	42
	5.3 Copy, cut and paste	45
6	**Structuring material**	47
	6.1 Standardize format	47
	6.2 Batch files	48
	6.3 32-bit files	54

7 Multitrack — 55
- 7.1 Loading — 55
- 7.2 Tour — 56
- 7.3 FX — 61
- 7.4 Automation — 64
- 7.5 Envelopes — 67
- 7.6 Multitrack for a simple mix — 69
- 7.7 Chequer boarding — 71
- 7.8 Track bouncing — 73
- 7.9 Fades and edges — 74
- 7.10 Prefaded and backtimed music — 76
- 7.11 Transitions — 76
- 7.12 Multitrack music — 77
- 7.13 MIDI and video — 78
- 7.14 Loops — 78

8 Post-production — 80
- 8.1 Timing — 80
- 8.2 Level matching — 80
- 8.3 Panning — 81
- 8.4 Cross fading — 82
- 8.5 Stereo/binaural — 82
- 8.6 Surround Sound — 83
- 8.7 Multi-layer mixing — 85

9 Audio design — 86
- 9.1 Dangers — 86
- 9.2 Normalization — 86
- 9.3 Basic level adjustment — 88
- 9.4 Compression/limiting — 90
- 9.5 Expanders and gates — 93
- 9.6 Multiband compressor — 95
- 9.7 Reverberation and echo — 99
- 9.8 Filters — 114
- 9.9 Restoration — 124
- 9.10 Special effects — 131
- 9.11 Time/pitch — 133
- 9.12 Spatial effects — 136
- 9.13 Basic effects — 138
- 9.14 Multitrack effects — 140
- 9.15 Real-time controls and effects — 142

10	**Reviewing material**		147
	10.1	Check there are no missed edits	147
	10.2	Assessing levels	147
	10.3	Listening on full quality speakers	148
11	**Mastering**		149
	11.1	Line-up and transmission formats	149
	11.2	Generate silence	150
	11.3	Generate noise	151
	11.4	Generate tones	152
	11.5	Changeovers	152
	11.6	Mastering process	153
	11.7	CD labelling	153
	11.8	Email and the World Wide Web	154
	11.9	Analytical functions	156
	11.10	Spectral editing	157
12	**CD burning**		170
	12.1	Types of CD	170
	12.2	Audio CDs	171
	12.3	Multi-session CDs	172
	12.4	Rewritable CDs	172
	12.5	CD recording software	173
	12.6	Recording codes	179
13	**Making programmes: acquiring material**		181
	13.1	Responsibilities	181
	13.2	Interviewing people	182
	13.3	Documentary	185
	13.4	Oral history interviews	185
	13.5	Interviews on the move	185
	13.6	Recorders	186
	13.7	Acoustics and microphones	187
	13.8	Multitrack	194
14	**Making programmes: production**		195
	14.1	Introduction	195
	14.2	Types of programme	195
	14.3	Talks	196
	14.4	Illustrated talks	197
	14.5	Magazine programmes	197
	14.6	Magazine item	197

			Page
	14.7	Simple feature	198
	14.8	Multi-layer feature	198
	14.9	Drama	198
	14.10	Music	198
15	**Archiving and backup**		**200**
	15.1	Selection for archiving	200
	15.2	Compact disc	200
	15.3	CD-ROM	200
	15.4	Digital Audio Tape	201
	15.5	Analog	201
	15.6	Computer backup	202
16	**Tweaks**		**204**
	16.1	Mouse options	204
	16.2	Hardware controller	204
	16.3	Keyboard shortcuts	206
	16.4	MIDI triggering	206
	16.5	SMPTE synchronization	207
	16.6	Favorites	207
	16.7	Preferences (F4)	208
	16.8	Icons	212
17	**Using the CD-ROM**		**213**
	17.1	Adobe Audition	213
	17.2	Bonus material	213
18	**Hardware and software requirements**		**215**
	18.1	PC	215
	18.2	Sound card	215
	18.3	Loudspeakers/headphones	216
	18.4	Hard disks	217
	18.5	USB	218
	18.6	Firewire	218
	18.7	Audio editor	219
	18.8	Linear editors	219
	18.9	Non-linear editors	219
	18.10	Multitrack	220
	18.11	Audio processing	220
	18.12	Mastering	221
	18.13	CD recording software	222
	18.14	DVD	223
	18.15	MIDI	223

18.16	Control surface	223
18.17	Automation	223

Appendix 1	**Clicks and clocks**	**225**
A1.1	Word clock in/out	226

Appendix 2	**MIDI**	**227**

Appendix 3	**Time code**	**229**
A3.1	MIDI	229
A3.2	Drop-frame time code	230
A3.3	Other systems	230
A3.4	What time?	231

Appendix 4	**Adding RIAA to FFT filters**	**232**

Glossary	**234**

Index	**247**

Foreword

At first glance . . . Adobe's Audition seems to be a fairly simple and straightforward audio editing program. That's because it is! But don't let that intuitive GUI (graphic user interface) fool you – it's an extremely powerful editor that has worked its way into the mainstay of the audio production industry, one heart at a time . . . simply by being able to do the job effectively and with as little confusion, muss or fuss possible.

Many of us will fondly remember Audition as Syntrillium's Cool Edit Pro. I actually had the fortune of working with the company for a number of years . . . I can honestly say that it was a caring and wonderful family. Lots of blood, sweat, tears and joy went into making this an innovative program that doesn't always tow the industry line. There are a lot of quirky applications, batch-processing sub-routines and convoluted processing functions that are so unique and different that they can't be found anywhere else. Some of these apps were quickly copied by other companies and made their way into the production mainstream.

. . . but the long and the short of it, is that Audition is a survivor because it's so good, so straightforward and so powerful that its user-base has become staggeringly large. Personally, I use it as my exclusive 2-channel editor and stereo mastering tool. Why? Because I've never found a program that does the job better, faster and more accurately. It's intuitive ability to cut, paste, fade and process a stereo file is simply superb. I also use it as a mastering tool for putting surround sound projects onto the web. It's little-known WMA surround-mastering tool is flat out the best in the biz. Honorable mention should also go to the "Marquee Selection" tool within the spectral analysis section that lets you alter selective frequencies within the audio spectrum. Want to cut out an annoying cough or unwanted sound from a live recording? Using this tool can nearly or completely eliminate it!

Within this updated edition, you'll learn some of the Audition's basic tools & tricks, as well as gain insights into some of its more advanced editing, multitrack and processing tools. I'd seriously advise you to load the Adobe 'tryout' time limited version on the book's CD-ROM or a full copy of the program and put it through its paces while you read through this book. Try not to overlook the quirky effects, tools and applications . . . a few of these might just be some of the really cool tricks and timesaving tools that you've been hoping for.

Foreword

Adobe's Audition is a deceptively simple program that has tons to power under its hood. With the help of this book, you'll soon be gaining valuable insights into computer-based audio editing and sound production. Spend time practicing on it! Play with it! Learn from it! Grow with it and most of all . . . Have Fun with it!

David Miles Huber
www.modrec.com
www.51bpm.com

Preface

For years PCs struggled hard to cope with the prodigious demands of handling audio, let alone doing it well, or at any speed. Nowadays 'entry-level' PCs can provide facilities that, in previous decades, professional audio technicians would have killed for. Indeed, there is no doubt that a contemporary 6-year-old could happily edit audio on a PC. But, as always with most skills, the trick is NOT doing it like a 6-year-old child.

This book is aimed at people who wish to make audio productions for issue as recordings or for broadcast, using a Windows PC with material acquired using portable equipment, as well as in studios. There are already plenty of books on Music Technology; this one is aimed at people who want to use speech as well; the making of CD sales promotions, radio programmes or, even, Son et Lumière! The book does not expect you to be a technician although, if you are, there will still be plenty to interest you, as the book includes production insights to help you use your PC for producing audio productions in the real world.

Even those who are experienced in editing and mixing with quarter inch tape can find the change to editing audio visually on a PC daunting. However, while this does bring a change of skills, most of the new skills are more easily learned.

Visual editing is done statically rather than having to use the arcane dynamic skills of 'scrub' editing, required by quarter inch tape. Hearing an edit while scrubbing the tape back and forth was a skill that took weeks, or even months, to perfect. Some were doomed never to be able to hear clearly. With the PC, making mistakes is far less of a problem; correcting errors is rapid, and safe, unlike unpicking bits of tape spliced with sticky tape.

The author' early career included being a specialist tape editor. He is now a convert to the use of a PC for editing. His career has included all the jobs that go to make a radio programme including production, reporting, management, studio and location technical operations (and fetching tea!). He can therefore offer advice and experience over the whole range of audio production.

This book is not intended as a manual but it is firmly based around Adobe Audition 2.0. This is the successor to Syntrillium' Cool Edit Pro which was bought by Adobe. Version 1.00 of Adobe Audition was largely identical to CEP 2.1. In the year following the take-over version 1.5 was developed. This continued the look and feel of the original but the combination of Cool Edit's original developers with Adobe's immense resources meant that it was already growing with important new facilities being added. With Adobe Audition 2.0 that growth has continued apace.The book attempts to show the basic principles of the new technology. While

Preface

each of the many audio editors that exist have their operational differences, basic principles remain constant. Most aspects of Adobe Audition are described in detail. Yet, audio is audio. Digits are digits. Other programs may implement any task differently, better or worse, but the principles behind those operations remain the same. As for Adobe Audition 2.0 itself, it is a large and complex program and I hope that this book helps clear a way through to the important core facilities.

This book contains a CD-ROM with a demonstration of Adobe Audition 2.0. However this will only run on Windows XP. Some users find that this operating system gives them compatibility problems with their existing software and hardware and have stayed with Windows 98 or even Windows 95. For these people the Windows XP installation CD gives advice on how to set up a 'boot' system so that you can carry on using older but valued software and hardware.

The CD-ROM has been especially made for this book and also contains various audio files; these are to give the total beginner something to edit. These should all be usable on any digital audio editor.

Roger Derry
May 2006

1
Visual editing

Anyone who is used to editing audio using quarter inch tape may approach editing on a PC with trepidation. PCs allow you to edit visually rather than using your ears to 'scrub' the audio back and forth to hear the edit point. Some Digital Audio Workstations provide a simulation of scrub editing, either using special hardware in the form of a search wheel, or an option to use the computer's mouse. This option is provided by some PC and Mac software such as *Pro Tools*, and now also by *Adobe Audition 2.0*. I find that these scrub options are usually unsatisfactory because of tiny, but noticeable, processing delays between the mouse movements and hearing the result. As a professional tape editor I used to find that the delay caused by having loudspeakers more than 2 metres away was sufficient to cause problems and that corresponds only to 6/1000ths of a second!

Editing on a PC is using a different medium and techniques change. After 25 years of quarter inch tape editing, I found that I took to editing visually like the proverbial duck to water. I did, however, have to overcome an emotional resentment at what seemed like a 'de-skilling' of the task.

My analogy is with word processing. In the days of manual typewriters, using carbon paper to make copies, there was a high premium on accuracy. It was also important to be able to type each key with an even pressure, so that the document had a professional look. This was a skilled thing to do and was not learnt in a day.

When word processors were introduced, the rules changed. It was soon discovered that it was more efficient to type as fast as possible and to clean up typos afterwards. Printouts take no notice of the key pressures that the typist has used. This has meant that even a person with no keyboard skills can – given a great deal of time – produce a document with a professional appearance.

So it is with audio editing; with plenty of experience and good training to become a skilled tape editor, it is possible to find the precise edit point very quickly, mark, cut and splice, to get a good edit every time. It is a skill. It has to be learned.

While this book concentrates on Adobe Audition, a great deal of its information applies to other audio editors. As with all technology there is a great deal of convergence and different editors begin to look and operate alike. At college, I offer my students the analogy of the time I had to drive my then boss from Cardiff to Swansea. The easy bit was the Cardiff to Swansea journey but first I had to ask him how to reverse his car so I could get out of the car park bay! If I had had to retune his radio presets, I might have had to look up the car's manual. Thus a good knowledge of Adobe Audition will see you in good stead for other programs.

The PC offers visual tools to aid your editing. It also has the potential to be much more accurate. With reel-to-reel quarter inch analog tape at 15 ips, using 60° cuts, the effective best accuracy is 1/120th of a second. 90° cuts can improve this but, in practice, not by much. Most edits are made using much lower accuracy.

In contrast, the PC can offer 'sample-rate' accuracy. This means that you can edit down to the resolution of a single number (sample) in the digital data. Using the standard professional sampling rates of 32 kHz, 44.1 kHz or 48 kHz, this gives a possible accuracy of 1/32 000th, 1/44 100th or 1/48 000th of a second! As I have already observed, most edits don't need this accuracy, so it is as well that you are not forced to work to this resolution! Most editors will provide a cross fade across the edit. This corresponds to the cross fade provided by slanting the cut on quarter inch tape. The digital editor has the advantage that it can do this with stereo recordings without the risk of image 'shimmying' caused by the right-hand channel being cross faded before the left-hand channel.

When viewing the sound file, as a whole, the PC offers you a static graphical display of your audio with level represented by the width of a line. For stereo, there are two variable width lines, side by side, representing the left and right channels. You can choose at what resolution you want to look at the audio. The width of the screen can encompass several hours or 1/1000th of a second.

Illustrated (Figure 1.1) is a 4-second chunk of mono audio. The text of what is being said has been added to the illustration. PC audio editors are not (yet) clever enough to transcribe speech from your recording!

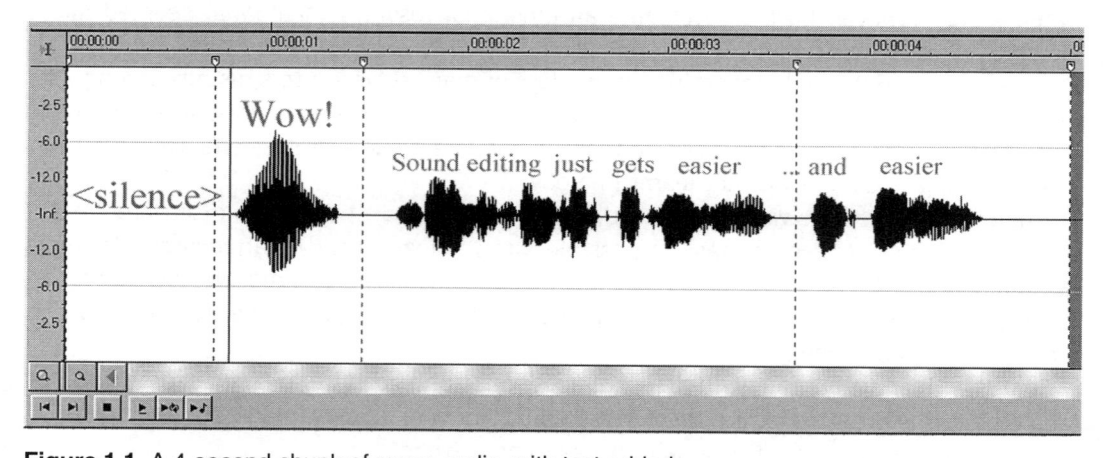

Figure 1.1 A 4-second chunk of mono audio, with text added

At this sort of resolution, you can easily see the rises and falls in level corresponding to individual syllables. Yet we can zoom in to even more detail. The usual way to do this is to select the area by dragging the mouse – just as you would to select text in a word processor. You then zoom to the selection. In this example, using the word 'and' (Figure 1.2) is enough to show the individual vibrations (Figure 1.3). The 'D' sound at the end is now extremely easy to see.

Visual editing

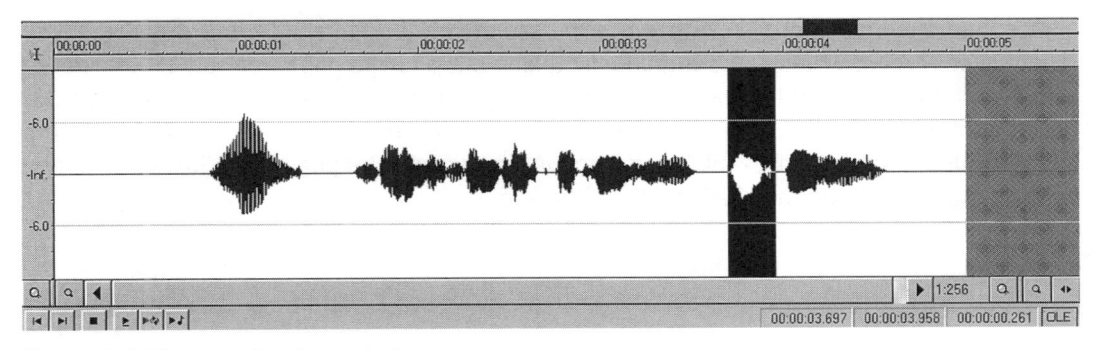

Figure 1.2 The word 'and' selected

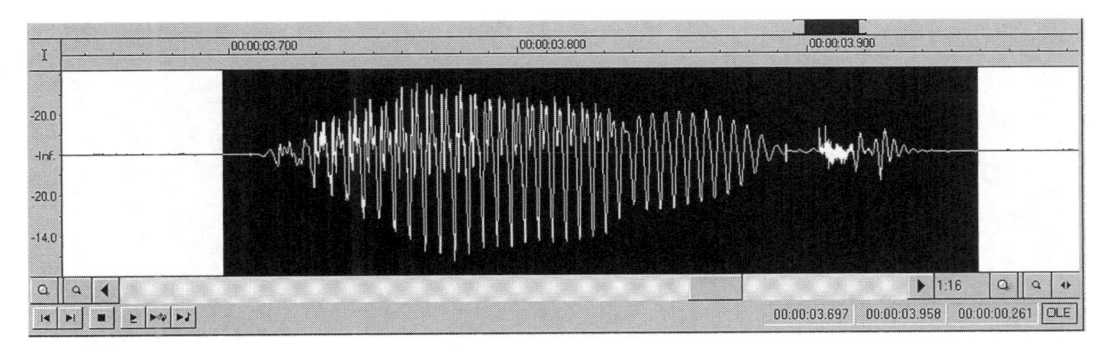

Figure 1.3 Zoomed in on the word 'and'

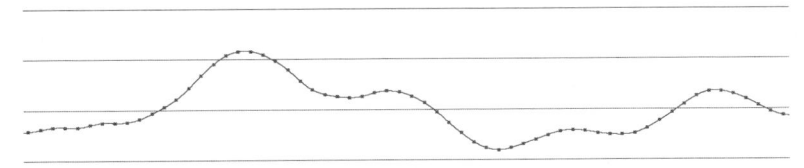

Figure 1.4 Audio zoomed in to showing samples as dots. The line joining them is created by the computer

Cutting a section of audio is as easy as selecting it with a mouse and pressing the delete key, just as in word processing. Also, like word processing, you can cut, or copy, audio to a clipboard and then paste it elsewhere.

There is usually an option for a 'mix' paste, where the copied audio is mixed on top of existing audio rather than being inserted. With several attempts and much use of the UNDO function, this can be a way of adding a simple music bed to speech but it lacks the sophisticated control available from using a non-linear editing mode giving you multitrack operation.

3

Most audio editors allow you to zoom to the sample level (Figure 1.4), but this is rarely of much value, except where you are manually taking out a single click from an LP transfer or some glitch picked up on the way. In general, noise reduction and de-clicking is more quickly done by software, although this will let through some clicks that have to be dealt with individually. Too high a setting on de-click software will 'over-cook' the audio introducing distortion where it starts attacking the sharp peaks of the audio.

2

Some technical bits

There are some technical terms that are much bandied about in audio. While it is possible to survive without knowledge of them, they are a great help in making the most of the medium. However, you may wish to skip this chapter and read it later.

2.1 Loudness, decibels and frequencies

How good is the human ear?

Sound is the result of pressure: compression/decompression waves travelling through the air. These pressure waves are caused by something vibrating. This may be obvious, like the skin of a drum, or the string and sounding board of a violin. However, wind instruments also vibrate. It may be blowing through a reed or by blowing across a hole; the turbulence causing the column of air within the pipe of the instrument to vibrate.

The two major properties that describe a sound are its frequency and its loudness.

Frequency

This is a count of how many times per second the air pressure of the sound wave cycles from high pressure, through low pressure and back to high pressure again (Figure 2.1). This used to be known as the number of 'cycles per second'. This has been given a metric unit name: 1 hertz (Hz) is one cycle per second and is named after Heinrich Hertz who did fundamental research into wave theory in the nineteenth century.

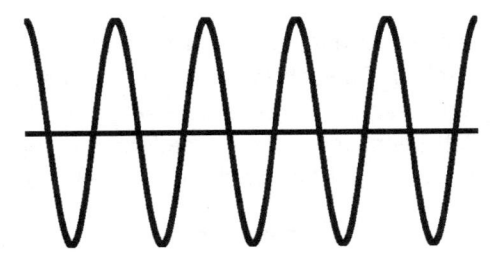

Figure 2.1 Five cycles of a pure audio sine wave

The lowest frequency the ear can handle is about 20 Hz. These low frequencies are more felt than heard. Some church organs have a 16 Hz stop which is added to other notes to give them depth. Low frequencies are hardest to reproduce and, in practice, most loudspeakers have a tough time reproducing much below 80 Hz.

At the other extreme, the ear can handle frequencies of up to 20 000 Hz, usually written as 20 kilohertz (kHz). As we get older, our high frequency limit reduces. This can happen very rapidly if the ear is constantly exposed to high sound levels.

The standard specification for high fidelity audio equipment is that it should handle frequencies between 20 Hz and 20 kHz equally well. Stereo FM broadcasting is restricted to 15 kHz, as is the Near Instantaneously Companded Audio Multiplex (NICAM) system used for television stereo in the UK. But Digital Radio is not, and so broadcasters have begun to increase the frequency range required when programmes are submitted.

The frequencies produced by musical instruments occupy the lower range of these frequencies. The standard tuning frequency 'middle A' is 440 Hz. The 'A' one octave below that is 220 Hz. An octave above is 880 Hz. In other words, a difference of an octave is achieved by doubling or halving the frequency.

Instruments also produce 'harmonics'. These are frequencies that are multiples of the original 'fundamental' notes. It is these frequencies, along with transients (how the note starts and finishes) that give an instrument its characteristic sound. This is often referred to as the timbre (pronounced 'tam ber').

Loudness

Our ears can handle a very wide range of levels. The power ratio between the quietest sound that we can just detect – in an impossibly quiet sound insulated room – and the loudest sound that causes us pain is:

$$1 : 1\,000\,000\,000\,000$$

'1' followed by twelve noughts is one million million! To be able to handle such large numbers a logarithmic system is used. The unit, called a bel, can be thought of as a measure of the number of noughts after the '1'. In other words, the ratio shown above could also be described as 12 bels. Similarly a ratio of 1 : 1000 would be 3 bels. A ratio of 1 : 1 – no change – is 0 bels. Decreases in level are described as negative. So a ratio of 1000 : 1 – a reduction in power of 1/1000th – is minus 3 bels.

For most purposes, the bel is too large a unit to be convenient. Instead the unit used every day, is one-tenth of a bel. The metric system term for one-tenth is 'deci', so the unit is called the 'decibel'. The abbreviation is 'dB' – little 'd' for 'deci' and big 'B' for 'bel' as it is based on a person's name – in this case Alexander Graham Bell, the inventor of the telephone and founder of Bell Telephones, who devised the unit for measuring telephone signals. Conveniently, a change of level of 1 dB is about the smallest change that the average person can hear (Table 2.1).

Table 2.1 Everyday sound levels

0 dB	Threshold of hearing: sound insulated room
10 dB	Very faint – a still night in the country
30 dB	Faint – public library, whisper, rustle of paper
50 dB	Moderate – quiet office, average house
70 dB	Loud – noisy office, transistor radio at full volume
90 dB	Very loud – busy street
110 dB	Deafening – pneumatic drill, thunder, gunfire
120 dB	Threshold of pain

3 dB represents a doubling of signal power. But human hearing works in such a way that 3 dB does not actually sound twice as loud. For audio to sound twice as loud requires considerably more than a doubling of signal power. A 10 dB increase in signal power is considered to sound twice as loud to the human ear.

2.2 Hearing safety

One of the hazards of audio editing on a PC is that it is often done on headphones in a room containing other people. Research has shown that people, on average, listen to headphones 6 dB louder than they would to loudspeakers. So already they are pumping four times more power into their ears.

When you are editing, there are going to be occasions when you turn up the volume to hear quiet passages and then forget to restore it when going on to a loud section. The resulting level into your ears is going to be way above that which is safe.

Some broadcasting organizations insist that their staff use headphones with built-in limiters to prevent hearing damage. By UK law, a sound level of 85 dBA is defined as the first action level (the 'A' indicates a common way of measuring sound-in-air level, as opposed to electrical, audio, decibels). You should not be exposed to sound at, or above, this level for more than 8 hours a day. If you are, as well as taking other measures, the employer must offer you hearing protection.

90 dBA is the second action level and at this noise level or higher, ear protection must be worn and the employer must ensure that adequate training is provided and that measures are taken to reduce noise levels, as far as is reasonably practicable.

The irony here is that the headphones could be acting as hearing protectors for sound from outside, but be themselves generating audio levels above health and safety limits.

If you ever experience 'ringing in the ears' or are temporarily deafened by a loud noise, then you have permanently damaged your hearing. This damage will usually be *very* slight each time, but accumulates over months and years; the louder the sound the more damage is done.

Slowly entering a world of silence may seem not so terrible but, if your job involves audio, you will lose that job. Deafness cuts you off from people; it is often mistaken for stupidity. Worse, hearing damage does not necessarily create a silent world for the victim. Suicides have been caused by the other result of hearing damage. The gentle-sounding word used by the medical profession is 'Tinnitus'. This conceals the horror of living with loud, throbbing sounds created within your ear. They can seem so loud that they make sleep difficult. Some people end up in a no-win situation where, to sleep, they have to listen to music on headphones at high level to drown the Tinnitus. Of course this, in turn, causes more hearing damage.

The Royal National Institute for Deaf People (RNID) has simulations of Tinnitus on its web site http://www.rnid.org.uk/information_resources/tinnitus/about_tinnitus/what_does_it_sound_like/. This can also be found on this book's CD-ROM.

If you are reading this book then you value your ears; so please take care of them!

2.3 Analog and digital audio

Analog audio signals consist of the variation of sound pressure level with time being mimicked by the analogous change in strength of an electrical voltage, a magnetic field, the deviation of a groove, etc.

The principle of digital audio is very simple and that is to represent the sound pressure level variation by a stream of numbers. These numbers are represented by pulses. The major advantage of using pulses is that the system merely has to distinguish between pulse and no-pulse states. Any noise will be ignored provided it is not sufficient to prevent that distinction (Figure 2.2).

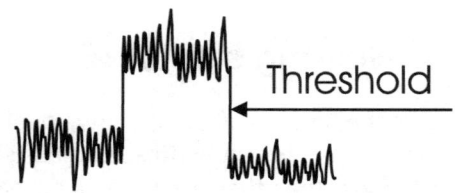

Figure 2.2 Distinction between pulse and no-pulse states

The actual information is usually sent by using the length of the pulse; it is either short or long. This means that the signal has plenty of leeway in both the amplitude of the signal (how tall?) and the length of the pulse. Well-designed digital signals are very robust and can traverse quite hostile environments without degradation.

However, this can be a disadvantage for, say, a broadcaster with a 'live' circuit, as there can be little or no warning of a deteriorating signal. Typically the quality remains audibly fine, until there are a couple of splats, or mutes, then silence. Analog has the advantage that you can hear a problem developing and make arrangements for a replacement before it becomes unusable.

Wow and flutter are eliminated from digital recording systems, along with analog artefacts such as frequency response and level changes. If the audio is copied as digital data then it is a simple matter of copying numbers and the recording may be 'cloned' many times without any degradation. (This applies to pure digital transfers; however, many modern systems, such as Minidisc, MP3s, Digital Radio and Television, use a 'lossy' form of encoding. This throws away data in a way that the ear will not usually notice. However, multi-generation copies can deteriorate to unusability within six generations if the audio is decoded and re-encoded, especially if different forms of lossy compression are encountered. Even audio CD and Digital Audio Tape (DAT) are lossy to some extent, as they tolerate and conceal digital errors.)

What is digital audio?

Pulse and digital systems are well established and can be considered to have started in Victorian times.

Perhaps the best known pulse system is the Morse code. Like much digital audio, this code uses short and long pulses. In Morse these are combined with short, medium and long spaces between pulses to convey the information. The original intention was to use mechanical devices to decode the signal. In practice it was found that human operators could decode by ear faster.

Modern digital systems run far too fast for human decoding and adopt simple techniques that can be dealt with by microprocessors with rather less intelligence than a telegraph operator. Most systems use two states that can be thought of as on or off, short or long, dot or dash

Some technical bits

(some digital circuits and broadcast systems can use three, or even four, states but these systems are beyond the scope of this book).

The numbers that convey the instantaneous value of a digital audio signal are conveyed by groups of pulses (bits) formed into a digital 'word'. The number of these pulses in the word sets the number of discrete levels that can be coded.

The word length becomes a measure of the resolution of the system and, with digital audio, the fidelity of the reproduction. Compact disc uses 16-bit words giving 65 536 states. NICAM stereo, as used by UK television, uses just 10 bits but technical trickery gives a performance similar to 14-bit. Audio files used on the Internet are often only 8-bit in resolution while systems used for telephone answering, etc. may only be 4-bit or less (Table 2.2).

Table 2.2 Table showing how many discrete levels can be handled by different digital audio resolutions

1-bit = 2	7-bit = 128	13-bit = 9192	19-bit = 524 287
2-bit = 4	8-bit = 256	14-bit = 16 384	20-bit = 1 048 575
3-bit = 8	9-bit = 512	15-bit = 32 768	21-bit = 2 097 151
4-bit = 16	10-bit = 1024	16-bit = 65 536	22-bit = 4 194 303
5-bit = 32	11-bit = 2048	17-bit = 131 071	23-bit = 8 388 607
6-bit = 64	12-bit = 4096	18-bit = 262 143	24-bit = 16 777 215

The larger the number, the more space is taken up on a computer hard disk or the longer a file takes to copy from disk to disk or through a modem. Common resolutions are: 8-bit, Internet; 10-bit, NICAM; 16-bit, compact disc; 18–24-bit, enhanced audio systems for production. Often, 32-bit is often used internally for processing to maintain the original bit-resolution when production is finished.

On the accompanying CD-ROM there is a demonstration of speech and music recorded at low bit rates; from 8 to 1 bits. There is no attempt to optimize the sound.

Sampling rate

A major design parameter of a digital system is how often the analog quantity needs to be measured to give accurate results. The changing quantity is sampled and measured at a defined rate.

If the 'sampling rate' is too slow then significant events may be missed. Audio should be sampled at a rate that is at least twice the highest frequency that needs to be produced. This is so that, at the very least, one number can describe the positive transition and the next the negative transition of a single cycle of audio. This is called the Nyquist limit after Harry Nyquist of Bell Telephone Laboratories who developed the theory.

For practical purposes, a 10 per cent margin should be allowed. This means that the sampling rate figure should be 2.2 times the highest frequency. The 20 kHz is regarded as the highest frequency that most people can hear. This led to the CD being given a sampling rate of 44 100 samples a second (44.1 kHz). The odd 100 samples per second is a technical kludge. Originally digital signals could only be recorded on videotape machines, and 44.1 kHz will fit into either American or European television formats (see box).

> From 'Digital Interface Handbook' Third edition, Francis Rumsey and John Watkinson, Focal Press 2004.
>
> In 60 Hz video, there are 35 blanked lines, leaving 490 lines per frame or 245 lines per field for samples. If three samples are stored per line, the sampling rate becomes $60 \times 245 \times 3 = 44.1$ kHz.
>
> In 50 Hz video, there are 37 lines of blanking, leaving 588 active lines per frame or 294 per field, so the same sampling rate is given by $50 \times 294 \times 3 = 44.1$ kHz.

The sampling process must be protected from out-of-range frequencies. These 'beat' with the sampling frequency and produce spurious frequencies that not only represent distortion but, because of their non-musical relationship to the intended signal, represent a particularly nasty form of distortion. These extra frequencies are called alias frequencies and the filters called anti-aliasing filters. It is the design of these filters that can make the greatest difference between the perceived qualities of analog to digital converters. You will sometimes see references to 'over-sampling'. This technique emulates a faster sampling rate ($4\times$, $8\times$, etc.) and simplifies the design of the filters as the alias frequencies are shifted up several octaves.

Errors

A practical digital audio system has to cope with the introduction of errors, owing to noise and mechanical imperfections, of recording and transmission. These can cause distortion, clicks, bangs and dropouts, when the wrong number is received.

Digital systems incorporate extra 'redundant' bits. This redundancy is used to provide extra information to allow the system to detect, conceal or even correct the errors. Decoding software is able to apply arithmetic to the data, in real time, so it is possible to use coding systems that can actually detect errors.

However, audio is not like accountancy, and the occasional error can be accepted, provided it is in a well-designed system where it will not be audible. This allows a simpler system, using less redundant bits, to be used to increase the capacity, and hence the recording length, of the recording medium. This is why a Compact Disc Recordable (CDR) burnt as a computer CD-ROM storing wave file data has a lower capacity than the same CDR using the same files as CD-audio. CD-ROMs have to use a more robust error correction system as NO errors can be allowed. As a result a standard CDR can record 720 Mbytes of CD-audio (74 minutes) but only 650 Mbytes of computer data.

Having detected an error a CD player may be able to:

1 Correct the error (using the extra 'redundant' information in the signal).
2 Conceal the error. This is usually done either by sending the last correctly received sample (replacement), or by interpolation, where an intermediate value is calculated.
3 Mute the error. A mute is usually preferable to a click.

While the second and third options are good enough for the end product used by the consumer, you need to avoid the build-up of errors during the production of the recording. Copies made on hard disk, internally within the computer, will be error-free. Similarly copies made to CD-ROM or to removable hard disk cartridge, will also have no errors (unless the disk fails altogether because of damage).

Multi-generation copies made through the analog sockets of your sound card will lose quality. Copies made digitally to DAT will be better, but can still accumulate errors. However, a computer backup as data to DAT (4 mm) should be error-free (as should any other form of computer backup medium. These have to be good enough for accountants, and therefore error-free).

How can a computer correct errors?

It can be quite puzzling that computers can get things wrong but then correct them. How is this possible? The first thing to realize is that each datum bit can only be '0' or '1'. Therefore, if you can identify that a particular bit is wrong, then you know the correct value. If '0' is wrong then '1' is right. If '1' is wrong then '0' is right.

The whole subject of error correction involves deep mathematics but it is possible to give an insight into the fundamentals of how it works. The major weapon is a concept called parity. The basic idea is very simple, but suitably used can become very powerful. At its simplest this consists of adding an extra bit to each data word. This bit signals whether the number of '1's in the binary data is odd or even. Both odd and even parity conventions are used. With an even parity convention the parity bit is set so that the number of '1's is an even number (zero is an even number). With odd parity the extra bit is set to make the number of '1's always odd.

Received data is checked during decoding. If the signal is encoded with odd parity, and arrives at the decoder with even parity, then it is assumed that the signal has been corrupted. A single parity bit can only detect an odd number of errors.

Quite complicated parity schemes can be arranged to allow identification of which bit is in error and for correction to be applied automatically. Remember that the parity bit itself can be affected by noise.

Illustrated in Figure 2.3 is a method of parity checking a 16-bit word by using eight extra parity bits. The data bits are shown as 'b', the parity bits are shown as 'P'.

The parity is assessed both 'vertically' and 'horizontally'. The data are sent in the normal way with the data and parity bits intermingled. This is called a Hamming code. If the bit in column 2, row 2 were in error, its two associated parity bits would indicate this. As there are only two states then, if the bit is shown to be in error, reversing its state must correct the error.

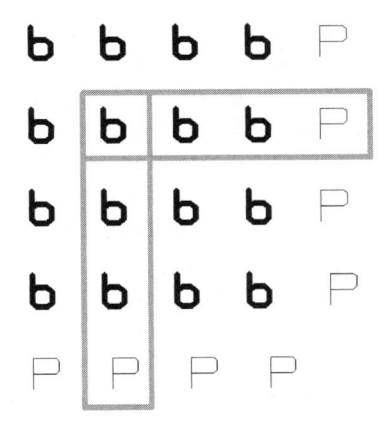

Figure 2.3 A method of parity checking a 16-bit word by using 8 extra parity bits

Dither

The granular nature of digital audio can become very obvious on low level sounds such as the die-away of reverberation or piano notes. This is because there are very few numbers available to describe the sound, and so the steps between levels are relatively larger, as a proportion of the signal.

This granularity can be removed by adding random noise, similar to hiss, to the signal. The level of the noise is set to correspond to the 'bottom bit' of the digital word. The frequency distribution of the hiss is often tailored to optimize the result, giving noise levels much lower than would be otherwise expected. This is called 'noise-shaping'. PC audio editors often have an option to turn this off. Don't do this unless you know what you are doing. Dither has the, almost magical, ability to enable a digital signal to carry sounds that are quieter than the equivalent of just 1 bit. There is a demo of dither on this book's CD-ROM.

2.4 Time code

All modern audio systems have a time code option. With digital systems, it is effectively built in, at its simplest level, it is easy to understand; it stores time in hours, minutes and seconds. As so often, there are several standards.

The most common audio time display that people meet is on the compact disc; this gives minutes and seconds. For professional players, this can be resolved down to fractions of second by counting the data blocks. These are conventionally called frames and there are 75 every second. This is potentially confusing as CDs were originally mastered from ¾ inch U-Matic videotapes where the data were configured to look like an American Television picture running at 30 video frames per second (fps).

In 1967, the *Society of Motion Picture and Television Engineers* (SMPTE) created a standard defining the nature of the recorded signal, and the format of the data recorded. This was for use with videotape editing. Data are separated into 80-bit blocks, each corresponding to a single video frame. The way that the data are recorded (*Biphase modulation*) allow the data also to be read from analog machines when the machines are spooling, at medium speed, with the tape against the head, in either direction. With digital systems, the *recording* method is different but the *code* produced stays at the original standards.

The SMPTE and the European Broadcasting Union (EBU) have defined a common system with three common video frame standards. 25 fps (European TV) and 30 fps (American Black&White TV). For technical reasons, with the introduction of color to the American television signal there is also a more complicated format known as 30 fps drop frame which corresponds to an average to 29.97 fps. Since the introduction of sound in the 1920s, film has always used 24 fps.

By convention, on analog machines, the highest numbered track is used for time code; track 4 on a 4-track; track 16 on a 16-track, etc. It is a nasty screeching noise best kept as far away from other audio as possible.

Time code can also be sent to a sequencer, via a converter, as MIDI data allowing the sequencer to track the audio tape. The simple relationship between bars, tempo and SMPTE time as shown by sequencers like *Cubase* is only valid for 120 beats per minute 4/4 time.

MIDI time code generators need to be programmed with the music tempo and time signature used by the sequencer, so that they can operate, in a gearbox fashion, so that the sequencer runs at the proper tempo.

People editing audio for CD will often prefer to set the Time Display to 75 fps to match the CD data. There is a small technical advantage to ensuring that an audio file intended for CD ends exactly at the frame boundary as the remainder of any unfilled frame will be filled with digital silence (all zeros). You will get a click on the frame boundary as audio silence will not be all zeros.

Right clicking on Adobe Audition 2.0's time display provides a popup with a wide range of options for the frame display (Figure 2.4) including being able to match to the bars and beats of a MIDI track.

See Appendix 3 for more information on time code.

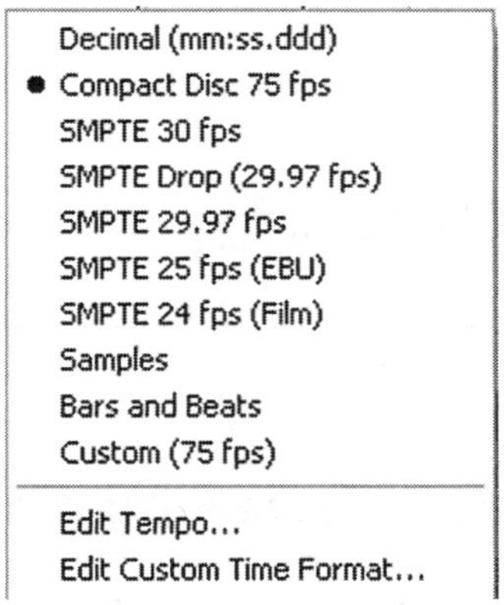

```
  Decimal (mm:ss.ddd)
● Compact Disc 75 fps
  SMPTE 30 fps
  SMPTE Drop (29.97 fps)
  SMPTE 29.97 fps
  SMPTE 25 fps (EBU)
  SMPTE 24 fps (Film)
  Samples
  Bars and Beats
  Custom (75 fps)

  Edit Tempo...
  Edit Custom Time Format...
```

Figure 2.4 Right clicking on *Cool Edit's* time display provides a wide range of options for the frame display

3

Transfer

3.1 Reviewing material

Some material will arrive already in a form readable by your computer (such as the audio material provided on the CD-ROM accompanying this book). Other audio will have been recorded on an analog or digital recorder that needs transfer to your computer. This may have been recorded by someone else or acquired by you. In Chapter 13 there is some advice on originating material but in a book called *PC Audio Editing* let's get on with that: having acquired all your material, how do you transfer it onto your computer?

There are recorders that plug in to a USB port and appear as a disk drive. Here the audio files will already exist and can be copied across to your work folder. This can often be done on the desktop but it can be useful to copy them via Audition so that the files can be 'topped and tailed' so that the file starts and finishes cleanly. The file format can be changed to .WAV if a different format is used by the recorder. This is particularly important if a compressed format has been used. Changing to a .WAV format will stop further quality loss due to data compression. Most material will have to be copied, in real time, through the sound card of your computer, preferably if the source is digital, using the digital inputs to preserve the best quality.

You should not waste this time, for it is now that you can review the material and make notes. You can save time if there are session notes by skipping the false takes. This is especially easy if you have a note of ident points if you were using DAT or Minidisc.

Logging speech recordings can save an immense amount of time. Your notes should have a convention so that they mean something to you several days, or weeks, later. Write down who the speaker is and what tape or disc they are recorded on – you did label the recordings didn't you?

Be consistent as to how you log where things are in the recording. With reel-to-reel or compact cassette, you are usually stuck with arbitrary numbers produced by the playback machine's counter. These can be wildly different on a different make of machine.

Digital machines log progress as time but there are different options. Quite a few portable machines actually record date and time of day continuously. This does mean that, on location, you can use your wristwatch and a notepad to identify points. However, a lot of mains machines do not handle this information.

DAT machines usually have different counter modes:

- Counting time from when last reset
- Counting absolute time from the beginning of the first recording on the cassette
- Counting time starting from 00:00:00 each time a new track marker is encountered.

Minidisc recorders usually offer:

- Time elapsed of track
- Time left of track
- Time left of disc.

Once you have established which timing system you are going to use, make your notes. Unless you have good shorthand you will not be able to transcribe what is said. Instead, log each question with the time and then list the major points made in the answer. There are many formats possible, one such is:

Joe Smith 29th March
0'00   ?Why has this organization been set up?
0'10   Fulfil need by public
0'20   Contact point for victims
0'35   Money not available from government
0'50   ?Should public money be provided?
0'55   Not a practical proposition
1'05   Get on with it
1'20   500 people already involved
1'30   ?How long before organization effective?
1'35   First effects within 6 months
1'45   90 per cent after two years
1'55   Complete after three years
2'10   OUT... 'Everyone needs this now'.

When you are editing something that is scripted, then most of the edits will be overlaps (going back to the beginning of a sentence at the time) or retakes (sections redone after the main recording). It should not be necessary to edit fluffs as they should be covered by overlaps and retakes. Pauses may need adjusting. Beware of misreads. 'These mighty heralds that make the Gods tremble' is what is intended and more effective than 'These mighty herons...'. Spell checker artefacts are a growing hazard where, during a spell check, an entirely correctly spelled but entirely wrong word is absentmindedly substituted.

Better results are often achieved by not cutting at the beginning of an overlap but cutting, instead, a few words into it, because speakers tend to overemphasize the start as they are angry with themselves for making the mistake. With this in mind when making notes, mark where, in the overlap, you think will be a good return point.

When marking an overlap, mark where it begins and how many times the speaker went back. There is nothing more irritating than editing an overlap and then discovering another one a few seconds later. Some people put brackets, the number of brackets indicating the number of repeats. I prefer to mark the point and place an 'E' in the margin. This I subscript with the number of repeats: E_1, E_2, E_3 etc. By E_3, the performer is usually quite annoyed with themselves and definitely tend to overemphasize the first few syllables, In this case I mark a likely late in-point with a Δ.

You can play out at speed by holding down the mouse button on the 'fast forward' button during playback. Easier – use a keyboard shortcut. Identifying the repeats should be relatively easy, even at speed. In practice professionals rarely listen at normal speed while searching for edits. About 150 per cent is usual.

One of the advantages of a visual display is that you can see small pauses and often breaks in rhythm. This means that it is usually quicker to 'bet' on an edit, make it and play from there. If there is another overlap then a click can select from the last edit to the new point. If the wagered edit is wrong then UNDO instantly restores.

It is amazing just how easily edits can be missed. The ear is sensitive to a break in the speech rhythm. It is always a mistake for the speaker to apologize or to swear at themselves when they fluff. This has the uncanny ability to maintain the speech rhythm and it is these edits that are most likely to be missed, especially in a news and current affairs situation. The professional way is to stop, give three-beats pause, and then go back to the start of the sentence without comment. This will sound more natural if the edit is missed but, ironically, that three-beat pause will usually mean that it is not missed.

3.2 Head alignment

If you are using an analog recorder, a compact cassette machine or a reel-to-reel quarter inch tape machine, then try to use the same machine that you used to record to dub your material onto your computer. This is because analog tape systems are very sensitive to head alignment differences between machines. This gives a muffled sound. On a cassette, Dolby C doubly emphasizes this sensitivity. In practice Dolby B noise reduction will reduce tape hiss to below the ambient noise in your recording environment without being hypercritical of head alignment.

Head alignment can be an issue with digital machines as well and can mean that a recording will play on one machine but not on another. Try always to have the original machine available, just in case.

3.3 Digital

In some ways, these recordings are the easiest to transfer. You do not have to worry about setting levels as you did that on the original recording. If you got them wrong then there is nothing that you can do to correct this until the digital audio data is on your computer. This is because the digital interface is merely transferring the numbers representing the audio that the original recorder laid down at the time of recording. Some sound cards allow you to alter the digital level but this leaves you vulnerable to digital overloads. Correcting within the audio editor is much more flexible.

There are several standards for digitally interfacing your computer to a recorder. The most common is known as S/PDIF which stands for Sony/Philips Digital Interface. This comes in three main variants, one electrical and two optical.

Electrical

The electrical connection uses standard RCA phono plugs, as used for audio on a lot of domestic equipment (Figure 3.1). A single connection is used for both the left and the right sides of a stereo signal. Unfortunately many domestic digital recorders, while having an electrical S/PDIF input, do not have an equivalent output.

Mains-operated machines will usually have an input and output in optical form but many portable machines have no digital output at all. A cynic might believe that manufacturers think that their machines will only be used for copying CDs, rather than for creative work.

For best results you should use phono leads designed for digital connections. While any old audio leads, you happen to have lying around, may get you out of a jam, they can lead to unpredictable problems by introducing errors. This is because they are not designed for the supersonic frequencies involved in digital transfer. Purpose-designed digital leads tend to be thicker than audio leads.

Professional set-ups will use balanced AES/EBU connections using XLR plugs. There are high end sound cards that can use this format. However, S/PDIF inputs and outputs are usually also available.

Figure 3.1 RCA Phono plug

Optical

Instead of electricity, optical connectors use modulated red light to transfer the data. This light is visible to the naked eye when a connector is carrying an output. This is extremely useful in avoiding confusion between input and output leads.

There are two physical standards. The original, usually found on mains-operated equipment, is known as TOSLINK (from TOShiba Link; Figure 3.2). Small portable machines use optical connectors the same shape as audio 3.5 mm minijacks (Figure 3.3). Some machines have dual function sockets where the socket can either be used for audio or for optical digital. Both connectors use the same signal, which is S/PDIF in light form, so machines using different connectors can be joined using a minijack to TOSLINK lead.

Some early DAT machines had a proprietary connector at the machine end and a lead terminating with a TOSLINK plug. This can lead you to find yourself wanting to copy from recorder to recorder to 'clone' a tape but are faced with joining two TOSLINK connectors. While not recommended, this can be

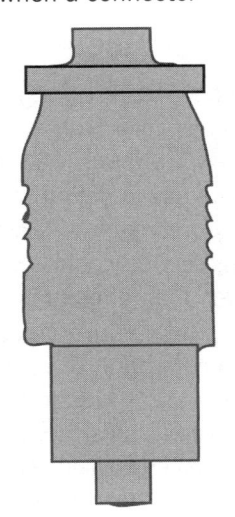

Figure 3.2 TOSLINK optical plug

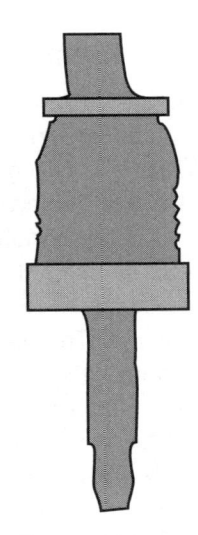

Figure 3.3 'Minijack' optical plug

Figure 3.4 Joining two TOSLINK plugs

done. Here the plastic tube containing the ink in a cheap ball-point pen can come to the rescue. Snipping the ink-free centimetre off the top will provide you with a suitable 'gender-changer'. Make sure that you trim it short enough for the ends of the TOSLINK connectors to touch (Figure 3.4). This technique can also allow you to join two optical leads together. Yes, of course, you should have bought a longer one and you will tomorrow but it can save the day today!

Optical leads are relatively fragile in that they will fracture if bent too far, as they are made of transparent plastic not copper. They are surrounded by opaque plastic. Because the signal is conveyed by light the connection is immune to interference from electrical equipment. However, because of the internal reflections in the fibre, the waveform received by the sound card is less pure than with an electrical connection which should, therefore be preferred. Inputs that 're-clock' the signal can correct the waveform distortion and remove the 'jitter' that otherwise results.

SCMS and copy protection

As well as the numbers representing the audio signal, the S/PDIF signal contains extra data including various 'flags'. Normally these need not concern the user, except for the flags indicating copy protection. Literally, the system is specified so that a digital flag can be raised that will tell a well-behaved digital recorder input not to accept the data.

The all-or-nothing approach of a single flag was thought to be too crude and the SCMS (Serial Copy Management System) was devised. This allows a single generation digital copy to be made. However, if an attempt is made digitally to copy that digital copy then that is prohibited. This is an attempt to reduce multiple digital cloning of commercial recordings. Annoyingly, it is made to apply to your own recordings as well. This only becomes a problem when copying from an SCMS recorder to another SCMS recorder. Digital interfaces on computer sound cards ignore the system. Analog copies always remain possible.

However, the original copy protect flag can be a problem when copying from a professional DAT machine to a domestic DAT. Professional machines use a variation on S/PDIF called AES/EBU which is almost, but not quite, compatible when fed to a phono plug like an S/PDIF signal. The flags are different; a professional machine will often set the flag which means copy protect to a domestic machine, but not to itself. This leads to a situation where you can copy from the domestic machine to the professional machine but not in the other direction!

Clocks

Stereo sound cards are relatively straight forward but If you are using a multiple input sound card you will need to set the card to be controlled by the external data by switching it to the external incoming data clock. (See Appendix 1.)

USB/Firewire

There are digital audio recorders that record onto their own hard drive or a memory card. The one illustrated (Figure 3.5) can transfer via the USB port giving all the advantages of digital transfer along with being able to do it at around 40 times faster than real time. Effectively, for the time of connection, the recorder becomes part of your computer and you are copying a files from one storage device to another. With USB or Firewire you can connect or disconnect the recorder without having to reboot.

Memory cards

Audio equipment is benefiting from the massive market for memory cards provided by the digital camera. Capacities of 512 Mbyte and 2 Gbyte are now easy to obtain at inexpensive prices. They take little space. Because the memory cards have no moving parts, microphones can be built into the recorders without the old motor noise problems. Many computers come with drives built in as well as the software. USB connectors to card drives are cheap and easily available. There are quite a few different types on the market, In practice, it is down to the customer to choose their cameras and audio equipment to use the same kind of card! (Figure 3.6)

Figure 3.5 Edirol R-09 by Roland; an example of a modern small audio recorder using SD (Secure Digital) memory cards of up to 2 Gbyte capacity to record MP3 or 16/24-bit stereo .WAV files using internal capacitor microphones or external feeds. Playback is by audio jack or USB connection to a computer

Figure 3.6 SD memory card shown actual size. The tag at the top is a write protect switch

Flash drives

These are 'key-ring' devices designed to be carried around and are more robust than bare memory cards. They plug directly into USB sockets. They appear as drives to the computer and need no adapter. Modern XP computers already have the drivers for this sort of device (Figure 3.7).

SCSI/IDE

Some recording systems allow direct connection to a computer via a SCSI bus or a proprietary connection of their own. Alternatively they may use removable hard disks. Contained in caddies, these can be unplugged from the recording device and slotted into the computer you use for editing. These often use the SCSI connection system but IDE systems also exist. In the best implementations the inserted hard drive is allocated a drive letter by Windows and can be edited directly by Adobe Audition 2.0 or any other editor. This disks can contain many gigabytes of data and hence many an hour of work. A backup copy should be a high priority. Carrying 20 hours of original audio across town on a single disk is quite frightening as it is vulnerable to so many possibilities of damage from being dropped to being stolen. For the first hazard carry them in a small foam filled flight case available from camera shops.

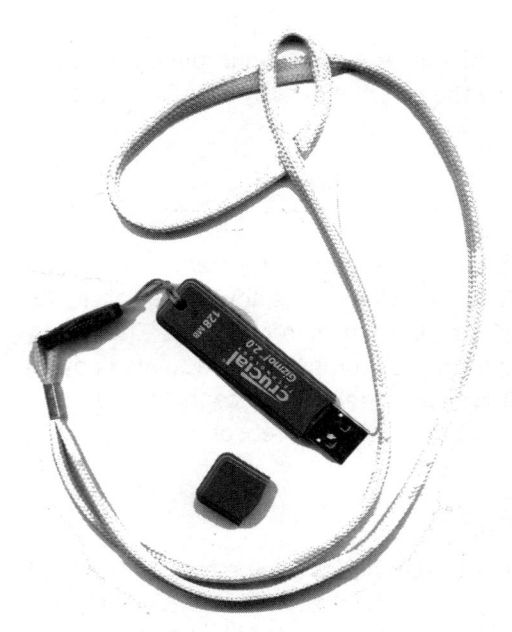

Figure 3.7 Key-ring device

CDs

CDs can be played by the Windows CD player. This is fine for listening to select material, but transfer is best done digitally. This is done either by using CD 'ripping' software, often supplied by the manufacturer of the CD-ROM drive, or by having your machine set-up so that Windows 'sees' the audio tracks as wave files (see 'Blue bar blues' Page 42). Adobe Audition 2.0 has CD ripping built in and can read audio CDs directly; but you may find that setting your CD-ROM to see CD audio tracks as wave files is much quicker and convenient as then every audio program you use can read CDs.

When they were introduced, there was a lot of hype that CDs were indestructible. Most people have discovered that they are not, except compared with LPs. Unlike LPs, which will always produce something no matter how crackley, CDs share with other digital media, the tendency to work perfectly or not all. While inevitably the best advice is to look after your CDs and to handle them by their edges, failed CDs can often be recovered.

CDs play from the centre outwards. This is to allow easy compatibility between 3-inch and 5-inch discs. The beginning of the recording contains a 'table of contents'. While this is particularly heavily error corrected, it means that if the table is covered by a mark, the CD will fail even to be recognized by the machine. If this happens examine the playing surface for a mark or marks at the inside of the playing surface.

CDs slow down as they track towards the outside. This gives the digital track a constant linear velocity. The LP had a constant angular velocity, and the speed of the playing surface past the stylus decreased as it went towards the centre. This means that half way through the playing time of a full CD is not half way across the disc but further out. Incidentally, the transfer speed of the CD-ROM is usually specified as a multiple of the standard playing speed of an audio CD. So 'x52' implies that the device can extract the data from a 52 minute in 1 minute. In fact it won't. The CD-ROM runs at a linear angular velocity and only reaches its specified speed at the outside of the disc.

If the CD is jumping and skipping a section, then this is likely to be a scratch or an opaque mark. If the CD repeatedly sticks at a section then this is likely due to a transparent mark (like jam). This bends the laser and thus confuses the player as to where it is.

Armed with this information, it soon becomes possible to work out where to look, and what to look for. Because of the nature of the digital track, scratches radiating from the centre are much less likely to cause problems compared with scratches running along the line of the track. This is why CDs should be cleaned with radial movements rather than the instinctive rotary action (Figure 3.8).

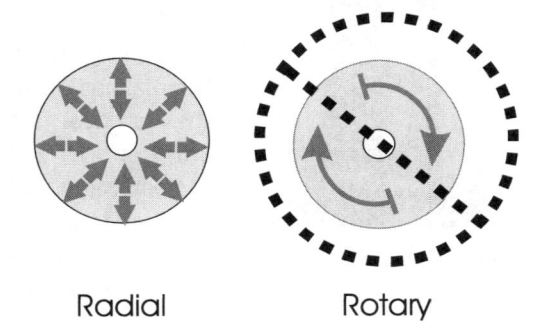

Radial   Rotary

Figure 3.8 Always clean CDs by wiping radially

The first emergency treatment is to wash the CD. Smear washing-up liquid over the surface and then run water over it from a tap, helping the washing-up liquid off with inside to out radial movements of your fingers. Cold or warm water is fine but it is probably best to avoid hot water: if it is too hot for your fingers then it is too hot for the CD. Shake the remains of the water off the CD and look to see if there are any obvious marks remaining. These may need individual attention, again with the washing-up liquid. Try the CD in the player (any slight remains of water will be spun off in the player).

A really persistent mark or scratch should now be identifiable. If rescue has still failed then the mark can be cleaned with brass cleaning wadding (*Brasso Duraglit* wadding in the UK). Remember to rub radially.

It may be that this still doesn't work for you. If your CD player's laser lens is dirty this will make it less able to correct errors. Regularly clean the lens with a special lens cleaning CD. This has little brushes on it which clean the lens as the track is changed – usually between 1 and 3. Cleaners using the same principle are also available for Minidisc machines.

If there is still a problem then try a different make of CD player, as different manufacturers have different error correction strategies. If you are 'ripping' the audio digitally straight off the CD, see if the CD-ROM has management software that will slow down the process. A CD that will not rip at x12 speed may be OK at x4, x2 or x1. If all fails try the audio output of the CD-ROM drive.

3.4 Analog

Provided care has been taken to get the levels right on the original recording, analog transfer can be entirely satisfactory. You should make sure that your system is connected satisfactorily and well earthed. The playback should be checked for hum pickup. This can be due to electrical equipment on the desk or to the computer itself.

Running the dubbing lead near to a power supply (including the playback machine's own) is likely to introduce hum. Running the lead near to the computer's monitor is likely to introduce not only hum but also buzzes and whistles.

Keep the dubbing lead as short as it practical, as very long runs will lead to loss of high frequencies unless you are using fully professional equipment using 'balanced' connections with tip, ring and sleeve jacks or XLR connectors (Figures 3.9 and 3.10).

Figure 3.9 Tip, ring and sleeve jack

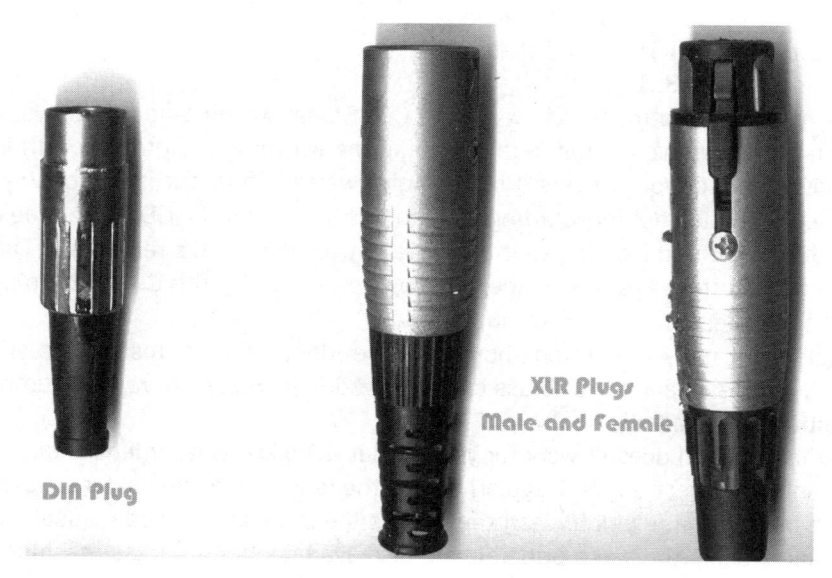

Figure 3.10 XLR connectors

Normally you will connect the 'LINE OUT' socket of the playback machine to the 'LINE IN' socket. The most common connectors at each end are 3.5 mm minijacks (Figure 3.11). However, the better quality sound cards have 'breakout' boxes. These either sit at the end of a thick flying lead or are mounted in one of the disk drive compartments of the computer. These will have phono, quarter inch jack or XLR connectors often with a balanced connection option for use with professional machines.

The LINE OUTs of playback machines are of a fixed level so level control has to be done on the computer. Using the headphone output instead is an option and can give excellent quality. It can also produce crackles and distortion if the volume control is dirty!

Figure 3.11 Stereo and mono 3.5-mm minijacks with 2.5-mm plug for comparison

Gramophone records

Despite the popularity of CDs, there are still occasions when old LPs are the only available source of a piece of music, or even actuality. It makes sense, therefore, to get the best possible sound out of them before resorting to software solutions within the computer.

Gramophone turntables cannot just be plugged straight to a computer sound card. Not only do they have an output much lower that the line level required by the sound card, but they also need what is called RIAA equalization. This is a standardized fixed top boost and bass cut made on recording that has to be corrected at playback.

In the past, this was done by the 'phono' input of a Hi-Fi amp. Nowadays, many do not even possess such an input. Maybe you have an old Hi-Fi amp that can be pressed into service. The TAPE output can be fed to the computer's sound card.

However, such a complicated device is not needed. Cheap battery-operated amplifiers are available; they deliver an equalized line level output suitable for your sound card. Being battery operated, they should have a lower hum level than conventional amplifiers. A typical example is illustrated in Figure 3.12.

Some can double as simple mic amplifiers, if required, by switching off the RIAA equalization.

It can be useful to bypass the RIAA top cut when transferring old recordings. The top cut softens clicks and de-clicking software can work better without it having been applied. You then correct the recording afterwards.

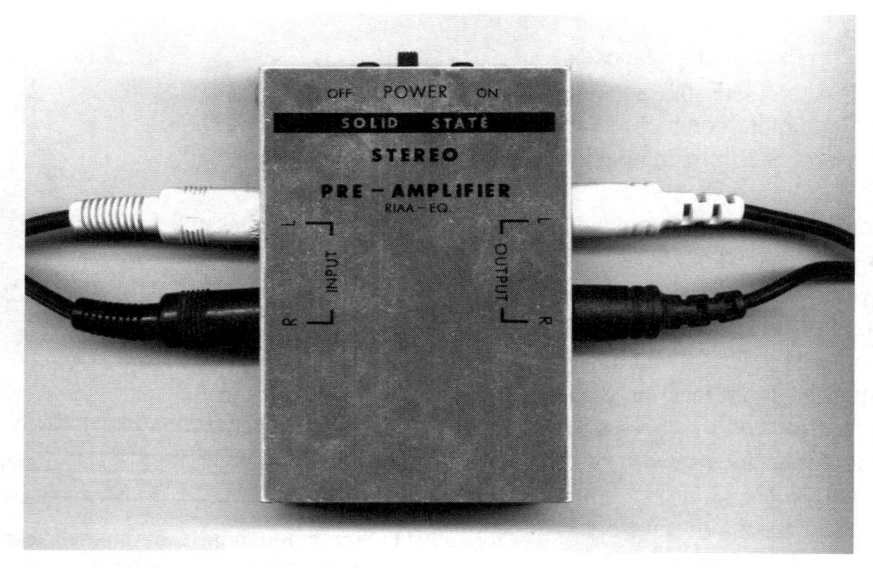

Figure 3.12 Typical record turntable preamplifier

Http://sound.westhost.com/project91.htm has a circuit board project for building your own RIAA amplifier.

Http://www.a-reny.com/iexplorer/restauration.html has very useful advice about transferring gramophone records.

If you are transferring 78 rpm recordings then RIAA should then definitely be left out as they did not use this system. They did use some form of equalization but there were more than a hundred different ones used by different companies over the half century reign of the medium. As a compromise the FFT filter on AA can be set to apply RIAA or Reverse RIAA setting. (See Appendix 4.)

3.5 Recording

Before making a recording you have to tell the program how you want to record. Some audio programs like *Pro Tools* will also ask you where you want to record before you start. Adobe Audition 2.0 uses buffer files and you can decide on file names and paths after the recording. You can set where these buffer files are by changing the temporary directories to be found in the system tab of the settings dialog. (When you first install Adobe Audition 2.0, it will suggest where to put them based on the disk space available). There is a speed benefit by having them on physically separate hard drives. This means that copying is achieved by two heads moving across their separate drives rather than one head clacking back and forth between different parts of the same disk.

Click on 'New' and you will see a dialog asking you to set various options (Figure 3.13). This will be based on what you set previously, so, for most people, this becomes a simple matter of clicking OK.

The sampling rate is how many numbers per second are used to define your audio. For high quality use, there are two main standards. 44 100 is used by CDs. If your material is going to end up on CD audio then this is the rate to choose (CD-ROM data can be any sampling frequency).

The other commonly used sampling rate is 48 000. This is required by some broadcasting companies that accept data files. If your material is sourced from, or is going to end up on DAT then this is the frequency to use. Many non-pro DAT machines will only record analog signals at 48 000 (or 32 000 at their Long Play speed). If you are making a digital transfer then it is vital that the sampling rate set matches the rate used on the recording. The sampling rate of a recording can be converted within Adobe Audition 2.0 but this is relatively a time-consuming process.

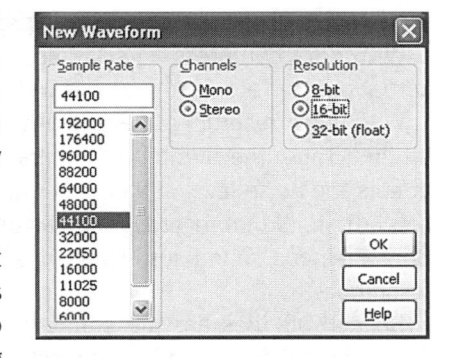

Figure 3.13 Setting recording options

The mono/stereo option is straight forward. Operationally it is much simpler to record everything stereo, even mono sources. A stereo transfer of a mono recording can help identify problems such as head alignment. Sometimes the left and right channels will be of different quality. You can choose the better one. Much of the noise in a gramophone transfer appears as the difference between left and right and will disappear if the left and right channels are summed together but being able to get at this difference signal can materially help noise reduction software.

However, mono files are half the length of stereo files and, if everything is in mono and good quality, then make the saving. Beware that CD audio has no mono mode so burning to CD may need stereo files. However, many CD writing programs have the ability to convert mono to stereo on-the-fly, thus removing this constraint on the use of mono files. Minidisc does have a mono mode and a 74-minute Minidisc can record 148 minutes of mono, an 80-minute disc 160 minutes at full quality.

Hi-MD has higher capacities. Old style minidiscs can be reformatted to have 305 Mbyte capacity. Hi-MD blanks have 1 Gbyte capacity. Taken from Http://www.minidisc.org/ the following capacities are possible.

305 Mbyte	1 Gbyte
Linear PCM 1.4 Mbps	28 m	1 h 34 m
ATRAC3plus 256 kbps 'Hi-SP'	2 h 20 m	7 h 55 m
ATRAC3plus 64 kbps 'Hi-LP'	10 h 10 m	34 h
ATRAC3plus 48 kbps	13 h 30 m	45 h
ATRAC3 132 kbps 'LP2'	4 h 50 m	16 h
ATRAC3 105 kbps	6 h 10 m	20 h 50 m
ATRAC3 66 kbps 'LP4'	9 h 50 m	32 h 50 m

ATRAC is Sony's own audio compression system comparable to MP3.

With data compression systems, the number of bps represent the quality possible. The actual output rate from a player will usually be 44.1 kHz.

With linear recording, the resolution sets how big the numbers that represent your audio can be and, thus, the number of separate discrete levels that can be represented. This directly affects the noise level of your recording. Eight bit only has 256 levels and is, as a result, very hissy. It should be regarded only as an end-user format. With suitable control and processing 8-bit files can sound surprisingly good. However, that processing has to be done at a higher resolution.

Sixteen-bit files are the standard, match CD and DAT and will be your normal choice. Thirty-two-bit files give the highest quality and, while taking up twice as much data space, give superior results when being modified. The floating-point format used by Audition gives almost complete protection against overload. If your acquisition source has higher than 16-bit resolution, or you are dubbing from an extremely high quality analog source (e.g. 15/30 ips Dolby reel-to-reel tapes) then, provided that your sound card has better than 16-bit analog to digital conversion, 32 bit will give extra quality. Similarly, if it can cope, doubling the sampling rate to 88 200 or 96 000 will further improve the quality but your resulting wave files will be enormous. The floating-point format also allows recovery from higher than 0 dB peaks.

The disadvantage of working at 16 bit is that this is what the domestic consumer has at home. A 16-bit recording will have to have some headroom left in case of unexpected loud peaks. Quiet passages may have to be amplified. This can mean that your final recording is, in parts, only equivalent to a 14 bit or less recording. Working at 32 bits allows you to correct levels, normalize and then reduce to a full 16-bit master.

Setting the level

Oddly programs for audio rarely have input gain controls. The job is usually done either by a utility provided by the sound card, or by the Windows sound mixer. Some installations have an icon on the task bar that will call this up (often via the playback mixer. Audition will call the windows mixer from the menu option *options/windows recording mixer*. You can then select which source you wish to record from. If you have a specialist card, maybe with several separate inputs, then it is likely to have its own manager for routing the audio. Here you will have to tell the editor which card to use.

In Adobe Audition 2.0, this is done using the Edit/Audio Hardware Set-up menu and clicking on the Edit View tab. This will provide information on the card's selection of input and output channels.

Multi-channel sound cards usually present themselves to Windows as a number of separate stereo sound cards. So a card with eight analog inputs and outputs plus a stereo digital in/out will appear as five separate cards. This is for the single wave view mode. In multitrack mode, each track can be separately selected to a sound card input or output (see Chapter 9, 'Multitrack', Page 000).

Because running level meters requires a fair amount of processing time, most editors leave them off except when actually recording (playback is less of a problem as the system already 'knows' what the levels are). It is this processing overhead that makes programmers avoid emulations of analog meter pointers like VU meters and peak programme meters.

In Edit View, Options/metering/monitor record levels (F10) selects the mono or stereo meter in Adobe Audition 2.0.

Recording a file

Having clicked the "New" dialogue, click the record button to begin recording. The waveform of the recorded audio will be drawn on the screen. Just as the waveform is getting towards the right of the screen there will be a redraw to move the window along to keep up with the pointer which will remain static (this can be switched off if you wish). You can 'crash' start into record from a blank window by clicking record and pressing ENTER to the dialog requesting the record parameters. This will use the settings that you last used. The recording is stopped by clicking the stop button.

| 44100 • 16-bit • Stereo | 292.22 MB | 00:28:57:07 | 8.56 GB free | 14:28:41.94 free |

Figure 3.14 Audition disk usage display

At the bottom of the screen can be listed your disk resources (Figure 3.14). In the illustration all the disk related boxes are enabled. The first box shows the sampling rate, bit rate, mono or stereo. The next box shows how many megabytes your recording has used, followed by how many hours, minutes, seconds and frames. These two increment in blocks, rather than by the second or frame, as Audition allocates the space this way. The next two boxes give you the equivalent figures but for how much space you have left (Figure 3.15).

Right clicking on the bar allows you to choose to show only what you are interested in. You have the additional options of showing the data under the cursor; the instantaneous level for that channel. You can see keyboard modifiers; Shift, Control, etc. The lock stability is available if you are using time code synchronization as is the display mode (waveform, spectral) which is probably quite obvious anyway (Figure 3.16).

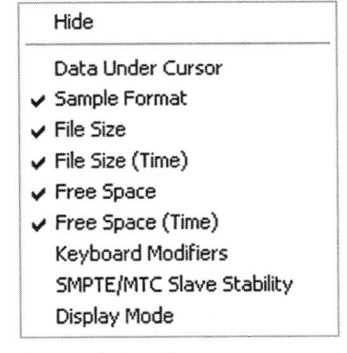

Figure 3.15 Disc usage selection

| R: -1.6dB @ 00:16:17:08 | Shift | 00.0% SMPTE/MTC | Waveform |

Figure 3.16 Time code synchronization

How much space is available?

This raises a vital issue. What is actually meant by how much space is left? Some editors like *Pro Tools* insist that you define where you want to record at the beginning of each session. In its Edit View, Adobe Audition 2.0 records to its own buffer. This is different from where you intend to put the finished recording. An ideal set-up system will put Audition's two buffers on separate disk drives. Finished recordings can be saved to different folders or partitions on the same disks (or, indeed, to other drives on the system). Personally I have two SCSI drives which are used as audio work files with IDE drives used for storage and non-audio programs. However, these days, with so much development the practical advantages of SCSI are probably marginal.

Topping and tailing

Once the recording has been finished, top and tail it by removing any extra material at the front and end. Ends usually sound cleaner if you use a fade out transform on the last half-second or so. This guarantees that your recording ends on absolute digital silence and removes a potential source of clicks when transferring to CD.

Mobile phones

Mobile phones have become almost universal. They are invaluable, especially in a business where gigs can be set up at short notice. However, you should be aware that they can cause severe interference with audio gear. Semi-pro equipment is particularly vulnerable as it uses unbalanced circuits, but even the most expensive professional products can be affected. This why they are banned in hospitals where they can interfere with medical equipment. When you are transferring your audio to the computer, you should avoid having your mobile phones switched on nearby.

Each phone has a radio transmitter and it is this that causes the interference. The interference shows up as burbling noises or even semi-musical arpeggios. Therefore, it is best practice to turn off mobile phones while in the same room as audio equipment in use. It is particularly impolite not to turn off your phone if the audio gear is being used by someone else! Interference can be caused even in standby mode although this will come in short bursts spaced minutes apart.

Essentially a mobile phone has five modes:

1 Off   This is the only safe mode.
2 Calling   The phone is transmitting to attract the attention of the cell transmitter.
3 Ringing   Although the ringing is incoming, the phone is transmitting data to the network to tell it where it is and that it is responding to the call.
4 Talking   The transmitter is sending your side of the conversation to the network.
5 Standby   Here the transmitter is usually off; but every now and then, it wakes up and sends a 'Hi there!' message to tell the network that it is still about. These messages can be even more frequent, if the phone is out of range of the network, as it will be trying to re-establish contact.

4

Editing

4.1 Introduction

Adobe Audition 2.0 is two editors in one, linear and non-linear plus a CD Burner (Figure 4.1). They are toggled between by clicking the appropriate icon at the top of the screen just below the menu.

We'll look first at the linear editor which physically changes the audio files so that data is changed when the files are saved.

When you first run Adobe Audition 2.0, it will open 'Audition-Theme.ses' in the multitrack editor (Figure 4.2).

Figure 4.1 Adobe Audition 2.0 Edit, multitrack and CD Burner icons

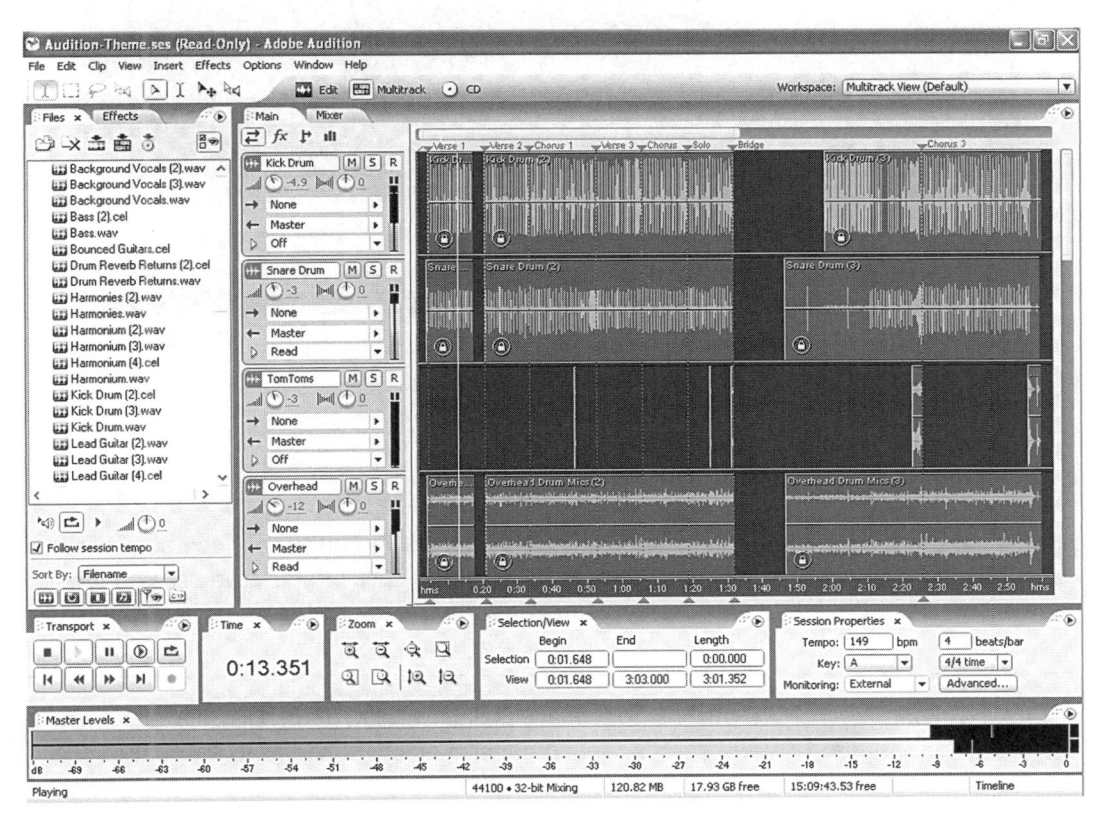

Figure 4.2 Audition-Theme.ses

Apparently this is because, in earlier versions, some customers had not noticed the multi-track editor was there!

If you click 'Edit' in the workspace panel at the top you will see a blank centre screen surrounded by a large number of icons, as shown in Figure 4.3. This should not be a cause for panic! The icons at the top can be ignored until you are familiar enough with the editor to find them useful. The most important controls are the transport and zoom icons at the bottom. We will come to them as soon as we have loaded a file.

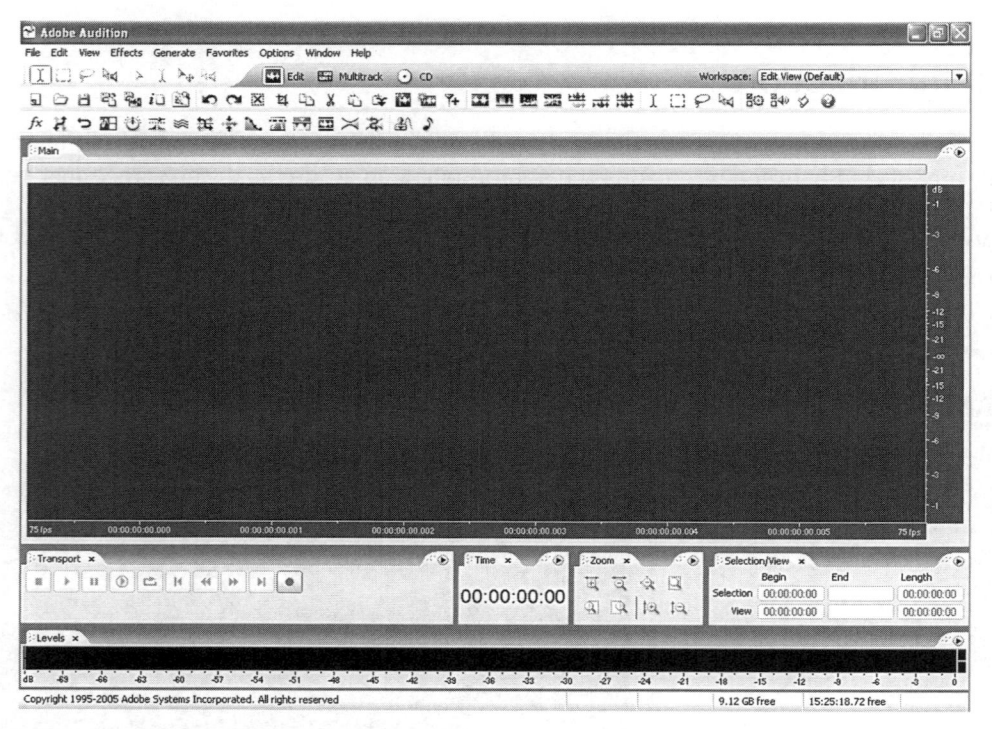

Figure 4.3 Entering Adobe Audition 2.0

4.2 Loading a file

OK, maybe you have not recorded anything yet, but you will find that your computer already has a number of wave files. They have the file extension WAV (if you have set Windows to show extensions). Click File\open... and you get the file selector.

> *To show Windows file extensions*: Open a desktop window (e.g. Disk C) and click tools/folder options. This will open a tabbed panel. Select the View tab and untick the 'hide extensions for known file types' box.

You will find a lot of them in your C:\WINDOWS\MEDIA directory. These are the sounds that Windows can make when it is asked to undertake various actions. You may have turned some, or all, of these off, but the wave files are still there. Indeed you may decide that your first exercise with the audio editor is to make some new sounds to personalize your computer. With Windows XP loading JUNGLE WINDOWS EXIT.WAV gives the illustrated display (Figure 4.4). This is a clap of thunder. (It is possible that it was not copied to your hard disk when it was installed. This depends on what sound scheme options were selected. It is part of the Jungle Sound scheme.)

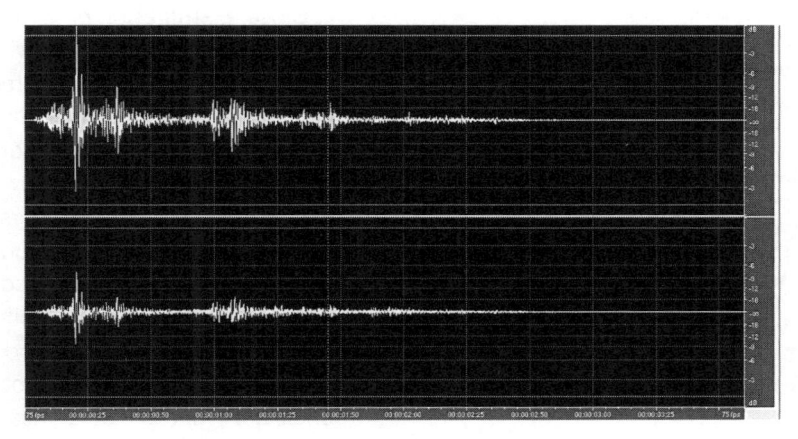

Figure 4.4 Display when JUNGLE WINDOWS EXIT.WAV is loaded

Figure 4.5 shows the status line at the bottom of the screen revealing that the recording uses 324k of disk space, is 3 seconds and 59 frames long, stereo, at 22 050 samples per second (Audition's display in the illustration is set to the CD standard of 75 frames per second. Don't confuse this with the European TV structure which has 25 frames per second).

Figure 4.5 Status line

This particular recording looks as if it is low level but, in fact, it is only 0.12 dB short of full modulation. How do I know this? Because I can get Audition to tell me!

Selecting Window/Amplitude Statistics (Figure 4.6) will give a dialog similar to Figure 4.7. This gives statistics for the left and right channels of the file. Clicking on the arrows will take you to the point, such as Minimum sample value, described. Halfway down the box are the entries for DC Offset. This should be very low – almost zero. Click the 'Account for DC' box and then 'Recalculate RMS'. You should now get the second table showing the peak amplitude of the left channel as being −0.12 dB.

PC Audio Editing with Adobe Audition 2.0

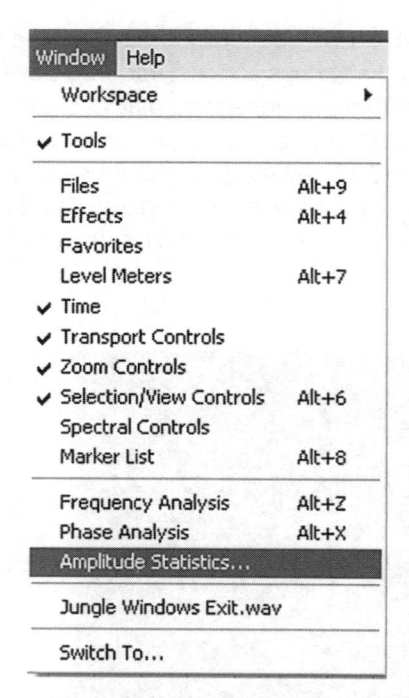

Figure 4.6 Select Window/ Amplitude Statistics

Figure 4.7 Selecting Window/Amplitude Statistics gives the Amplitude Statistics dialog

The Normalize option – Effects/Amplitude/Normalize (Figure 4.8) – allows you to set the normalization level, choose whether to apply DC Bias correction and, for stereo files, choose whether you want left and right channels separately normalized or to keep the level change the same for both (this preserves sound images positions in the sound stage).

Close examination of some waveforms can reveal that they have what is called a DC offset. I have simulated a 10 per cent offset in Figure 4.9. If you look at the detail, you can see that the whole waveform is offset slightly below the centre line. Offsets like this are highly undesirable as they will produce a click every time the track is started or stopped and can cause clicks on edits.

A good quality sound card should not introduce an offset, but a lot of cheaper ones do. This is not a disaster, as the offset can be removed by the editor. There is an offset on the thunder file supplied with Windows.

In Adobe Audition 2.0 this can be corrected as part of the normalization process when the level is corrected to produce full level. It used to have an option to remove DC on-the-fly during recording. This had the disadvantage that if you left this switched on 'just in case' it prevented you from doing bit-for-bit digital clones via Audition as it acts as a low frequency filter.

If you actually want a DC offset you can change the value from 0 per cent to the value you require. This may seem an unlikely need but could be useful to correct an offset manually.

The Decibels format option sets the values entered into the 'Normalize to' box to show decibels rather than percentages. It's conventional to set the level to −0.5 dB rather than 0 dB as, if two samples are either side of a peak the Digital to Analog converter may produce an instantaneous level higher than 0 dB causing distortion in apparatus downstream of the converter.

The reason the audio looks so low level is that there is a brief moment at the start of the thunder where the level virtually reaches the peak. By zooming in on the detail we can see that it is literally one cycle of audio from negative to positive. As the stereo sound is not coming from the centre, the right-hand channel (bottom display) is even lower in level (Figure 4.10).

Short peaks are a common problem with level adjustment. It is one of the prime skills of mastering engineers in the record industry to modify these so that the resulting CD can be made louder.

In this case, we can make a louder file by highlighting the section, reducing it by 6 dB and then renormalizing. We can also, if we wish, untick the Normalize L/R Equally box, to bring up the right-hand channel. Using the Edit/Zero crossings options can ensure that our selection does not stop, or start, in the middle of a cycle, which can introduce a click (Figures 4.10–4.13).

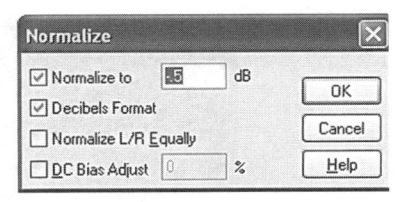

Figure 4.8 The normalize option

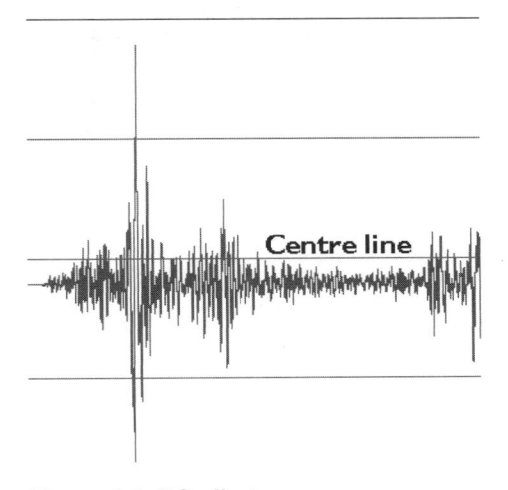

Figure 4.9 DC offset

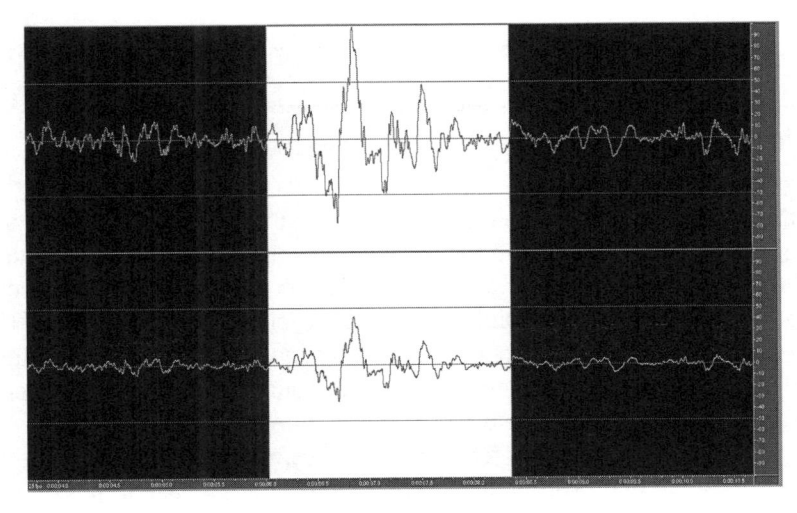

Figure 4.10 Thunder peak selected

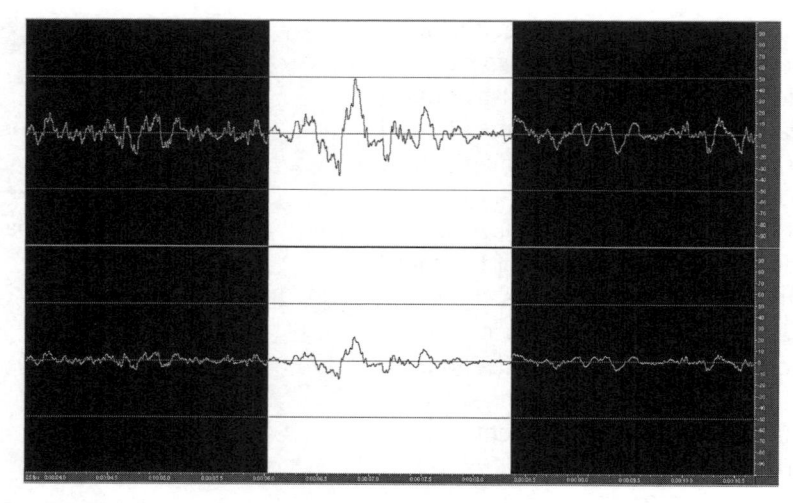

Figure 4.11 Thunder peak reduced by 6 dB

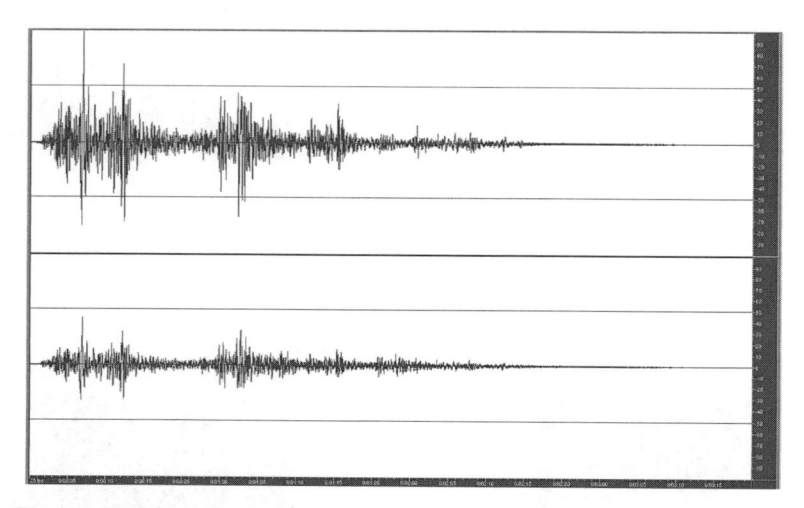

Figure 4.12 Thunder, renormalized with 'normalize L/R equally' set

By doing this manipulation we achieve a louder recording, although the peak level remains the same. However, we have brought up the background noise as well, which may or may not be a good idea. You may like to try this, if you have this file on your hard disk. The process has now revealed two more peaks that are holding down the level. When do you stop? That's where skilled ears come in. A similar process is gone through when music is mastered for CD. Removing a few peaks can make a commercial disc sound louder. This has become rather competitive as no company wants its discs to sound quieter than their rivals.

While a full level recording is ideal for production use, this may not always be the case for the version heard by the end user. Windows sound effects are a case in point; it is generally desirable to set these 6–12 dB below maximum recording level. This allows the PC speakers

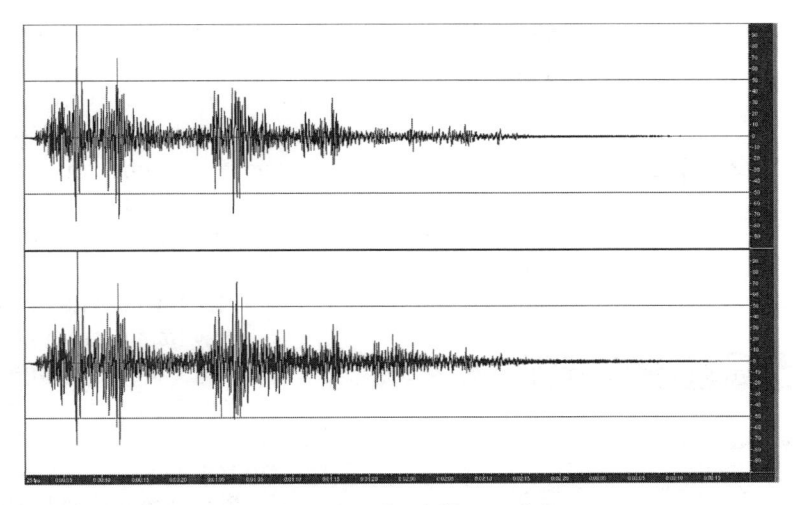

Figure 4.13 Thunder, renormalized with 'normalize L/R equally' not set

to be set at a good level for when they are reproducing music from a CD, or even your editing session, without the Windows sound effects blasting you out of the chair!

4.3 Transport

To modify recordings, we need to be able to navigate round a sound file. Adobe Audition 2.0 presents the whole of the file when loaded. Personally, I find this reassuring and have never got on with other editors that only show the beginning of a loaded file.

Figure 4.14 Transport icons

The keys to navigation are the transport and zoom icons at the bottom left of the screen (Figures 4.14–4.17).

The left-hand block is for the transport and is reassuringly like a cassette recorder, with a few extra options. Stop, Play and Pause are followed by the triangular play icon in a circle – a second play button!

Clicking the standard play triangle symbol plays the file, as you would expect. The triangle in a circle play button will also start the playback but, optionally, in a different way.

The spacebar will "press" the last play button you used which, as they can be given different start and stop points, is useful or confusing

Play View
- Play from Cursor to End of View
Play from Cursor to End of File
Play Entire File
Play Preroll and Postroll (Skip Selection)
Play Preroll and Selection
Play Postroll
Play Preroll, Postroll, and Selection

Preroll and Postroll Options...

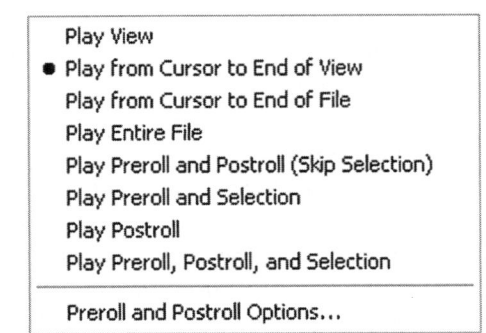

Figure 4.15 Playback options

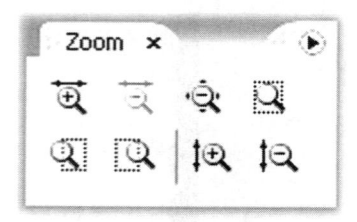

Figure 4.16 Zoom icons

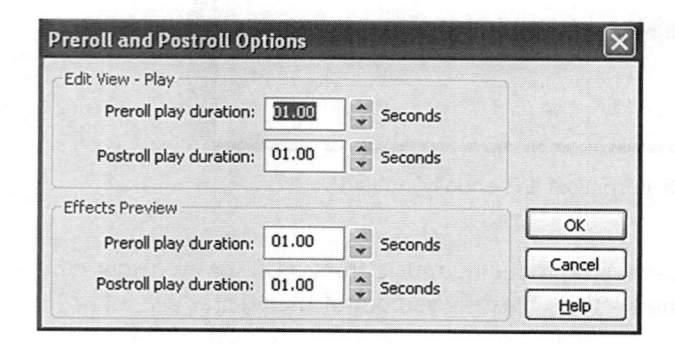

Figure 4.17 Preroll and Postroll options

- Variable (3x, 5x, 10x)
- Variable (2x, 4x, 8x)
- Constant 2x
- Constant 3x
- Constant 4x
- Constant 6x
- Constant 8x
- Constant 2x (reverse 1x)

Figure 4.18 Options

depending on your point of view. The next icon – an arrow pointing to its own tail – sets up a loop mode so that playback never ends but loops round, either to the start of the file, or to the start of the selection. In fact, right clicking on the play buttons gives you a number of options (Figure 4.15) most of which are self-explanatory.

Play View: play the visible part of the file from beginning to end.

Play from Cursor to End of View
Play from Cursor to End of File
Play Entire File: wherever cursor is the playback starts at the beginning of the file and plays through to the end.

Play Preroll and Postroll (Skip Selection): This rehearses an edit by playing from one side of a selection to the other skipping the selection.

Play Preroll and Selection: play from just before selection of end of selection.

Play Postroll: play from end of selection for a few seconds.

Play Preroll, Postroll and Selection: Do not be confused by the word order the playback is from just before the start of the selection to just after the end of the selection.

The double arrow buttons 'spool' the cursor along the file (audibly if in the play mode). Figure 4.18 shows the options revealed by right clicking the buttons. The other spool buttons take you to the beginning or end of the file, or to the next marker if you are using them. The final button, with the 'red LED', is the record button.

The playback 'head' is represented by a vertical line. This can be moved to where you want, merely by single clicking the point with the mouse. If you are already playing the file then a cursor line will appear where you clicked but playback will not be interrupted. Clicking a play triangle will instantly move the playback to the newly selected point. The space bar toggles playback mode on and off. If you are using this, then it toggles through stop, before going to the new location on the next press.

You can select an area by dragging the mouse over the area you want. The alternative is to click one end, with the left mouse button, and then click the other end with the right mouse button. The selected area can be modified with right mouse clicks. The selected edge nearest to the mouse click is the one modified. By default, Adobe Audition 2.0 requires you to press control to change the selection with a right click, otherwise it gives you a popup menu.

This can be reversed by clicking the Extend Selection option within the Edit View Right clicks section in the General Tab of the Edit/preferences dialog. This reversed operation was how the original Cool Edit worked and is, in my view, preferable.

You can also select the edit area 'on-the-fly' as you play the file. There are two keyboard shortcuts called 'Selection anchor left when playing' and 'Selection anchor right when playing'. 'Out of the box', Cool Edit sets these to the '[' and ']' keys. This means that if you press th'e '[' key while you are playing a file the start of the selection will be set to the point playing at that moment. At the same time, the point where Cool Edit starts playing from when play is pressed is moved there as well. Pressing the ']' key marks the end of the selection. Stop, then by pressing 'Delete' you will delete that selection. Pressing the spacebar will restart play from your edit. You can then either 'spool' back or use a preroll option and check from before your edit. The spool controls are not allocated a keyboard shortcut by default but can be (see Section 16.3 on Keyboard shortcuts in Chapter 16 'Tweaks'). Being able to use the keyboard, rather than the mouse, can vastly improve efficiency of a session where a great deal of edits are made. This is more typical of a speech editing session than a music one. This could transform the use of the program for blind users who can now edit by ear rather than being frustrated by the visual nature of PC audio editing, but, unfortunately Audition breaks some existing conventions.

Zooming

The Zoom panel has eight icons controlling how the audio is seen (Figure 4.19):
From left to right on the top row:

- + zoom in horizontally
- − zoom out horizontally
- Zoom out full both axes
- Zoom to selection.

From left to right on the bottom row:

- Zoom in to left edge of selection
- Zoom in to right edge of selection
- Zoom in vertically
- Zoom out vertically.

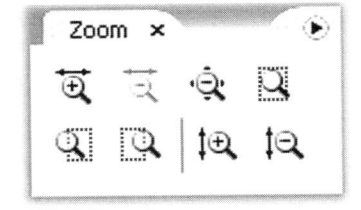

Figure 4.19 Zoom Panel

It is very easy, quickly to examine a section of audio. Click on the point you are interested in. Define a selected area by dragging or using the right mouse button and click on the 'Zoom to selection' icon. You can repeat this as many times as you like to get into really fine detail. You can even set the magnification to be so large that you can see the individual samples (Figure 4.20).

These are represented as dots. They are joined by a line generated by the computer. This line is not simply dot-to-dot for cosmetic reasons, but will show what will happen to the audio after the output filter.

If necessary you can move the dots, literally to redraw the waveform. This can sometimes be the only way to eliminate a very recalcitrant click that has slipped through declicking software.

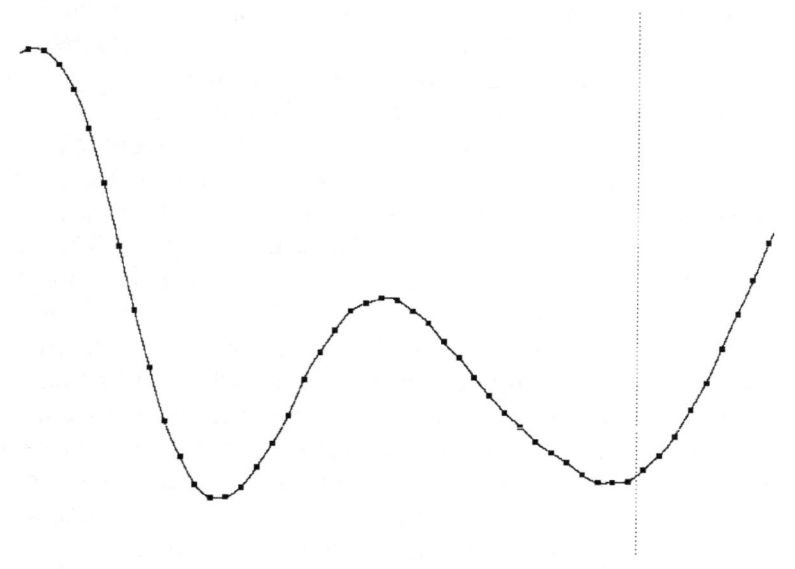

Figure 4.20 Waveform zoomed-in to show samples

In Figure 4.21, I have drawn a five sample click this way, to illustrate that the line joining the dots is more intelligent than it first appears; the wavy lines show what is known by audio engineers as 'ringing'.

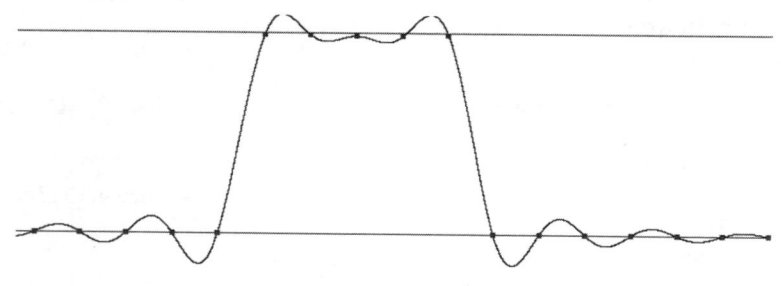

Figure 4.21 Five sample square pulse drawn onto waveform showing computer's interpretation of the result

4.4 Making an edit

Editing on a PC is best done visually. Some editors (including Audition 2.0) have a 'scrub' editing facility where the sound can be rocked backwards and forwards as on a reel-to-reel tape machine. People used to quarter inch tape ask for this, probably out of fear, or sentiment. In practice, nearly everybody edits visually. The technique is very simple and the eye and the ear work together easily. It is much easier to learn, from scratch, than is scrub editing. Very soon you will be able to read the waveform. For example, you may know a particular point is overlapped three times. It is often possible to spot the individual retaken sections and select the last one without listening to the intervening audio. If, on performing the edit, you find that you are wrong then UNDO quickly restores the cut. By making 'bets' like this you can considerably speed up the editing process, particularly with scripted speech editing.

Editing audio is surprisingly similar to editing text in a word processor. If you use a word processor, then you already know the basics of editing audio. You use your mouse to select a start and finish point. The space between is highlighted by reversing the colors. Pressing DELETE removes the highlighted section. Alternatively clicking Edit/Trim (or Control/T) will eliminate everything except the selected area (ideal for 'topping and tailing').

As already described, the selected section can be created in many ways. The most obvious is to place the mouse at the start of the section you wish to cut and drag the pointer to the end of the cut section. Audition also allows you to select the other end of the selection from the cursor by right clicking on the point (while pressing control unless you have reversed the action of the right click). This works forward or backward.

4.5 Visual editing

This is all very well but how do you find a phrase within an item? Here you use the zoom facility already described.

Figure 4.22 shows a short section of speech 'The um start of the sentence'. The text has been added to the picture and is not a feature of current audio editors!

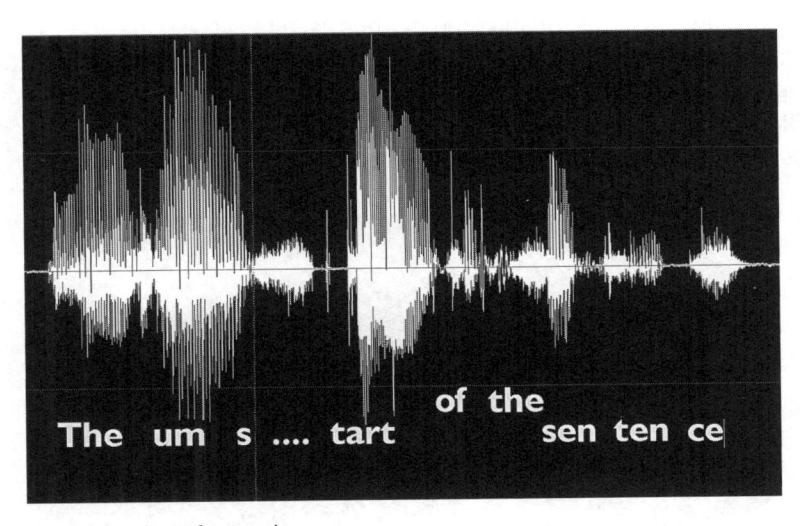

Figure 4.22 A short section of speech

How do we go about removing the 'um'? Clicking the play button with the mouse (or pressing the space bar) will play the section. A vertical line cursor will traverse the screen showing where the playback is coming from. Your eye can identify the section that corresponds to 'um'.

With the mouse click at the start of this section. The cursor will move to this point.

Press play again and the playback will start from where you clicked. You can hear if you have selected the right place as, if you have not, there will be something before the 'um', or it will be clipped. Reclicking to correct is very fast.

Now identify, not where the 'um' ends, but where the next word starts – the 's' of 'start'. This time click with the RIGHT mouse button. This sets the end of the section to cut (Figure 4.23).

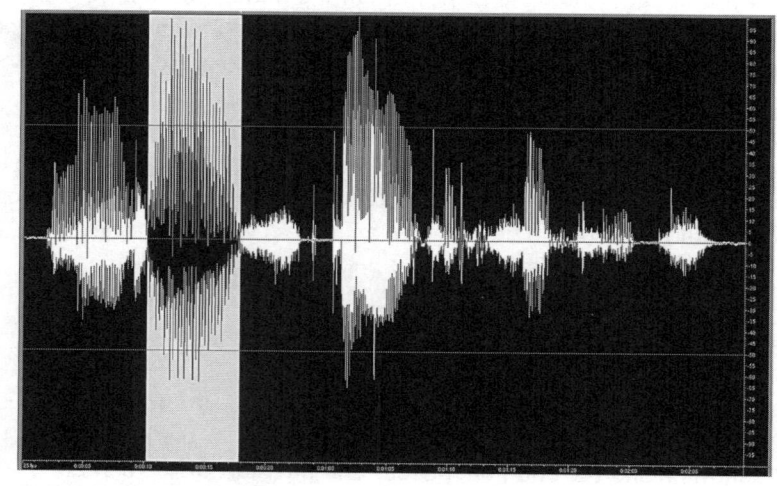

Figure 4.23 'Um' selected for removal

Hit play again and just the highlighted section will be played. Is the end of the 'um' clipped? Can you hear the start of the 's'? If not then the edit is probably OK. Within Audition 2.0, pressing shift/R will rehearse the edit by playing before the edit (preroll) and after the edit (Postroll). Please note that it is the key presses that are required not just capital 'R'.

Now press the delete key. The highlighted section is removed (Figure 4.24).

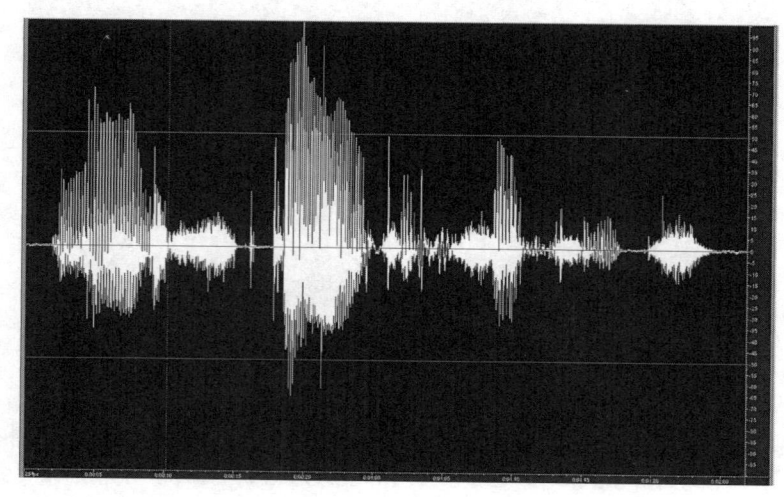

Figure 4.24 'Um' removed

Click a little before the join and play the edit. Is it OK? If so go on to the next edit.

If it is not, then click UNDO and you will be restored to the previously highlighted set-up. You can adjust the edges of the selected area by right clicking near them.

At first, you may well have to cycle round several times, but soon you will be getting edits right first time. Remember even redoing the edit three or four times is still likely to be quicker than it used to take with a razor blade.

5

Quarrying material

5.1 Introduction

Returning to base with material of exactly the right duration and needing no editing is a rare occurrence. If your recording is of a continuous event then simply cutting it to time is all that is required.

With unscripted speech-based items, you can return with substantially more than you will ever use, especially if you have recorded a series of interviews which you plan to 'quarry' out as extracts, rather than use them complete.

However, I cannot stress too much that the very best way to edit an interview is to have asked the right questions in the first place! A well-structured interview will also be easier to edit for duration. However, circumstances can force returning with far too much material, especially if vox pop (in the street interviews) have been recorded.

Once upon a time there was no alternative; quarter inch tape had been used and the only way to edit was with a razor blade, wax pencil and sticky tape. Copy editing added extra time and was avoided.

5.2 Blue bar blues

No matter how fast your computer and its hard drives, audio files consume a vast amount of data compared with a word processor, and take time to be saved. Most editors have a very useful UNDO facility. This works by saving a copy of the section of the file you are changing onto hard disk.

All this takes time and slows down the editing process. This means that you should structure your material so as to avoid long audio files of more than 3–5 minutes. Put plainly a single edit within a 30-minute file takes much longer to do because of the time taken to save that long file. A 5-minute file takes one-sixth as long to save. A strategy of one or two files per item within a longer piece will speed things along nicely. Having said this, by using clever programming, Adobe Audition avoids many of the delays. Even so you can be caught by the 'flushing buffer' alert when saving files. This is where Audition is rescanning the file and reorganizing itself before actually saving the item and can double the time you were expecting it to take.

Extracting sections of a file to new separate files is most easily achieved by highlighting the section and using the 'File/Save Selection' menu option.

Where you just want to copy and paste, then use the clipboard where, as with all modern operating systems and programs, Control/C will save the selected area. Control/X will cut it out and Control/V will paste it to the current cursor position. Adobe Audition, as well as using the Windows clipboard, also provides you with five clipboards of its own. These can be selected by using Control/1 to Control/6 inclusive with Control/6 being used to select the Windows clipboard. This is slower but allows cutting and pasting to other programs. They can be set to keep the material when the program is closed and, for example, could be used to store commonly used audio 'stings'.

However, there will be things you will want to do to long files which will take time, such as noise reducing an archive recording. It is inevitable that you will spend some time staring at the screen as the blue progress bar slides slowly from 0 to 100 per cent. Ironically, the time is much less than used to be taken with the mechanics of razor blade editing. However, it feels much longer because you, personally, are not doing anything while the process is taking place.

You can improve your efficiency by having a number of other tasks that you can do while the processing is going on; telephone calls to make, facts to check, or even letters to write, as the processing can carry on in the background while you use your word processor on the same PC. You can push Audition into the background by clicking the triangle symbol in the top right corner of the process alert (Figure 5.1).

Some routine operations, such as normalization, can often be done using 'batch' files. These allow you to leave computer to get on with the task while you do something else.

You will very often want to load many files at the same time. Please remember that the Windows file selector allows you to select many files which will all be loaded when you click OK.

Figure 5.1
Process alert background icon

One way is to use the 'rubber band' method of multiple selection (Figure 5.2).

The mouse point is placed by a track and the left mouse button pressed and held down. With the left button still held down, moving the pointer will produce a rectangular 'rubber band' which will select any files included within it. You can include all the available files by starting the rubber band at the first file and then dragging downwards and rightwards. The file selector window will scroll the list so that you can include everything.

If you want to select a number of individual files then click on each file you want (Figure 5.3).

Normally selecting a new file will deselect the previous selection, but if you hold the control key down while you do this then this will not happen. The control key also introduces a toggle action so that clicking on a selected file will deselect it. This is particularly useful if you want to make a minor modification to a number of files as, once loaded, switching between them is rapid. Once you have made all the changes you can SAVE ALL and all your changes will be preserved.

Obviously, loading and saving a large number of files takes an appreciable time but by triggering this with one action you can be doing something else that is useful while this is happening. If you are saving 10 files each taking 15 seconds to transfer you have given yourself 2½ minutes to make a phone call or whatever.

Many programs will include material from CD. This can be commercial music but also sound effects, or even from your own personal archive. Here considerable time can be saved

PC Audio Editing with Adobe Audition 2.0

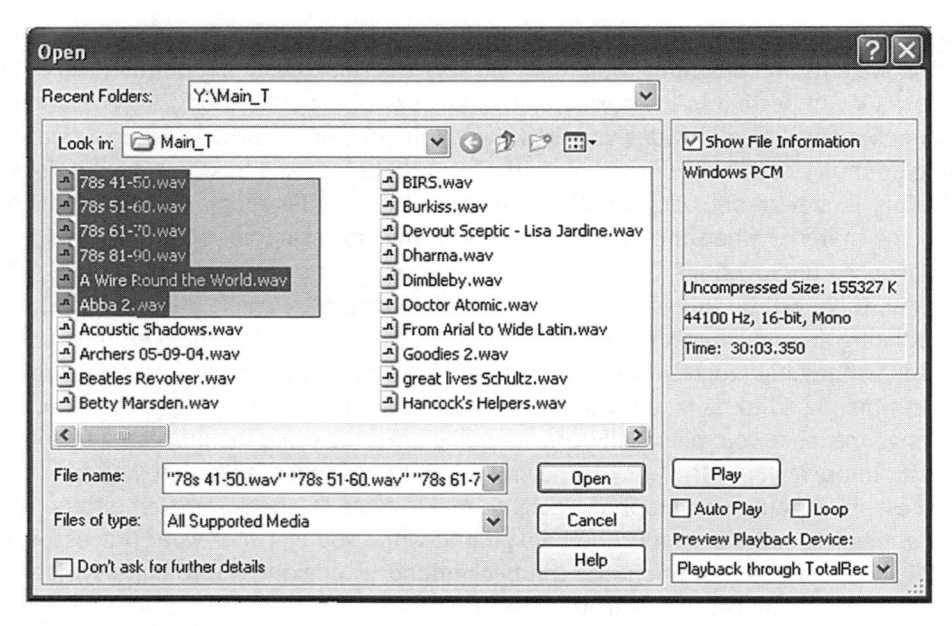

Figure 5.2 File selection using 'rubber band'

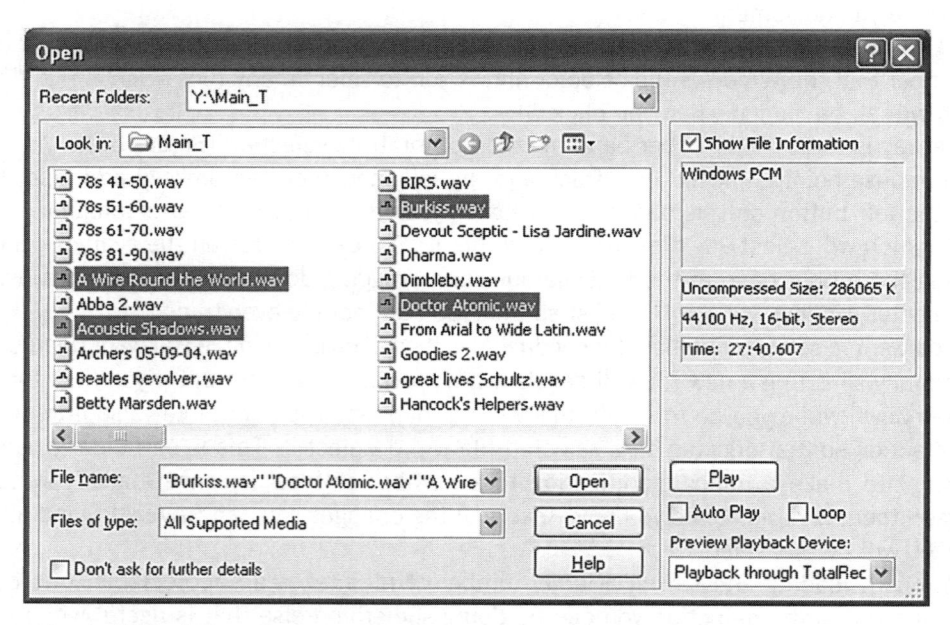

Figure 5.3 Individual file selection using mouse with control key held down

as a properly set-up computer will be able to 'rip' the audio data from the CD much faster than real time. This means that 10 minutes of material can be transferred in less than 1 minute. You don't even have to listen to it! The quality will be better as the audio is not converted back to analog and then back to digital again by your sound card.

The easiest way is to use a background program that modifies the Windows desktop so that audio CDs appear in the desktop window with their tracks showing as .WAV files. These are usually named 'track1.wav', 'track2.wav', etc. This allows you to copy the audio files as if they are normal data files. Desktops without this feature will open a CD drive containing audio, but the tracks will be shown as 'TRACK1.CDA', etc. They are not normally directly accessible as audio. Adobe Audition has CD Ripping software built-in and can access the CD audio directly using the .CDA files as if they were the actual audio files.

5.3 Copy, cut and paste

Adobe Audition gives you the option of six separate clipboards, as mentioned. It has five of its own, plus the standard Windows clipboard. These are selected, whether through the 'Edit/Set current clipboard' menu option or by using control/1 to control/5 for Adobe Audition's own clipboards and control/6 for the Windows one.

This means you can work with multiple pieces of audio 'in memory' at the same time; so you can, for example, copy different jingles or link music sections to each clipboard, and place them in your file at chosen locations. The current clipboard can also be set to be the Windows clipboard. This is available to other programs and is a convenient way to copy audio from Adobe Audition to another program or vice versa.

These internal clipboards save audio in your temporary directory as wave files, and they can be retained even after Adobe Audition closes. The 'Delete clipboard files on exit' setting in the Edit/Preferences/System tab switches this on or off.

As well as using the normal paste function to insert material you can create a new file from the clipboard using Edit/Paste To New (Shift/Control/N). Save selection will usually be quicker as you do not have to copy to the clipboard first. Edit/Copy to New will create a copy of the entire current file with (2) or (3), etc. added to the end. This is in memory and makes no assumptions about the format. It is down to you to choose when you save it.

Edit/Mix Paste (Shift/Control/V) will add material to the file on top of existing audio; the length is not changed except if the insert option is used. If the format of the waveform data on the clipboard differs from the format of the file it is being pasted into, Adobe Audition converts it before pasting.

There are four mix paste modes (Figure 5.4):

- *Insert*, inserts the clipboard at the current location or selection, replacing any selected data. If no selection has been made, Adobe Audition inserts clipboard material at the cursor location, moving any existing data to the end of the inserted material.
- *Overlap*, the clipboard wave does not replace the currently highlighted selection, but is mixed at the selected volume with the current waveform. If the clipboard waveform is longer than the current selection, the waveform will continue beyond the selection.

- *Replace*, will paste the contents of the clipboard starting at the cursor location, and replace the existing material thereafter for the duration of the clipboard data. For example, pasting 1 second of material will replace the 1 second after the cursor with the contents of the clipboard.
- *Modulate*, modulates the clipboard data with the current waveform. I suspect that this option is here because they can do it. It is yet another way of producing weird noises. This has the potential to do magical things. The classic use for this sort of facility is to make every day sound talk by modulating them with speech. However, the Dynamics Effects Transform can generate a modulation envelope. This can be used to modulate the level of a file as if it were being compressed using the audio from another file.

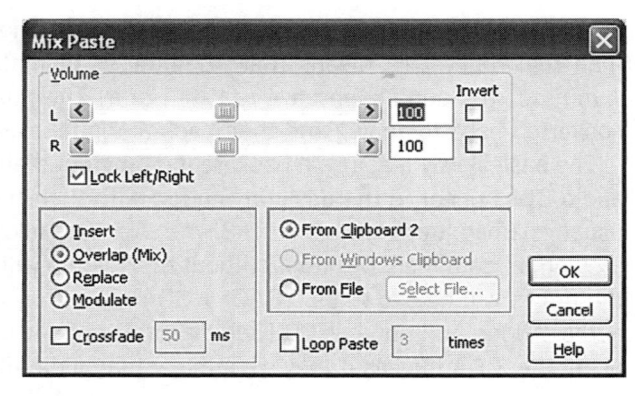

Figure 5.4 Four mix paste modes

Loop paste allows you to multiply paste the same clip. This can be useful for music samples or for extending backgrounds and atmospheres for documentary work. You can choose to copy from the current Adobe Audition clipboard, Windows or a file. The cross fade allows you to smooth the transition of the mix.

The invert tick boxes turn the waveform upside down. Sometimes mixes work better if this is done. You can use this to compare nominally identical copies. By making sure that the starts match, mixing the two, with one file inverted, *should* give total silence. Any noise or audio that is left corresponds to errors. It can be fascinating to do this when one of the files has been created using a lossy method, Minidisc for example. What you end up with is the audio that was 'thrown away'.

6

Structuring material

6.1 Standardize format

A simple way to split a large file is to mark it with regions corresponding to the separate files you want to obtain. This is done by using the mouse to select a section of audio and then pressing F8. This will create a 'basic' region (Figures 6.1 and 6.2). F8 combined with shift or shift/control produce different types of regions used when burning CDs. There is also a 'Beat' and 'phrase' marker that can be used when generating music loops using options available under 'Edit\ Auto-Mark\…' (Figure 6.3).

You can trim each end of a region by dragging its marker to the right place. When you select View Marker List you will then see your selections named 'Marker 01' upwards. Select the markers you want to turn into files. This is done in the same way as the Windows file selector. Hold down control individually to select a marker or hold down shift to select all the files from an initial selection to where you click. Press the batch button and you will see a new dialog.

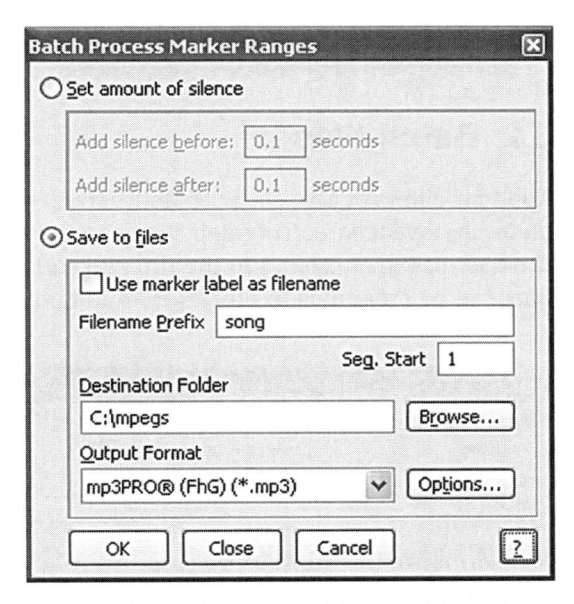

Figure 6.1 Batch process dialog used to split a long file into many small ones

Figure 6.2 Marker list of basic regions

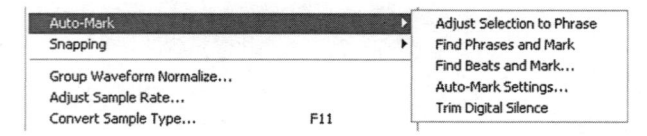

Figure 6.3 Edit\Auto-Mark\...

There are two options. The first will insert a set amount of silence before and after each marker. This can be useful to pad out the 0.7 seconds that are often missed at the start of MP3 files. In this case, we want the second option 'Save to files'. Here you can set the type of file that will be output and to where on the hard disks on your system. The 'filename prefix' is what they will be named with '01', '02', etc. added on to the end. In the illustration, I have called the prefix 'song' and selected 'C:\mpegs' as the output path. The resulting files will be 'C:\mpegs\song01.mp3', 'C:\mpegs\song02.mp3', 'C:\mpegs\song03.mp3', etc.

The files marker names can be edited and these changed names can be output as file names by ticking the 'Use marker label as filename' box. For example, I will transfer several students work from Minidisc in one go and then divide up the different songs as regions naming each marker with the student's name.

6.2 Batch files

Audition allows you to set up batch files to automate some activities. Using batch files is quite an advanced activity and they can take a while to get right. This means that, in the short term, they may add to the time taken to edit an item. However, once they are sorted, they can be great aids to efficiency. Audition's batch files contain sets of scripts which can be grouped into three types (Figure 6.4).

The first type 'Script starts from scratch' creates a new file. The first command has to be File/new (Audition will impose this if you forget). This would be useful if you had a standard setup you wanted to create regularly. An example might be to create a file at a particular sampling rate and bit resolution. Generate 20 seconds of line-up tone followed by 10 seconds of silence.

The second type 'Script works on current wave' is where the same sequence of treatment is applied to the whole file and you can select a series of files

Figure 6.4 Scripts and batch processing dialog showing typical range of scripts

to be actioned. This means that you can run a series of effects transforms that takes several hours to do in total but, instead of several hours of blue bar blues, you can be off doing something more interesting, such as lunch.

The last type 'Script works on highlighted selection' is one that is going to be applied to part of a file and consists of a series of treatments, say, equalization followed by normalization followed by 6 dB of hard limiting.

Scripts and batch processing can be found under the 'File' menu heading.

The scripts are, literally, text files containing instructions in a sort of programming language. But don't panic! You create scripts, not by writing them, but by getting Audition to record what you do to a file. You can save this and recall it in the future.

To use an existing script, click on its name and then click 'Run Script' (Figure 6.2).

To record your own script is simply a matter of entering a new name in the New Script title box. This will enable the record button. Which type of script you generate is controlled by your starting point in the editor. If there is no file then the 'Start from Scratch' mode is assumed. If a section of the file is highlighted then a script for action on a highlighted section is created. A file with no selection creates the sort of script that can be applied to a series of different files.

Clicking 'record' starts the process. You now perform the actions you want to be able to repeat each time the script is run. If you had ticked the 'Pause at Dialogs' button then when the script is run, the dialogs will appear for you to enter settings (or press Enter to OK them). With the box not ticked then the values you enter as you record the script are used. It makes sense to use a short test file to create scripts, so you do not have long to wait for each action to be processed during the record process.

The 'Execute Relative to Cursor' is an advanced option and is used when operating 'Works on Current Wave' script types. With this, you can have all script operations performed offset from your cursor position. If you record the script with the cursor set 10 seconds in from the beginning, then, when the script is run it will operate 10 seconds after your current cursor position. Recording with the cursor at zero is a way of producing a script that works from the current cursor position, rather than using a highlighted section. This might be used for generating line-up tone sequences, for example.

As you would expect, the script recording is stopped by clicking 'Stop current script'. The '≪Add to Collection≪' button is now enabled so that you can save what you have done, or you can cast it to oblivion by clicking "Clear'. This is the best option if you have made a mistake.

For most people who are not Techies, redoing the recording is preferable to clicking the 'Edit Script File' button, as this dumps you into Notepad and a raw text file that is the actual batch file of commands for the whole collection of scripts.

Where this can be useful is when you want a set of near identical scripts. For example if you want to add a bit of silence to the front, normalize and then add a variable amount of hard limiting. In the illustrated case (Figure 6.5) this is a script I have used for preparing speech recordings of audio books that are going to be transferred to MP3 files. I have four files that change the amount of hard limiting to give values of 6, 8, 12 and 16 dB of hard limiting. The scripts are identical except for the value on the line starting '2:' under 'amplitude\hard limiting'. Rather than record four times, with different values, I copied the section

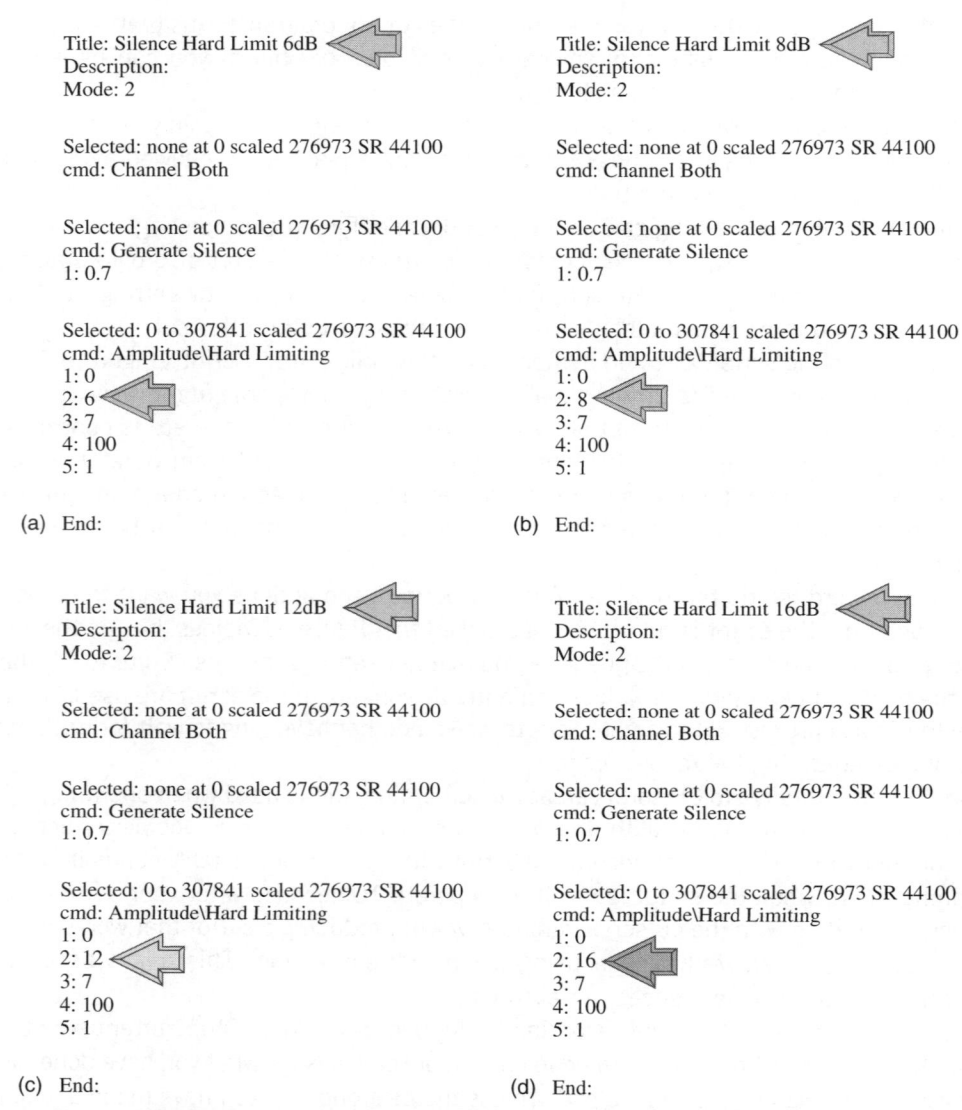

Title: Silence Hard Limit 6dB
Description:
Mode: 2

Selected: none at 0 scaled 276973 SR 44100
cmd: Channel Both

Selected: none at 0 scaled 276973 SR 44100
cmd: Generate Silence
1: 0.7

Selected: 0 to 307841 scaled 276973 SR 44100
cmd: Amplitude\Hard Limiting
1: 0
2: 6
3: 7
4: 100
5: 1

(a) End:

Title: Silence Hard Limit 8dB
Description:
Mode: 2

Selected: none at 0 scaled 276973 SR 44100
cmd: Channel Both

Selected: none at 0 scaled 276973 SR 44100
cmd: Generate Silence
1: 0.7

Selected: 0 to 307841 scaled 276973 SR 44100
cmd: Amplitude\Hard Limiting
1: 0
2: 8
3: 7
4: 100
5: 1

(b) End:

Title: Silence Hard Limit 12dB
Description:
Mode: 2

Selected: none at 0 scaled 276973 SR 44100
cmd: Channel Both

Selected: none at 0 scaled 276973 SR 44100
cmd: Generate Silence
1: 0.7

Selected: 0 to 307841 scaled 276973 SR 44100
cmd: Amplitude\Hard Limiting
1: 0
2: 12
3: 7
4: 100
5: 1

(c) End:

Title: Silence Hard Limit 16dB
Description:
Mode: 2

Selected: none at 0 scaled 276973 SR 44100
cmd: Channel Both

Selected: none at 0 scaled 276973 SR 44100
cmd: Generate Silence
1: 0.7

Selected: 0 to 307841 scaled 276973 SR 44100
cmd: Amplitude\Hard Limiting
1: 0
2: 16
3: 7
4: 100
5: 1

(d) End:

Figure 6.5 Four files that change the amount of hard limiting to give values of (a) 6 dB; (b) 8 dB; (c) 12 dB and (d) 16 dB

from 'Title:' to 'End:' and pasted it in three times to the text file. I then edited each section to have a different title for the script and changing the hard limiting value.

File/Batch processing gives you access to copying and changing files with or without running scripts.

The Batch Processing dialog will appear. It has five tabs along the bottom. The first is 'Files'. Click 'Add files' to get the file selector and add the files you want to the list (don't forget that the Windows file selector allows you to select more than one file at a time).

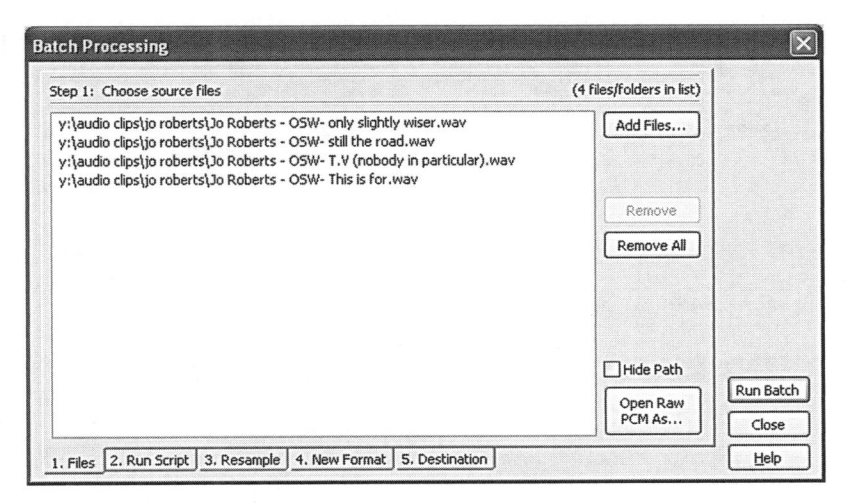

Figure 6.6 Adding and deleting files from the batch process

If, having loaded a number of files, you see some that you realize you should not have included then you can highlight them with the mouse and remove them by clicking the 'Remove' button (Figure 6.6).

Figure 6.7 Selecting a script

The second tab 'Run Script' allows you insert a script by ticking the 'Run a script' box. The browse button allows you to select a script collection file and then the drop down box gives you access to the actual script (Figure 6.7).

The third tab allows you to change the sample rate, etc. when you have ticked the 'conversion settings' box. The 'Change Destination Format' button produces the comprehensive

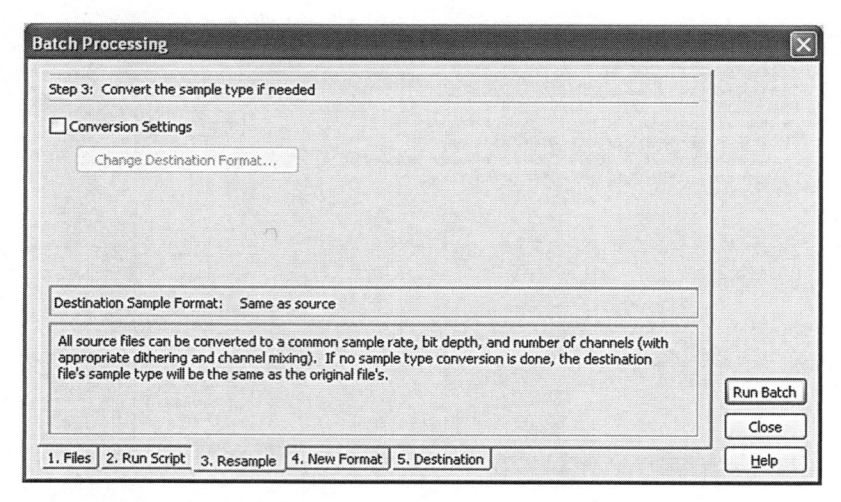

Figure 6.8 Selecting a sample format

'Convert Sample Type' form which is also produced in the Edit view by F11 or 'Edit/Convert Sample Type' (Figure 6.8).

A table gives you a choice of standard sampling frequencies or you can type in a nonstandard one. Normally resampling will involve pre and/or post filtering but this can be switched off (probably to demonstrate why you need it!). You can convert between stereo and mono, controlling how left and right channels are mixed. Dither settings can be altered and a bit depth of 8, 16 or 32 bit can be selected. The Low/High Quality slider adjusts the quality of the sampling conversion. Higher values retain more high frequencies (they prevent aliasing of higher frequencies to lower ones), but the conversion takes longer. Lower values requires less processing time but result in certain high frequencies being 'rolled off', leading to muffled-sounding audio. Adobe advise that values between 100 and 400 are fine for most conversion needs (Figure 6.9).

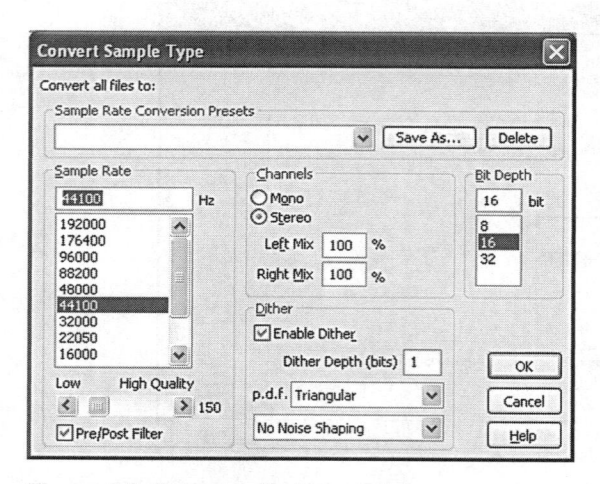

Figure 6.9 Selecting file format dialog

The fourth tab selects the new file format (MP3, .WAV, etc.). If you have not set the previous tab to resample the files to a common format then this tab will scan the selected files and produce a list of the different formats used. By selecting from this list you can change the option for each list (Figure 6.10).

The fifth tab controls the destination of the batch copies files. The simplest option is simply to copy them into the source file's folder. This can be messy and as like as not it is more

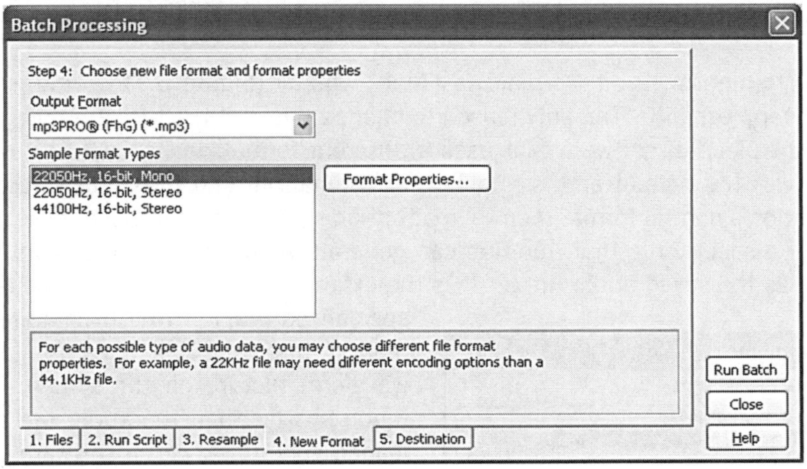

Structuring material

Figure 6.10 Selecting file format options

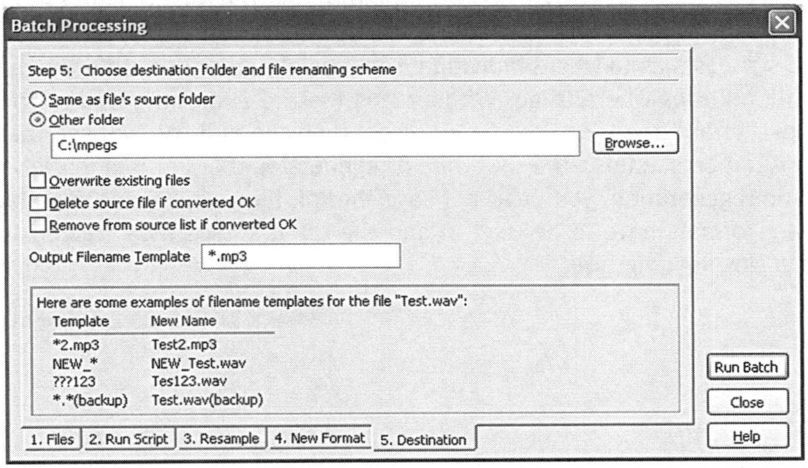

Figure 6.11 Selecting file names

convenient to create a folder specifically for the batch copied files and select it with the browse button. Ticking the 'overwrite existing files' is safe in these circumstances and more convenient when you decide to redo the sequence. Personally I would never tick the 'Delete Sources file if converted OK' option; it is far too dangerous. The 'Remove from the Source list' can help you keep track of things in a busy session (Figure 6.11).

Finally the 'Output Filename Template' is a powerful way of adjusting the output filename. You can force another extension, or alter the filename itself (the portion before the '.'). There are two characters to use when altering the Output Filename Template: a question mark '?' which will signify that a character does not change and a star '*' which will denote the entire original filename or entire original file extension. The destination panel gives some useful examples.

6.3 32-bit files

The normal format is 32-bit Normalized Float (Type 3) (Figure 6.12). This is the format Audition is happiest with. The only reason to change this is if the files are going to be used by a different piece of software that uses a different format. In its Cool Edit days '32-bit 16.8 float' was the default and is retained for compatibility. If the other software will not accept Audition's normal format then try that option next. This is because these two formats make use of a second file that Audition can generate when saving audio. This file has the same name as the saved audio file but has the extension .pk. This 'peaks' file is actually a specialized graphic file. It contains the information that Audition uses to generate the waveform picture on the screen. This file is much shorter than the audio file and can be loaded very quickly. The software checks that the .pk file was actually generated at the same time as the audio file. If it is valid then the 'picture' of the audio can be flashed onto the screen extremely rapidly. It also provides the software signposts as to where individual sections of audio are within the real audio file so that Audition knows where to go. Without this file the audio will load much slower. Try deleting a .pk file and then loading the corresponding file and you will see the difference. You will see Audition scanning the file while it regenerates the file. Incidentally, because of this automatic regeneration, you need not save the .pk files when you are archiving a project. The other formats have to be read in and the .pk graphics files created from scratch which gives a slow loading time.

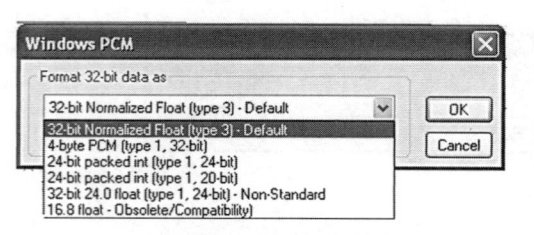

Figure 6.12 32-bit options for .WAV files

7

Multitrack

7.1 Loading

When you first install Adobe Audition 2.0 it loads the 'Audition Theme'. This is a short demonstration of what it is capable of and will reward study of how it is set up. Figure 7.1 is how it shows on a 1024 × 768 laptop. Given a larger resolution screen more information can be shown. Figure 7.2 shows the same opening screen but as shown on a 1600 × 1200 screen.

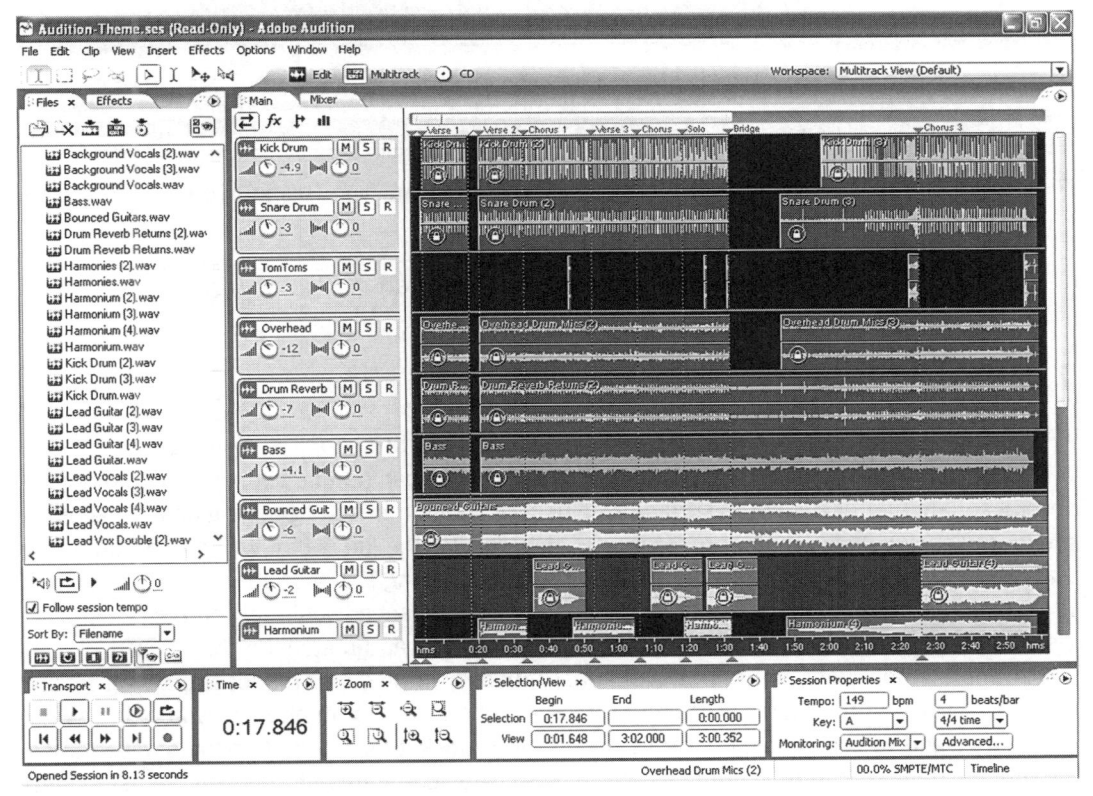

Figure 7.1 The start-up screen after installation showing the 'Audition Theme'

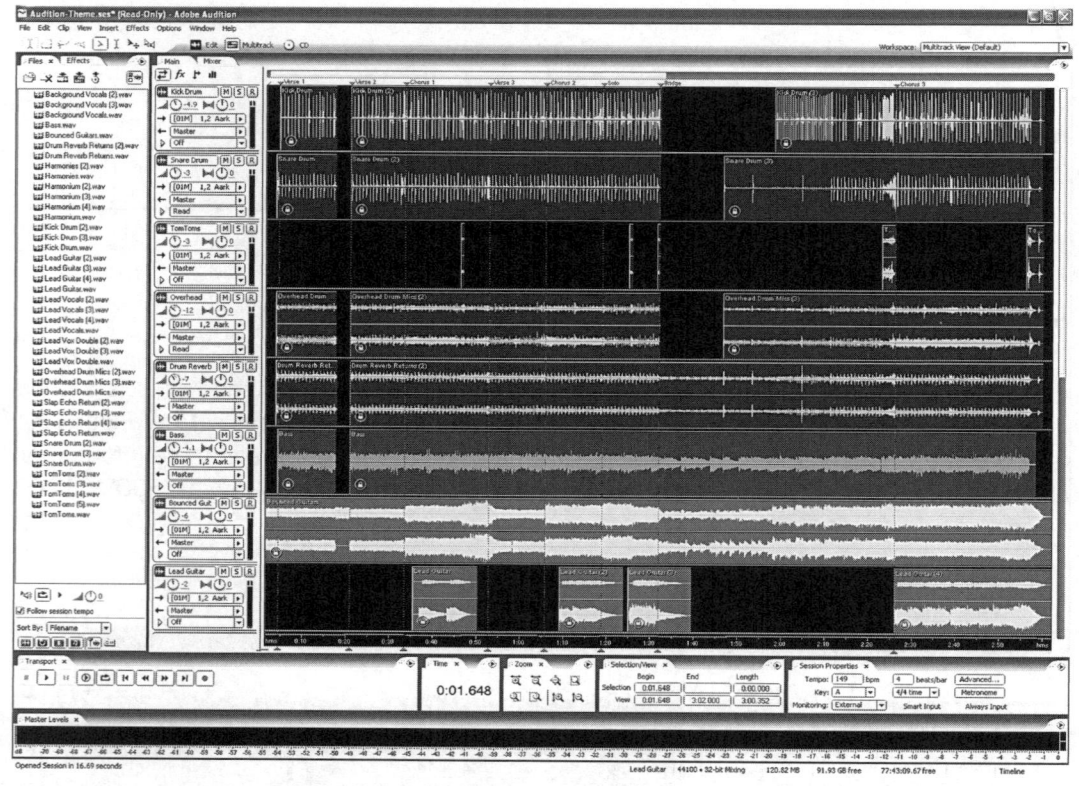

Figure 7.2 The start-up screen after installation showing the 'Audition Theme' 1600 × 1200 screen

But, at first, even a blank new session can look intimidating (Figure 7.3).

7.2 Tour

A simple music balance such as in Figure 7.4 can look intimidating at first but, fact, it is a minimalist portrayal of what is happening. Adobe Audition can handle a mix that would require several square metres of control surface area using a traditional analog mixing desk. One of the great problems with a computer realization of the tools for the job is that you are limited by the size and resolution of the screen and, as like as not, a single mouse to control the process in conjunction with the keyboard. An experienced music balancer would rely on their peripheral vision to detect problems from hundreds of lights, displays and meters on their mixer.

Fortunately not every job is a 100-channel mix of a worldwide broadcast giga-gig involving the glitterati of the music world and Adobe Audition is adept in allowing you to hide what you don't need to see with its controls divided into dockable, resizable panels and frames. On the one hand, it can make full use of the two-screen capability of Windows XP using large high-resolution monitors and, on the other, fit a single laptop screen.

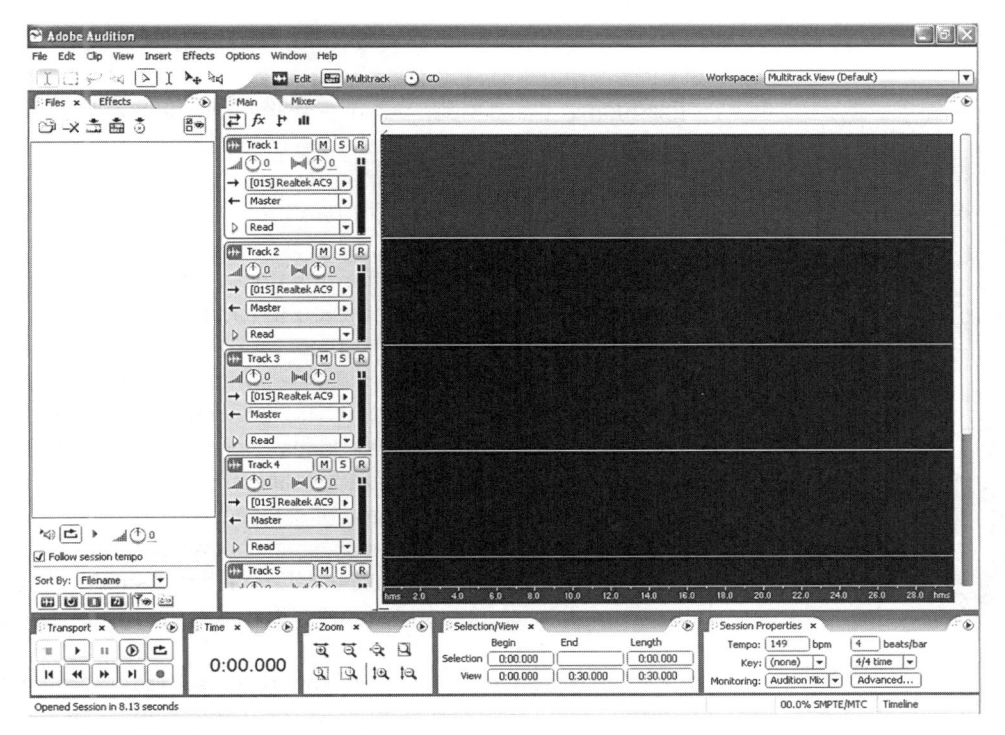

Figure 7.3 Multitrack new session view

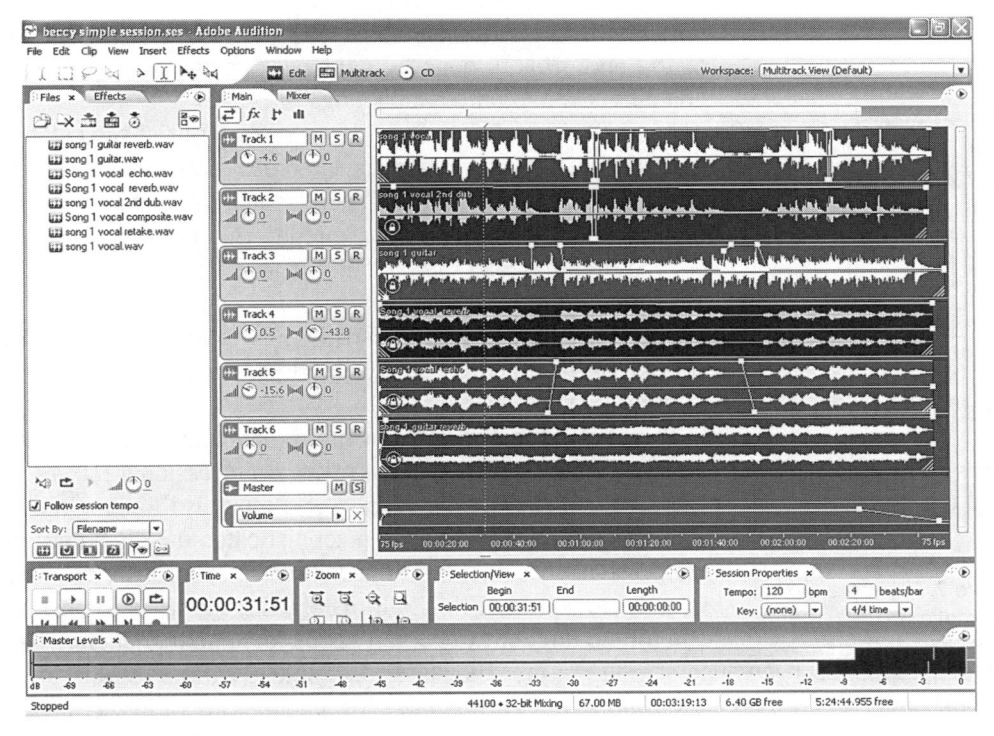

Figure 7.4 Multitrack simple balance

57

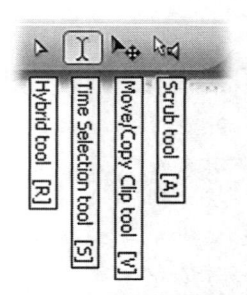

Figure 7.5 Mouse tool icons

Figure 7.6 Clip grab handle for trimming start of audio

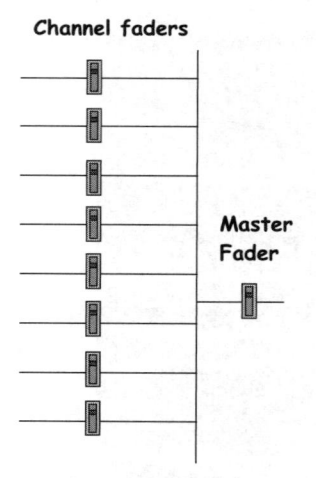

Figure 7.7 Simple mixer schematic

Examining the simple balance using the default multitrack view, we can see the menu at the top of the screen. Below this on the left are eight icons which control the mouse tools for the edit and multitrack views. The right hand four are for the multitrack view (Figure 7.5).

The hybrid tool combines the functions of time selection and move/copy. The scrub tool allows you to audition individual clips.

Below, the left panel lists the files in use. The icons at the top are for manipulating the file list. Those at the bottom are playback options for auditioning the files in the list. Followed by options as to which files are shown: audio, loops, video, markers (show cue and track ranges as separate entries) and show full paths. Tabbed behind is a list of effects for ease of selection.

To the right is the main panel which shows six source tracks and a master track. In this case each track has a single file on it but any number can be added either as whole tracks or sections of track. Audition calls these 'clips'. Each clip can be non-destructively 'topped and tailed' by dragging the bottom corner of either end – a grab handle is revealed when the clip is selected by clicking on it (Figure 7.6).

The master track represents the final mix volume. Normally hidden, I have revealed a volume envelope which shows a fade placed at the end. This is done by right clicking the master track and enabling 'Expand automation lanes'.

Audition is capable of more than a straight mix; it can create and use 'busbars'. 'Bus' is short for 'omnibus' and derives from early electricity power stations where individual generators were combined together onto a common circuit. This is the same word as derived from the road bus 'Omnibus transport' 'Transport for all'.

As the currents were so high this was a copper bar rather than a wire. The term has been carried over to audio. In the simple set-up the master output is sometimes called the 'masterbus' (Figure 7.7).

This is all simple mixers have; but more sophisticated mix set-ups with multiple busbars are used by professional mixers.

Figure 7.8 shows a schematic of a mixer with a single busbar with its own associated fader. A sophisticated desk will have more than one busbar and even busbars can have busbar sub-mixes. Why would you want this? Ultimately, the answer is the 20 finger and thumb limit of two human hands for handling faders. There is also the practical convenience of group mixes. The classic example is a drum kit with perhaps five or six microphones on it. You will want to lift or lower the whole drum kit in the mix; moving that many faders is likely to disrupt the internal mix on the drums. Having a single group

Multitrack

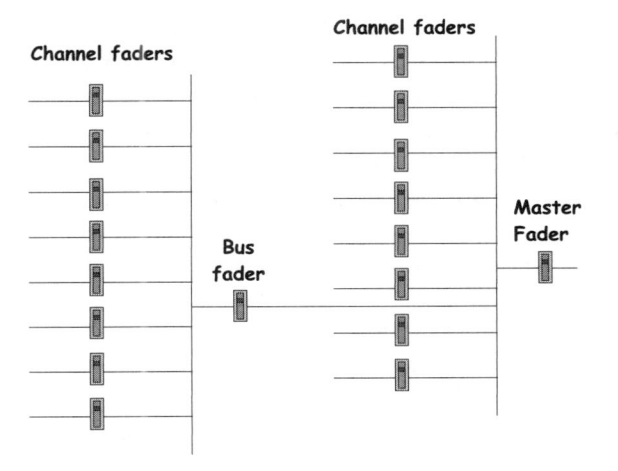

Figure 7.8 Simplified bus mixer schematic

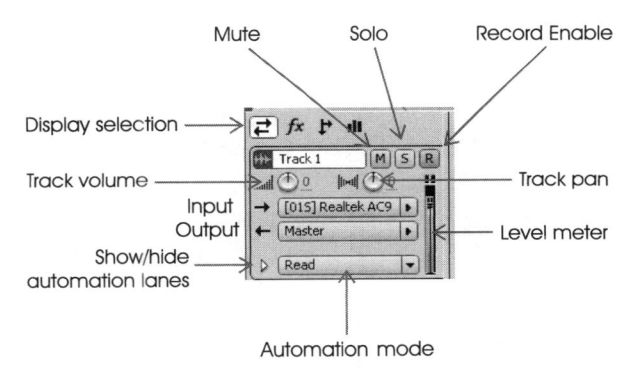

Figure 7.9 Track description panel

Figure 7.10 Track volume control

or bus fader to represent the drums is a boon.

When you are mixing using computer software there is an additional reason for busbars and that is processing. While it is technically possible, say, to feed each drum mic to reverb, that will be five or six lots of intensive processing. Whether your computer is up to this will depend on how powerful it is and what other tasks it is performing. A single set of processing on the drum bus will save a lot of processing power.

Returning to our simple mix, the tracks have a panel on their left. Beside each of the six tracks shown are the individual controls for each track, as shown in Figure 7.9

The first thing to realize is that Audition hides a great deal of information as, much of the time, you don't need it. If you are not using the effects why show them? At other times information is concealed because there is no room on the screen. This track control panel is the most obvious example of this. For a start, there are four icons at the top which select between four sets of display (Figure 7.9). But whatever you select you will get the vital controls. These are the default 'Volume' and 'Pan' controls. Personally, I find rotary controls operated by a mouse very irritating but they perhaps have an advantage in terms of displaying their setting efficiently. Compared with some audio software Audition is very straightforward. Placing the mouse over the control gives a finger cursor with horizontal arrows (Figure 7.10).

Briefly a tool tip will show the current setting of the control. Pressing down the left hand mouse removes the finger symbol leaving just the horizontal arrows (←→). Moving the mouse cursor left/right OR up/down will change the setting. Holding down SHIFT while you do this will make the change faster while CONTROL will make it slower. These shift and control modifiers work for most of the mouse-controlled variable controls except for the faders on the graphical mixer. The drag works on the knob, its icon and also the numerical display. Additionally you can double click a numerical value and type in a new one.

59

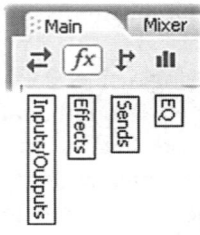

Figure 7.11 Facility view selector icons

Also shown are the editable track names, Mute, Solo and Record enable buttons. In Figure 7.9 you can also see input source and where the output is fed (to the master output, to a busbar or to a physical sound card output. The last could be to feed artist's headphones or external 'fairy-dust' hardware). At the bottom is the selection for real time automation. On the right is a level meter. Incidentally, if you are tight on processing power, you can switch these track meters to look at the inputs only or you can shrink the height of the track so that they are removed.

The default is the input/output set with a choice of showing the effects in use, the sends to the effects and the EQ setting (Figure 7.11).

If a lot of tracks are being shown on the screen then an even more truncated display will be shown (Figure 7.12).

But an individual track can be widened by dragging the bottom of the panel; the mouse cursor will change to a ⇕ symbol. Figure 7.13 shows track 1 in a minimalist configuration and track 2 in expanded.

Figure 7.12 Truncated track display

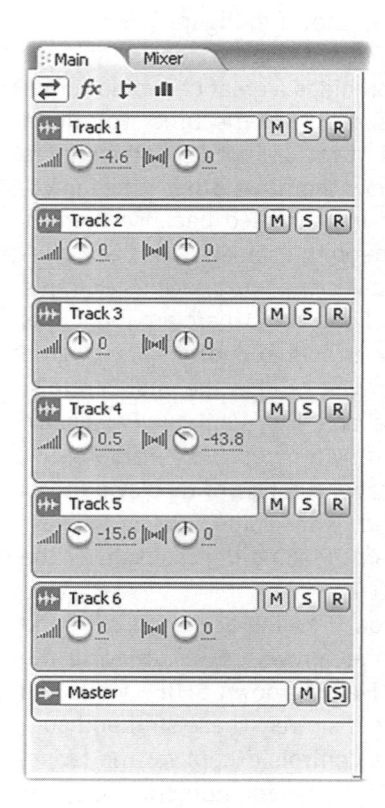

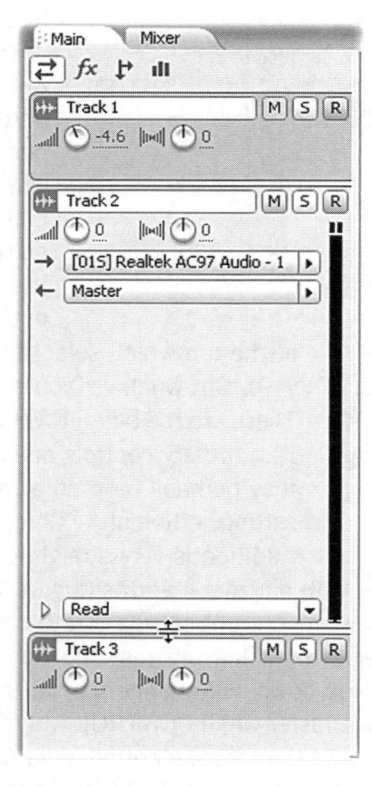

Figure 7.13 Track 2 expanded with outputs selected

The essential ability of a multitrack recorder is that it can record on all of its tracks simultaneously, on only one, or on any number in between. So, in addition to the transport record button, each track has a red record enable button marked 'R'. If this is not selected then the track is in 'safe' mode and will not go into record when the main record button is selected. Additionally, already recorded tracks are be played back as a guide to the new recordings.

Sometimes you want to listen 'inside' the mix; the green 'M' button when pressed and illuminated will mute the track. This can be very useful for tracking down where an unwanted noise is coming from. The yellow 'S' solo button has the opposite effect. It mutes all the other tracks and plays only the selected track. Industry practice is for solos to be deselected when a new solo is pressed. This is what happens with Audition, provided CONTROL is pressed, otherwise a simple toggle action takes place. This is the reverse to how Audition 1.5 worked.

With multiple tracks, you are not restricted to stereo recording. Adobe Audition 2.0 is limited only by the processing capacity of the computer as to the number of simultaneous tracks you can have. Much multitrack recording is built up layer by layer, so a 60-track recording does not need 60-record inputs they can be recorded one at a time through a single record input.

Unlike some audio editors, Adobe Audition 2.0 can handle both mono and stereo files. Others can only provide stereo by linking mono tracks together. If they import a stereo file then that is split into two separate mono tracks. With Adobe Audition 2.0 you can use a mixture of mono and stereo tracks within the same session.

7.3 FX

Figure 7.14 shows the effects. Three are shown, inserted prefader. There is room for 16 different effects and you can see how a scroll bar on the right has appeared so that the slots off the screen can be selected. Each effect can be switched on or off using the 'power' button on the left or they can all be switched off using the 'power' button at the top of the panel. Effects, inevitably use processor power and even the most powerful computer is going to flag eventually. The symptoms are that it cannot keep up with real time and there are skips and mutes in the output. Fortunately there are ways round this. These centre round the fact that a straight play back of a file takes less processing than an effect. One answer is to use the edit view to generate, say, reverb (a particularly processor intensive task), and create a new file which only has reverb and no direct sound. This laid alongside the original and mixed as required. The disadvantage is that you have to do it all again if you want to change the character of the reverb rather than merely its volume. With Adobe Audition 2.0 it is altogether more efficient. Above the scroll bar is a 'freeze' button ; setting this to on (gray turns to blue) will produce a progress bar while a file is produced and will substitute for the real time processing of the effect. The waveform display will turn blue.

Figure 7.14 Effects display selected

Deselecting it will instantly return to 'live' processing. Thus you can freeze all the effects you are happy with; unfreezing them as required. A minor disadvantage is when you load a session with frozen files, you have to wait until they are regenerated. No, only joking, this is an option if you are short of hard disk space. Provided 'Save frozen audio tracks with session' is ticked on the Edit/preferences/multitrack tab (F4) the frozen data are saved into a folder named after the session file name with '(frozen tracks)' appended to the end.

Figure 7.15 shows the 'sends' panel. This shows a single send with its volume and pan control (as always, the track volume and pan are shown above). The 'send' is the destination of the output of the selected effect. This has to go to a bus; it cannot go straight to the master output. Clicking the right arrow allows you to select or create buses as destinations (Figure 7.16). You can then allocate buses as specialized outputs with their own faders and controls. An obvious one is for reverberation. You also have the choice of feeding the effect from before or after the track volume/fader. With reverb, a pre-fader setting will mean that if you fade out a vocalist between verses and they cough, then this will be fed to reverb and be audible where a post-fader setting will not be as the feed will fade out with the track volume. However if you are using a bus to feed the vocalist's headphones then they will not want to hear what you are doing to their track; just a constant feed.

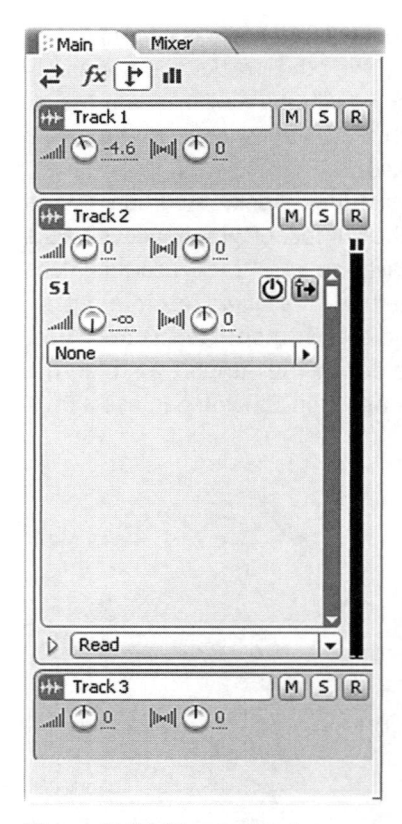

Figure 7.15 'Sends' display selected

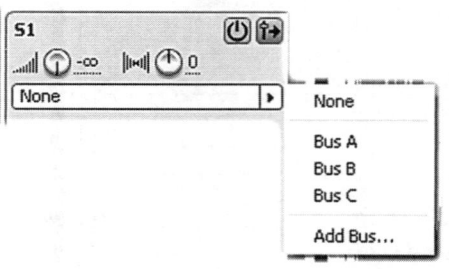

Figure 7.16 'Send' bus selection

Figure 7.17 shows the last panel which control the real time EQ available for the track. The parameters before the three equalizers are shown only as text – no graphical knobs here. EQ1 is a low-frequency shelf control. EQ2 is a variable 'Q' mid-range control and EQ3 is a high-frequency shelf equalizer. The 'power' button inserts the EQ. Remember that power-off saves on processing power compared with just setting the controls to zero. Clicking the EQ button brings up the EQ panel complete with a graphical display of the setting. EQ1 and EQ3 can be changed to variable Q controls like EQ2 by clicking the 'shelf' symbols on the right of the horizontal sliders (Figure 7.18).

The horizontal sliders control the turnover frequencies for the equalizers with the vertical sliders controlling the amount of lift or cut. In fact there are two equalizers which can be switched between by clicking the EQ button which will toggle between reading 'EQ/A' and 'EQ/B'. This is so that you can compare two settings; typically EQ/B will be set to flat however it can be useful to store your best yet setting so you can compare it with your latest attempt.

'Q'

'Q' is a term to describe the steepness of a filter. The boost or cut is defined as being at a particular frequency. For a shelf filter this frequency is the point where there is 3 dB change at the full setting. For a peaking 'presence' filter this is the centre frequency. 'Q' is a measure of the width. In the diagram, EQ1 is set to a very low Q (0.06) and has a very broad effect. EQ3 is set to a very high 'Q' (20) and has a very narrow width.

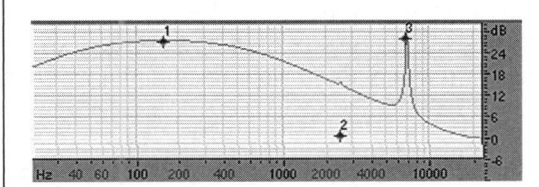

Tabbed behind the main panel is a full-scale graphical mixer which has the same controls rearranged as a mixer. There are some additional controls. Selecting ⌀ to Ⓠ will reverse

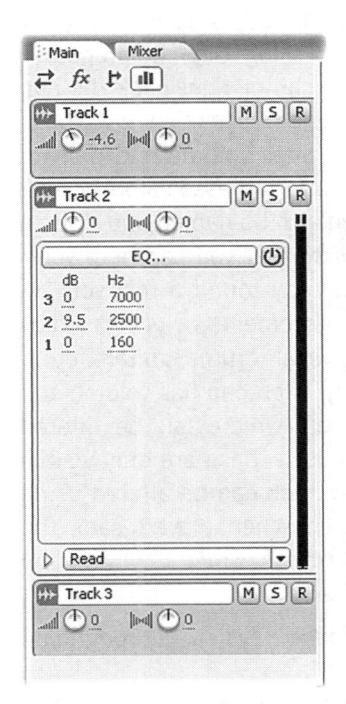

Figure 7.17 'EQ' display selected

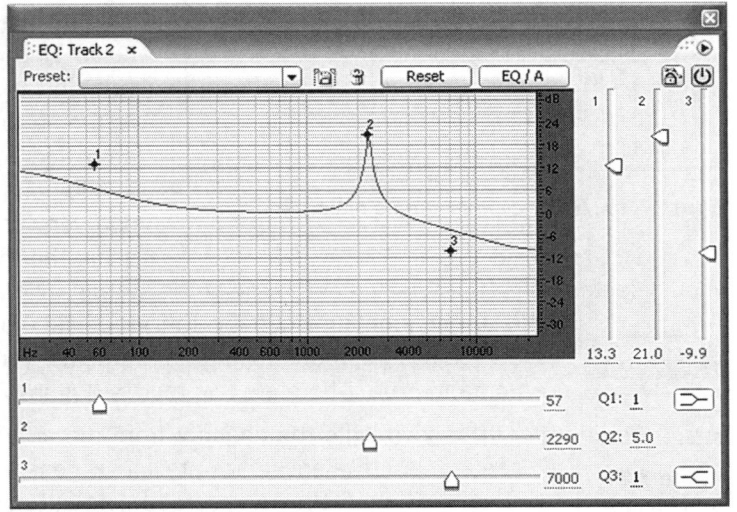

Figure 7.18 Equalizer controls

the phase of the CHANNEL; both left and right outputs are reversed, not, as you might expect just the right output. This is to handle a mono or stereo microphone on, say a drum kit, that is reversed in phase with the other mics on the same kit. Just above the fader is an unconventional control: selecting [((•))] to [•))] will sum the left and right outputs to mono. This would be useful where a domestic stereo mic is being used as a mono pickup.

Potentially confusing is that each channel has two volume controls. The top volume control is a preset for the volume to be set with the others for a 'straight-line mix'. The second volume control is the fader which corresponds to the control at the left of the main track display. Both of these are used to alter the mix during the run of the item and are both available to automation. Care should be taken not to confuse the 'Send' volume and pan controls with those just described.

7.4 Automation

All these settings represent an overall control of whatever is on the track; it may be just a single clip or many. While the overall volume is set by the track's 'vol' and 'pan' controls; as we have seen, this is not the only way the volume of a clip can be altered. Firstly the volume and pan setting of the clip itself can be adjusted by using the clip properties dialog: Clip/clip properties for a selected clip or right click the clip for popup. This produces Figure 7.19.

The sliders alter Volume and pan (or type a value into the associated boxes). The track can also be named and given a color. This is so that you can help yourself find your way round a mix you are returning to movements by giving different colors to vocal, drums, guitars, etc.

Additionally, each clip has volume and pan 'envelopes' which can be altered using the mouse. These are only visible when selected using the 'View' menu. Finally the setting for the track can be altered using the automation provided by the graphical mixer. When played back the faders move reproducing the movements you made while recording an automation run. There are five modes (Figure 7.20):

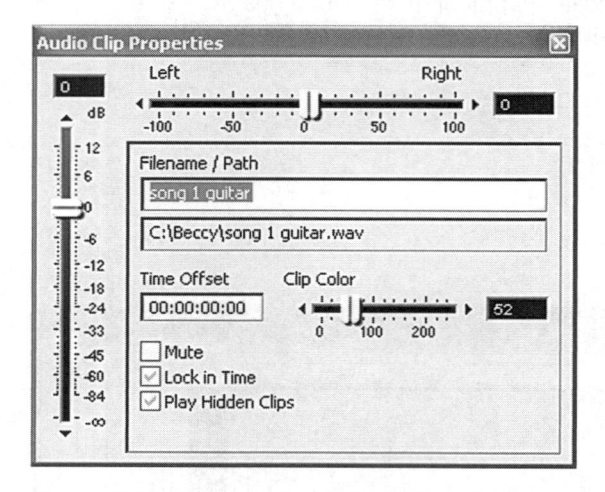

Figure 7.19 Audio clip properties dialogue

Figure 7.20 Automation settings

1 'Off' disconnects the channel from the automation system although you can still edit the envelopes previously generated.
2 'Read' will read already recorded data but not allow it to be changed. You can, however, rehearse some changes but they will not be saved.

3 'Latch' begins recording when you first adjust a setting, and continues to record new settings until playback stops.
4 'Touch' is similar to Latch, but returns settings to previously recorded values when you stop adjusting them.
5 'Write' Similar to Latch, but records current settings when playback starts, without waiting for a setting to change.

An alternative to overriding recorded fader movements is to reveal the track automation lanes which will graph the fader movements. You can, if you wish enter a mix directly to the automation lane and the mixer faders will follow your envelope.

AUTOMATION LANES AVAILABLE

Volume
Mute
Pan
Track EQ power on/off
Track EQ: Band 1 frequency
Track EQ: Band 1 gain
Track EQ: Band 1 Q
Track EQ: Band/Low shelf

Track EQ: Band 2 frequency
Track EQ: Band 2 gain
Track EQ: Band 2 Q

Track EQ: Band 3 frequency
Track EQ: Band 3 gain
Track EQ: Band 3 Q
Track EQ: Band/High shelf

Insert FX: Power on/off
Insert FX: Wet/Dry mix
Insert gain: Gain amount

Most of the mixer parameters are automated and can have their individual automation lanes revealed. Figure 7.21 shows the complete list revealed for one track. As you can see, unless you have a very large screen this is unlikely to be practical.

For playback, each track, be it mono or stereo, is fed to either a busbar or to an audio sound card output (Figure 7.22a, b).

When starting a 'New session', you can also select the sampling rate which is fixed for the entire multitrack session. You can mix 8-, 16- and 32-bit files but those with a different sampling rate

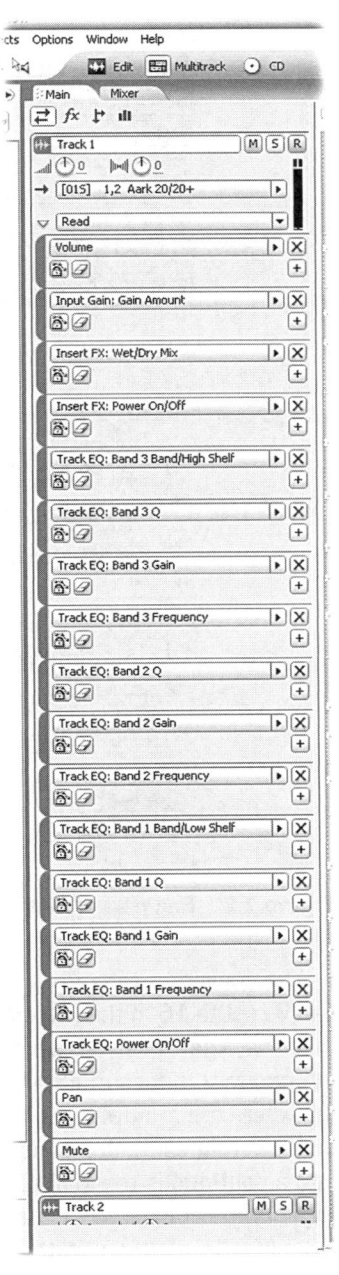

Figure 7.21 All the automation lanes for one track revealed

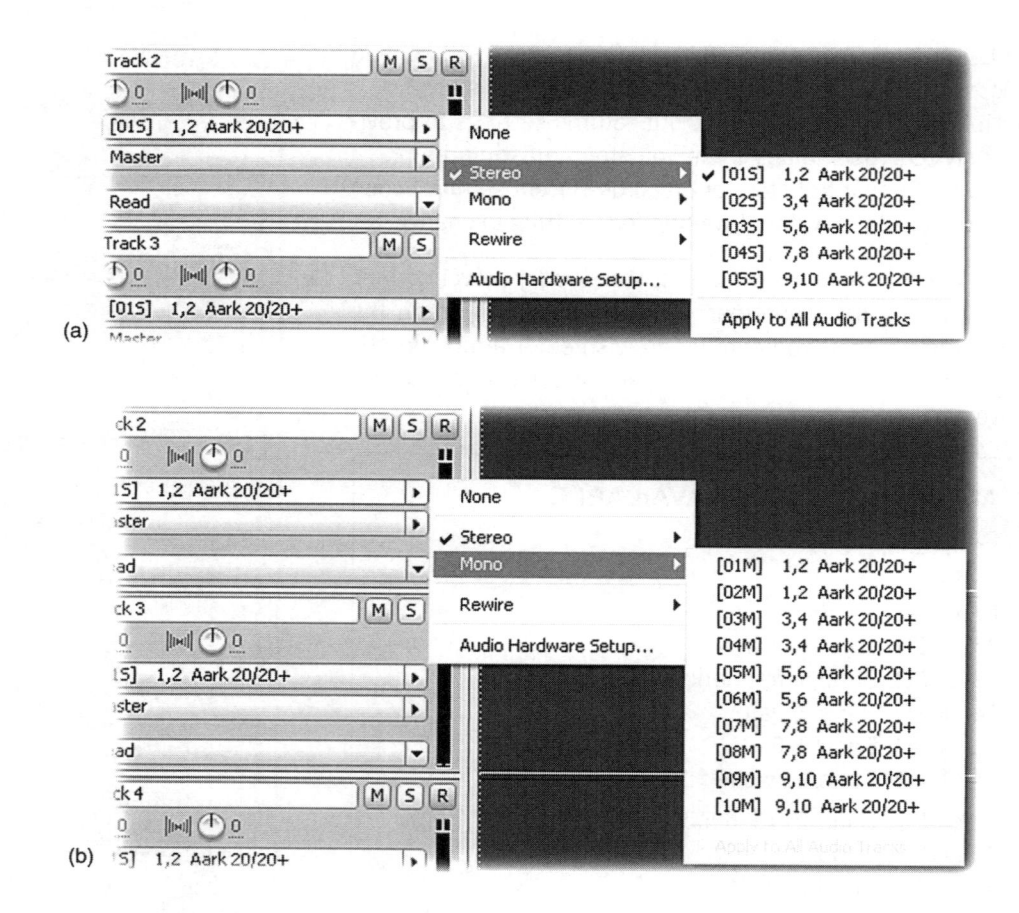

Figure 7.22 Example of audio output switching

will be converted when they are imported. The session normally records at 32-bit resolution but can be set to 16 in the Preference/multitrack tab (F4). The 16/32-bit option in the Export/Audio mix down dialog selects what the mix down output will be.

Important though multiple ins and outs are, they are likely to be little used outside music mixing where the multiplicity of outputs are useful mainly for feeds to an external mixer. The power of the non-linear editor is also the ease with which it can merge and mix mono and stereo audio into a continuous mono or stereo final mix. Audio items are added and laid out, as required, on any number of mono or stereo tracks. They can be regarded as a large number of play-in machines in a studio. Without our intervention, all the material will be mixed together at full level with no panning.

Clips

You can also set the volume and pan for individual clips on a track. These may represent separate wave files or they may represent wave files that have been split by the multitrack editor. They look like separate wave files but are, in fact, sections of a wave file. This is done by right clicking on a clip. Within the popup menu, Clip Properties gives you a vertical slider that adjust the volume of just that block and a horizontal slider to adjust the pan. The settings are shown in text at the start of the block with the letters 'V' and 'P' standing for volume and pan. These setting are in tandem with the track settings and do not override them (Figure 7.23).

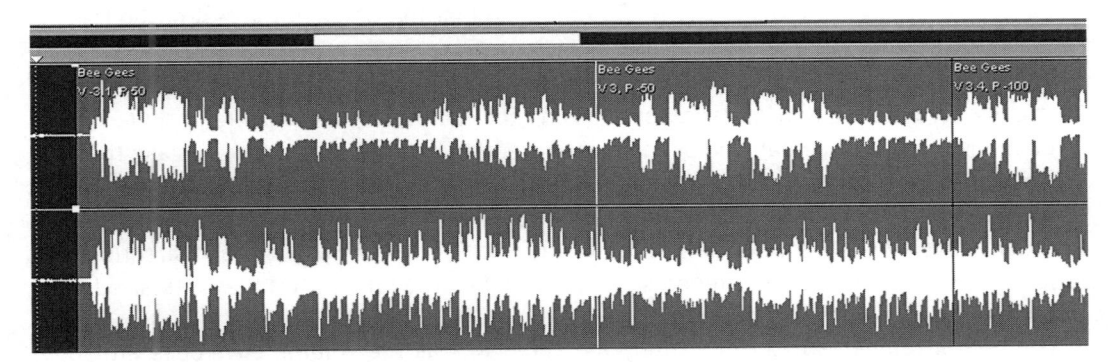

Figure 7.23 Three-wave blocks with different volume and pan settings indicated by text at the top left of each block

A good use for this is to make successive wave blocks consistent in level so that the master track level works for all the blocks. The block pan can be used to create consistent pan setting for individual speakers in a documentary feature. If you split a wave block (select track(s) and CONTROL/K or 'Split' from right mouse click popup) the two sections carry the same volume and pan settings.

7.5 Envelopes

As well as these 'global' settings for the whole track or block, you can control Volume and Pan on a moment-by-moment basis. This is done using 'envelope' controls within the track. To be able to see the envelopes, you have to set the 'View/Show clip pan envelopes' and 'View/Show clip volume envelopes' menu options so that they are ticked.

Until changed, the Pan envelope is a straight line along the centre of the track, and the Volume envelope is a straight line along the top of the track. These represent Pan Centre and Full Volume (as modified by the track settings on the left). Moving the mouse cursor onto an envelope line produces a '+' symbol. Clicking will produce a small blob where you clicked. This is a 'handle' which can be dragged by the mouse to change the setting. You must select the track by clicking it before this works.

To produce a simple fade-out (Figure 7.24), create a handle where you want the fade to start, and another one to the right of it. Drag this second handle to the bottom of the track (representing zero volume) and left or right to the point you wish the fade to finish. (You will

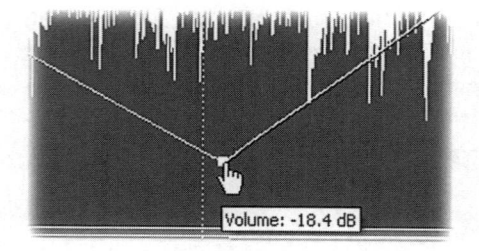

Figure 7.24 Moving envelope handle

not be allowed to drag it to before the start of the fade handle. However, this can be a useful 'end stop', if you just want to switch off the track.)

You will discover that the level immediately begins to increase after the fade, because the end of the envelope line is attached to the handle at the end of the track. This needs to be dragged to the bottom, so that the whole of the rest of the track is faded out.

Now you have your basic fade you can adjust it for timing by listening and, if necessary, moving the handles. A simple straight-line fade is often inadequate but you can add as many intermediate handles as you like to make the fade curved, or to have several sections (see Section 7.9 'Fades and edges' on Page 74).

You only need to use the pan envelopes where you want the sound from a track actually to be heard moving. You should beware of doing this to speech, as it is a distraction and reduces communication; the listener is likely to react on the lines of 'Oh they've started moving. Why? ... er ... what did they say?'

The pan envelopes are set in entirely the same way as the volume envelopes although, much of the time, a single pan setting for the overall track is all that is needed. In both cases, a handle can be removed by dragging it off the top or the bottom of the track.

Figure 7.25 shows a short section of a mono track. This starts at full volume and ends with a fade. The pan setting starts at centre. Then there is a sudden pan full left followed by a slow pan from full left to full right. The next two sections are panned fully right and then fully left with the track ending panned centre.

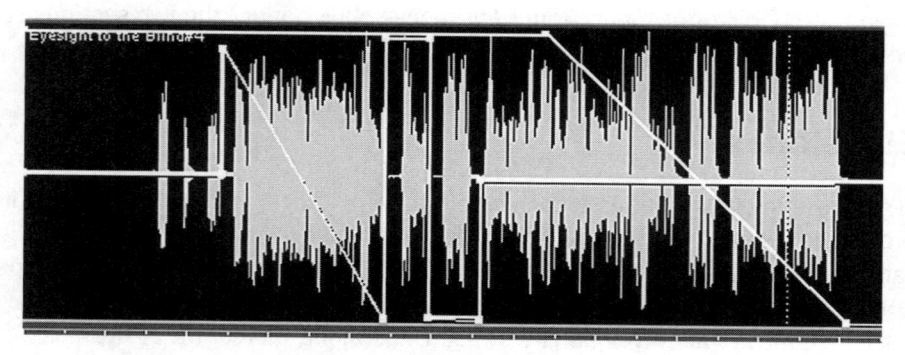

Figure 7.25 Pan and volume envelopes being used together

When you want to move a handle, you position the mouse cursor over the handle until the cursor becomes a pointing finger. You can now drag the handle to where you want it, positioning it anywhere between the previous and next handles. You can see the value which will popup beside the cursor. It takes a little while to get used to this, and it is very easy to create new handles rather than moving an existing one. UNDO can correct this, or dragging the unwanted handle off the track will remove it.

This use of envelopes is a common way of overcoming the imposition of 'one-finger' operation by using a mouse. While you can only change one thing at a time, all the changes are remembered exactly. While simpler and less impressive than a screen filled with a pretty picture of a sound mixer, it can be much more effective.

To get rid of one-finger operation you need either an external mixer or an external control device that may look like a mixer, but controls the software settings with the program. Many Digital Audio Workstations work this way, but at a price.

You now have complete control of the audio levels, and timing, of your material. Like many editors, Adobe Audition 2.0 allows real time processing of audio, applying equalization and special effects such as reverberation. This requires a lot of processing and a correspondingly fast machine, but some sound cards have Digital Signal Processors which take the load off the computer's processor and have software to allow effects to be added in real time. Where this is not possible, the technique is to make additional tracks which contain the effects.

7.6 Multitrack for a simple mix

It is very easy to see the non-linear multitrack mode as a complication to be avoided. For simple cut and paste editing, this is often true; however, as soon as mixing becomes involved, it is invaluable. It gives you the ease of multi-machine mixing with quarter inch tape, combined with absolute repeatability and flexibility.

Due to the integrated nature of the editor it can be convenient to 'pop in' to the multitrack editors when all that is required is to make a simple mix.

The following example takes a section of Berlioz's *Damnation of Faust* which has one bar repeated in it. A simple cut edit does not work as a retake has been cut in, which does not match the original. The answer is to crossmix between the two takes during the one bar overlap.

This is simply done by entering the multitrack mixer. If the wave file is already in the linear editor, it will already be in the editor's organizer list (Alt/9). All that needs to be done is to right click it, and click the 'Insert into Multitrack' option or just drag the file to a track. (To make the illustrations clearer, I have zoomed in vertically to show just two tracks.)

Next, the end of the first take of repeated section is selected with the mouse, and the track split at that point. This is done by right clicking on one side of the cursor and selecting the option to split the track or, if the playback cursor is already in position using the keyboard shortcut CONTROL/K. You will see the file name appear as well as the edge of clip drag handles (Figure 7.26).

Now that the audio is in two parts, the second half can be dragged, using the right mouse button, down to track two. In Figure 7.27 I have highlighted the first take of the repeated section.

The next thing to be done is to slide the second section to the left until the overlapping section matches musically that of the track above (Figure 7.28).

At this stage, the rhythm matches but the level bumps and a cross fade needs to be implemented. This can be done manually with the volume envelope but it is much easier to use the automatic cross fade facility.

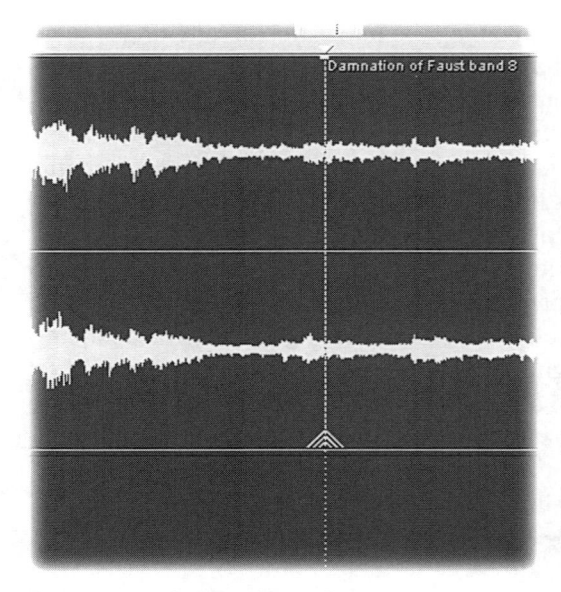

Figure 7.26 Splitting "Faust" file

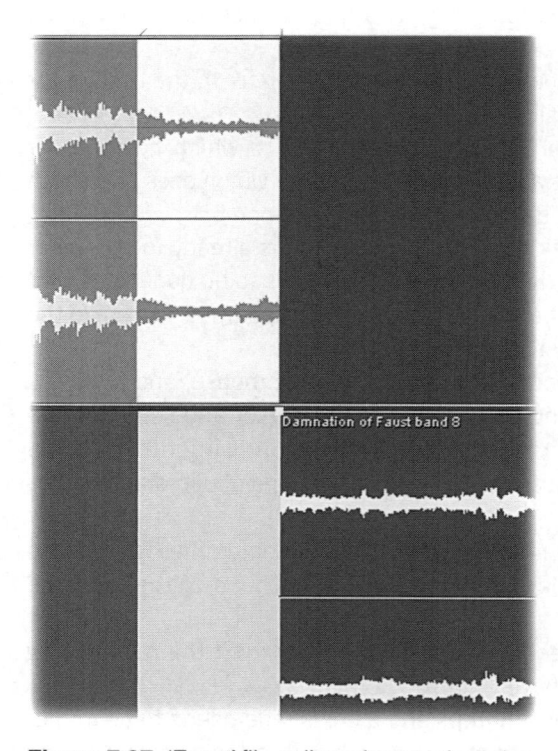

Figure 7.27 'Faust' file split and second section moved to track two

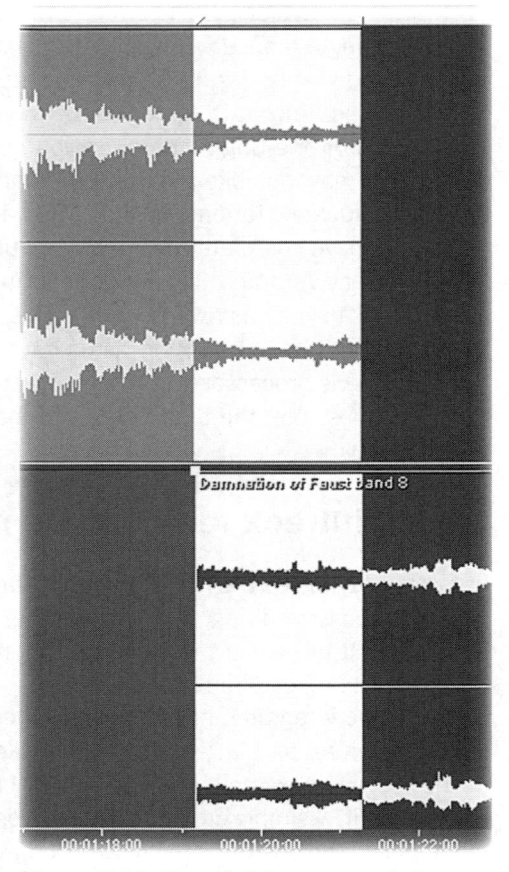

Figure 7.28 'Faust' slid to correct timing, ready for cross fade

First the cross fade area is highlighted. Then both clips are highlighted by clicking with the control key held down. Now select the cross fade from the right-click context menu.

There are four types of cross fade to choose from (Figure 7.29a, b, c, d). They all have their uses. However, linear cross fades tend to give a boost to the level in the middle of the fade. Logarithmic fades should give a better approximation to equal power throughout the transition. Due to their asymmetrical shape, there is a choice of 'direction'. Sinusoidal is another equal power option. There is no substitute for trying the

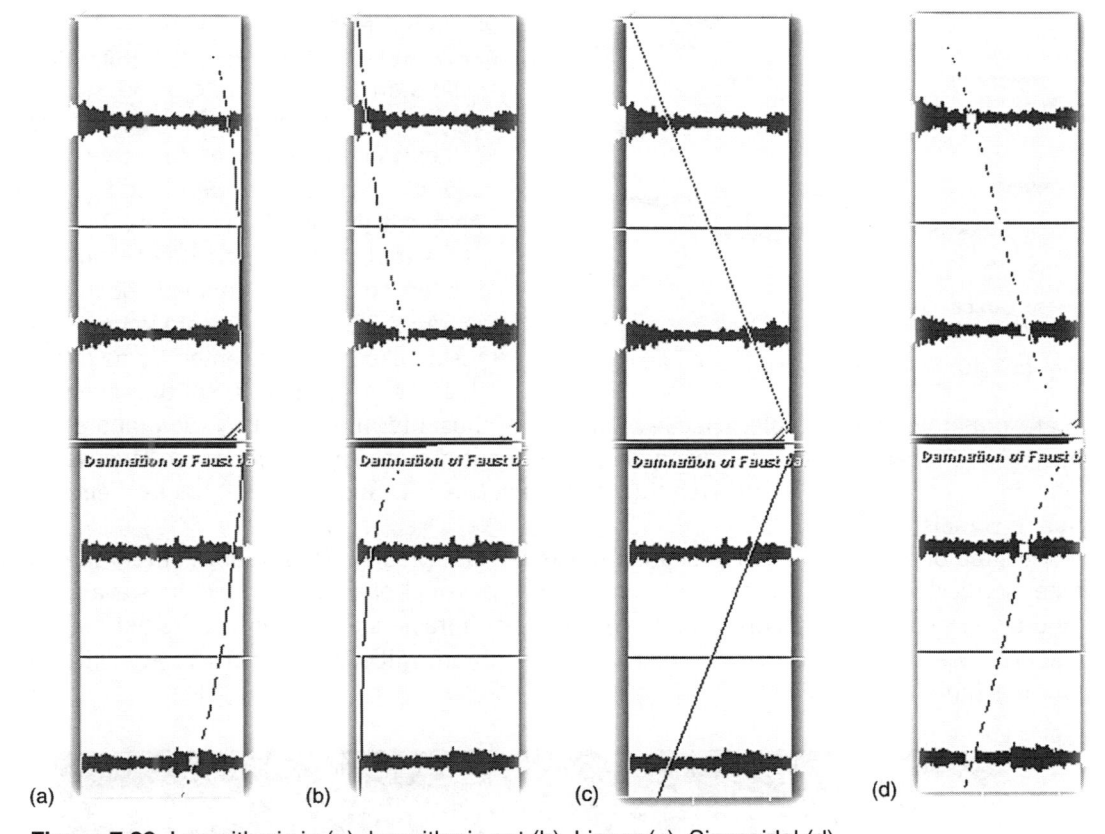

Figure 7.29 Logarithmic in (a), logarithmic out (b), Linear (c), Sinusoidal (d)

different options. The UNDO option is one menu click away (or CONTROL/Z). Playing the different transitions will soon convince you as to which is best for any edit.

To transfer this edit back to a file in the linear editor you merely mix down your tracks (File/export/audio mixdown and then save the file). If you are confident with your editing, you will select a short section with a little each side of the cross fade and mix down just that section. This can be copied and pasted into your original in seconds.

7.7 Chequer boarding

Once a programme item becomes much more than a simple interview or talk, then mixing becomes necessary. While mixing is possible with linear editors using Mix Paste, it usually involves a lot of UNDO cycles to get it right. The multitrack non-linear editor makes it so easy.

In the days of reel-to-reel tape, programmes of any complexity would be split into banded 'A' and 'B' reels with odd numbered inserts on the 'A' reel and even numbered inserts on the 'B' reel. Each insert could then be started over the fading atmosphere at the end of the

PC Audio Editing with Adobe Audition 2.0

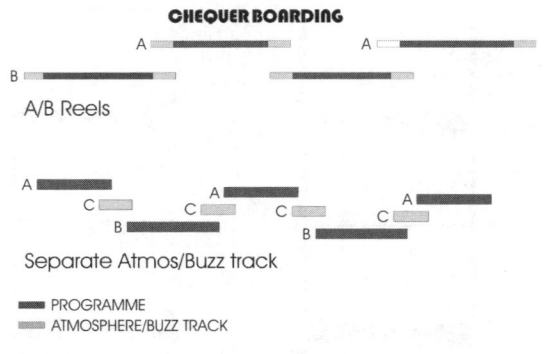

Figure 7.30 'Chequer boarding' on tape

preceding insert. Where both items had heavy atmosphere the incoming band would start with atmosphere and a cross fade made. Wherever necessary a third 'C' reel would be made up to carry continuity of atmosphere over inserts with tightly cut INs and OUTs (Figure 7.30).

In a multitrack editor, this technique is often known as chequer boarding. Figure 7.31 shows a programme of music linked by a presenter. Some of the music is segued and so the chequer board principle is invaluable. (Segue is an Italian music term adopted by radio and show business to mean one item following another without a break. It is pronounced 'seg-way'.) For convenience, the links are given their own track (track 1) and the music tracks their own 'A and B reels'.

A degree of extravagance in allocating tracks is not a problem. In the days of reel-to-reel tape, studios were often limited by the number of physical play-in tape machines available (and by how good the tape operator was!). Here, each track is effectively a different play-in machine. We can have as many of them as our hardware will support – rather better than a four-machine radio studio.

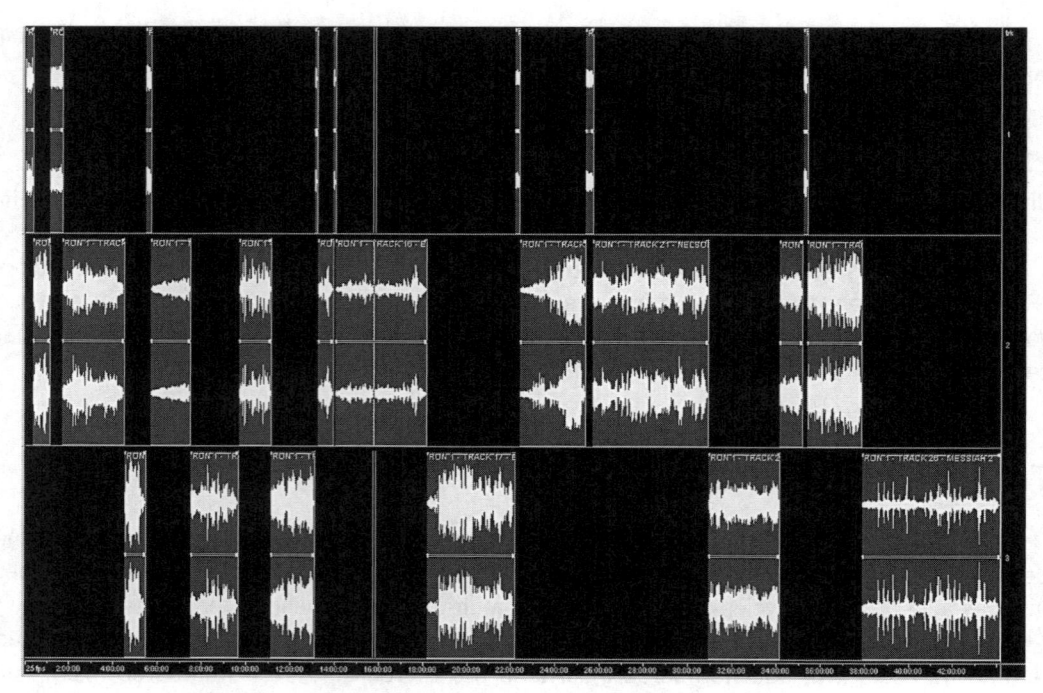

Figure 7.31 'Chequer boarding' on *Cool Edit Pro*

Even better, we are no longer reliant on the tape operator. Each 'play-in' is automated; fixed until we change it. No longer does one muffed play-in require a whole take to be redone. If an item comes in slightly late or early, it only takes a moment with the mouse to slide the clip until it is right (drag using the right mouse button).

While totally dominant in music recording, reel-to-reel multitrack was never a very useful device for drama and features because of this one limitation; you could not slide tracks relative to one another. There are ways to emulate this, but only by adding considerable complexity to the process.

7.8 Track bouncing

There are times when you want to undertake a quick mix to make a composite track. For example, you have two versions of a vocal track and you want to mix the good bits together into a single track. You can 'track bounce' selected tracks by using the 'Edit/Bounce to new track' menu option. If you enable the 'Lock in time' option on the newly created wave block right click menu, then it will stay in exact synchronization if you move it to another track.

The three bounce options (Figure 7.32) are straightforward: 'All Audio clip in session' does what it says; 'Audio clips in selected range' will mix down all the audio clips, horizontally as selected by the normal time range selection (Figure 7.33a) and 'Selected Audio clips' are

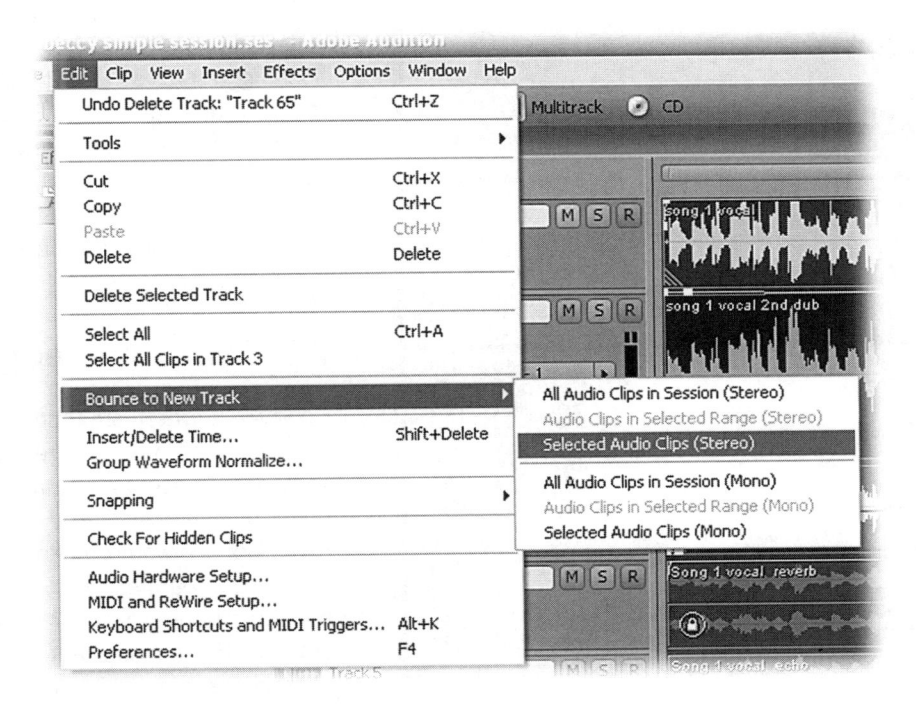

Figure 7.32 Track bounce options menu

clips you have selected wherever on the track layout. (Hold down control to select more than one clip.) Figure 7.33b shows the middle clip of each group of three not selected and the resulting Mixdown.

This speeds up background mixing and helps clear up your work area as you can now delete the originals from your session (although not, for safety, from your hard disk). The contents of all or selected enabled (unmuted) tracks are combined, with track and waveform properties (such as volume and pan) affecting the way the final mix sounds. Session elements such as looping, images and envelopes are all reflected in the mixed waveform.

As track bounce also picks up your real time FX it is also a way of 'locking' a track so that it is saved to hard disk and so is quickly reloaded, rather than having to be regenerated, when you reload a session. It is not possible to select a locked FX block, so you should unfreeze it first.

7.9 Fades and edges

However, a little more than simple chequer boarding is required. Together, a studio manager on the mixing console and a good tape operator would make material merge seamlessly by skilled and instinctive use of faders, both on the console and on the play-in machines. Often inserts would be started, not with an

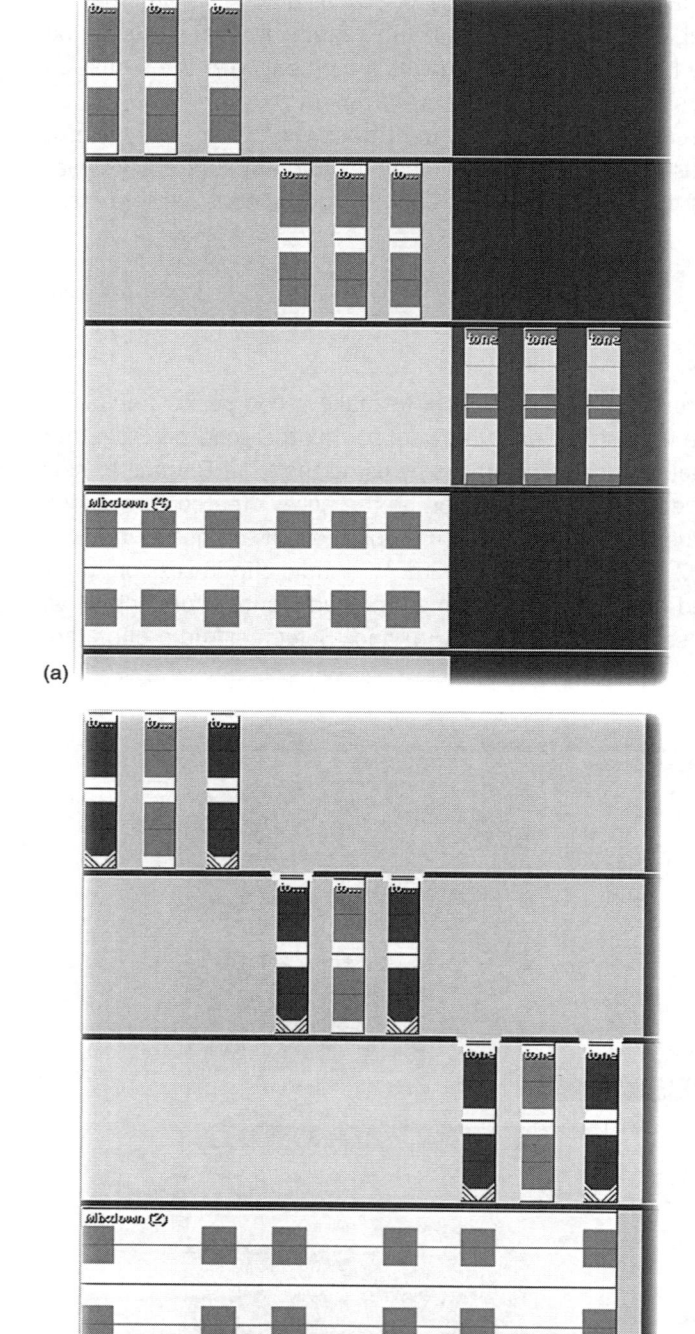

Figure 7.33 Results of mixdown options

actual fade-in, but with edge-ins, where the item was started with a sharp fade-up to 'half fader', and then a fade-in over a second. Sound balancers have also learned that most items sound better if they are started 2 dB to 4 dB low and edged up, over a second or so.

On a PC, all these instinctive skills have now to be defined. This is why many people, with the necessary money, will pay for an external mixing device to control the computer's mixer. This preserves the power minutely to adjust levels within an item merely by having your fingers on the faders. This 'pinky-power' can control up to ten sources simultaneously. A computer mouse can only control one. However, because of the automation provided by a PC, each change can be made individually and cumulatively.

It makes sense to preserve the old split of duties between the sound mixer and the tape operator. When preparing your material, pay attention to how each file begins and ends; put a rapid fade-out on the atmosphere at the very end of the file. Put an edge-in on the front. You can set this up as a 'Favorite'. Starting values of −6 dB and −3 dB fading up to zero are useful here.

Edge-ins should be inaudible to the listener. They are there to smooth the seams. Actual fade-ins and fade-outs, ones that are going to be perceived as such by the listener, are best done using the envelope controls of the multitrack mixer or the automated fader on the mixer.

Fades usually have the role of establishing a passage of time. The old convention, from the 'golden age of the wireless', was a slow fade-out followed by a slow fade-in. These days, the fades are much faster and the fade-in is more of a steep edge-in.

Fades also have the function of allowing voice-overs to be heard over music. Very rarely should the listener be aware that this is being done. What will sound particularly horrible will be a mix where the music is audibly dipped to make a hole for the voice to enter, with another hole following the end of speech before the music is faded back up.

Most important is that the voice-over should fit the music. The music should return at the start of a musical sentence. The dip-down is less critical in the context of pop music but, with more formal features, the voice should pick up from a cadence, or the end of the musical sentence.

When mixing voice-overs 'live', I used to say to presenters that I would 'go on the whites of your tonsils'. By this I meant that the music would dip on the very first syllable of their link. Equally it sounds better if the fade back up starts about a syllable before the end of the link. This is less critical if the presenter has hit the beginning of a musical phrase as you are fading up in the micro-pause between notes.

Music fade-outs are often done badly. A vital thing to realize is that any fade-out consists of three separate parts (Figure 7.34):

1 warning the listener they are about to lose the music,
2 fading down,
3 fade-out.

This applies as much to a clean fade as to one where the presenter comes in over the end.

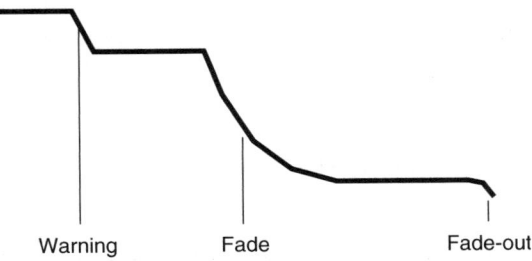

Figure 7.34 Notional structure of a fade-out

The classic example of the latter is the opening signature tune, a feature of so many audio items, from broadcast programmes to cassette promotions. The music starts full level and then towards the end of a musical phrase, the level is dipped slightly to warn the listener and then almost immediately after faded down so that the presenter can start their link at the end of the phrase with the music running under. The music is then lost under the presenter, and brought out on a cadence, or end of phrase.

Fading classical music badly can cause real offence, but pop music deserves equal care. Again the three-part fade applies; a slight dip at the start of the last phrase and then a fade towards the closing cadence, with the music taken out at the end of the cadence. A common mistake, especially when the fade is made while looking at the score, is to leave the start of the fade too late so that the listener is not prepared for the music to end.

Where the fade has to be quick, say, because the illustration is about the words of a song, then, rather than lose the last few words on a fade to cadence, a 4–6 dB dip at the beginning of the last line can provide the psychological warning and make the quick exit far less offensive.

7.10 Prefaded and backtimed music

A common technique to bringing a programme out on time, especially a live one, is to have 'prefaded' closing music. A piece of music is dubbed off, say of 1 minute duration, and started with the fader shut exactly 1 minute before the end of the programme. As the presenter makes their closing remarks, the music is faded up behind them, to bring the programme to a rousing and punctual close.

A technique used by some producers is to end the last item with music which is then run under the presenter's close and then faded up so that it ends. This effectively offsets the prefade back to the last item. However, it sounds so much better if that last piece of music actually begins at a musically satisfying place, rather than just emerging arbitrarily. A cheat that works most of the time – although I remain offended that it does – is to dub off the item's closing music to the end, and run that separately as a backtimed prefade closing music. The two recordings will not be in synch, yet, because they are the same music, they can usually be cross faded under the presenter's closing link without being audible to the listener.

These techniques may seem unnecessary for a recorded programme, as it can be edited to time. However, time to edit can be a scarce resource and it may be much more cost-effective to use a prefaded closing music than to spend another hour looking for cuts. Of course, you do not have to do any actual back-timing. All you have to do is to slide your closing music until its end is at the duration that you want. You then use the amplitude envelope control to fade it in at the right point.

7.11 Transitions

Transitions are the stuff of dramatic effect, not only in drama itself, but also in documentary. Cross fading from one acoustic environment to another or from one treatment to another can be extraordinarily effective.

The first type of transition, going from one location to another, is simply managed. Two tracks are used and the incoming track slid along to overlap the end of the exiting track. You can use Adobe Audition 2.0's automatic cross fade generation, as described earlier in this chapter, or you can manually create the envelopes if this does not produce the effect you want. Always remember that you can modify the cross fade that Adobe Audition 2.0 did for you by moving the envelope 'handles' with the mouse. With a transition, you are rarely going to use a perfectly engineered equal loudness cross fade.

The second type of transition is from one type of treatment to another on the same material. Examples include an orchestra playing in another room. The music is treated to remove all the high frequencies. The characters move into that room, and you want the high frequencies to return as they make that move, or they open the door. Sometimes you want to go from music or speech apparently heard acoustically – a jukebox playing in a coffee bar – to music heard directly off the recording. Here you need to take the good quality original and treat it, literally by playing it on a loudspeaker and recording it with a microphone (or, in a studio, playing the music on the foldback loudspeaker). A small room effect may be used instead.

With the example of the orchestra, your treated track is identical in length with the original and is inherently in synch with it. Within the multitrack editor place the cursor roughly where you want the treated track and, without moving the cursor, insert it. They will both line up on the cursor. Now they need to be grouped together so that you can move them without losing synchronization. Click the first track so that it is selected and any other selection lost. Now, holding down the control button, click the other. They should now both be selected. Right click on one of the waves and click 'group clips' on the popup. You can do this from the 'Clip/Group clips' (CONTROL/G) menu item as well. Now, when you move them about they will move together, until you match the dialog in the best way.

You now take the handles at each end of the volume envelope on the good quality track and move them to the bottom, so that there is no output from the track. Listening to the mixed output, adjust the track gain for the treated track so that it fits behind the dialog. If it is supposed to be coming from behind a door, then you may wish to pan the track to one side or the other.

At the point in the dialog where you wish to make the transition, create a cross fade by adjusting the envelopes on both waveforms. You can adjust the gain of the incoming untreated track with the track volume control, or by using the volume envelope.

In the second example, the music recorded off a loudspeaker, the two files will be of different lengths. However, it is very easy to slide one file with respect to the other until you can hear that they are in synch. Now you can group the two files and treat the transition in the same way as before.

7.12 Multitrack music

Music recorded on a multitrack recorder can be transferred to your computer very easily, if you have a multitrack sound card with enough inputs. This can even be done digitally with the right gear. A common interface is a variant on the optical TOSLINK connector. The same plug is used, but an eight-channel digital signal is sent instead. A common type of multitrack recorder is the video cassette- based machine, which records eight digital audio tracks instead of a picture. These can be linked together to become 16-, 24-, 32-track, etc. recorders.

In cases of desperation, short pieces of music, say up to 5 minutes, can be transferred using only stereo equipment. The multitrack tape must be prepared by having a synchronizing 'slate' recorded simultaneously on all of the tracks just before the music. A convenient way of doing this, in a studio, is to use the slate talkback: the talkback that goes to the multitrack and not to the artists. A short ident, and a sharp sound like a coin tapping the talkback mic, should be adequate. At the end of the music, a similar slate is recorded. The tracks can now be copied two at a time, either directly to the computer, or to a stereo recorder, preferably digital. When the tracks have been transferred to your computer, they should be split into their corresponding mono files. This will have been done automatically if you have used the multitrack facility to record them as two mono files rather than a single stereo file. Each wave file will begin with identical recordings of the slate. Each file is edited so that it starts with the most recognizable part of the 'sharp' sound. Sample accuracy editing is advisable, and practical, here.

Once loaded into the multitrack editor and lined up at 0:00:00:00 the files should now be back in synch. How much error has been introduced can be ascertained by examining the end slates. Usually, in my experience, the error, even with an analog multitrack, is better than two European Broadcasting Union (EBU) frames (2/25th second) which for many types of music (and performance) is good enough. Better accuracy can be obtain by using Adobe Audition's effects transforms to varispeed tracks digitally so that the error is corrected. The mix down can now be undertaken within Adobe Audition 2.0.

7.13 MIDI and video

MIDI and video files (AVI, WMV, ASF and QuickTime) can be loaded into the multitrack file and can be used as references for audio tracks. The picture is represented by a graphic of a series of frames like a sprocketless 35 mm film on the multitrack display itself (Figure 7.35). The display has three options; blank frames, a thumbnail of the first frame or selected frames throughout the video. Its track has no operational controls at the left. The actual picture can be seen in a separate window by clicking 'Window/Video'. The audio has the full range of operational controls.

A MIDI file's notes are represented graphically but has only the mute and solo buttons available at the left along with a volume control. There is an additional 'Map' button that allows you to adjust the MIDI device and channel assignments. The MIDI part also has a volume envelope for adjusting it within the mix.

7.14 Loops

WAV files can be configured as loops. Right click the block and select 'Loop properties'. Once so configured they can be dragged out to repeat for as long as required. The prime function for this is to use samples of drum beats, etc. to make up a rhythmic background for

Figure 7.35 AVI video and a MIDI file open in Adobe Audition 2.0

music balances. For speech and drama work this can be very useful for looping small sections background effects or actuality to extend them. However, beware that short loops soon set up their own rhythm. Longer loops can also become recognizable if run for a long time. For example, a horn sounding in a traffic noise loop will become startlingly obvious the third or fourth time around! Adobe Audition also uses .CEL files which are essentially .MP3 files with a .CEL extension. Each .CEL file has a header that contains loop information, such as the number of beats, tempo, key, and stretch method. The .CEL format avoids a potential problem with .MP3 files. During encoding, a very small amount of silence is added to the beginning, end, or both of an .MP3 file. The silence is very short – often only a few samples long. When you work with a loop, though, it's enough to throw off the entire loop. As it saves a .CEL file, Adobe Audition calculates how much silence will be added to the .MP3 file and writes this information into the .CEL header. Then, when Adobe Audition opens a .CEL file, it reads this information and automatically removes the silence from the file so that it loops smoothly.

8

Post-production

8.1 Timing

This covers the period between the compilation of an item – often in a studio – and the final mastering of the finished item. In some cases, it will be difficult separately to distinguish this phase, but it is rare for it not to exist at all.

With a 'live' transmission, it is unusual for adjustments not to have to be made after a rehearsal. Only in live news and current affairs programmes do adjustments have to be made on-air. An overall studio producer/director will ask colleagues to cut arbitrary amounts of time out of items. Hellish though this can be, it usually works.

Typically, when all the sections of an item have been put together, whether in a studio or with your audio editor, the item will be too long, poorly paced, with further editing of sections required either for style and pace or 'fluffs' and overlaps. It is always a waste of time to 'de-um' speech recordings until this stage as otherwise you waste time on edits that are thrown away.

Don't regard it as a 'mistake' when you find your post-production material is too long. It's difficult to judge the worth of individual items until they are in their final context. Something that seemed relevant within an individual item becomes a drag within a programme. Ten per cent over-recording is about right. Now is also the time to be rigorous with yourself.

A good traditional policy when looking for something to cut is 'Cut your darlings'. That piece which was particularly difficult to get, that fact that is so fascinating to you, will, as likely as not, turn out to be supremely indulgent and weaken an otherwise tightly crafted item.

The post-production phase is also where you apply your audio design, which is covered in the Chapter 9.

8.2 Level matching

Make sure that levels are matched. The human ear is extremely sensitive to quick changes in audio levels. It is much less sensitive to slow changes of level. This is, at once, a danger and a boon.

The danger is that, unless you are constantly keeping an eye on levels, they can slowly sink. There is a tendency for the ear to want successive sounds to be about 2 dB quieter for the level to sound matched. This is why the edge-in is so important in smoothing transitions. You can have the level the ear expects and, by slowly increasing the level, restore it to normal.

An edge-in involves a small change in level but sometimes original material has far too much dynamic range. An obvious example might be classical symphonic music where the range from *ppp* to *fff* can easily exceed 50 dB. The traditional BBC guideline for dynamic range was a mere 26 dB. That is still valid for an item that someone is going to sit down and listen to. However, it is extraordinarily rare for anyone to do that. Radio broadcasting and audio presentations have the advantage that they can be absorbed by people who are busy, or on the move. In the modern noisy environment a dynamic range of 12 dB is about as wide as you can go without audibility suffering. Real life has a dynamic range of 120 dB, a power ratio of 1 000 000 000 000 : 1. How can we represent it with 12 dB, or only 16 : 1? The answer is by manipulating the ear's strength and weakness.

A fully modulated speech programme item will consistently peak to full level. Dynamic range is implied by cheating levels at transitions. A slow change of level is not perceived; a sudden change is. You can make an entry sound loud, be it a fortissimo passage or a shout, by dipping the level before the transition.

Say we have a presented item which is back announcing a previous item on Mozart, followed by music played by *The Who* or *The Rolling Stones*. This is meant as a surprise. It is meant to sound loud. The presenter then voices over the music and must now sound louder than the music. Here, the answer is to creep down the level of the back announcement and start the music at full level. This is crept down and then dipped conventionally for the presenter's voice over which is now at full level. This happy jig is kept up throughout the programme giving a sense of dynamic range yet the programme is audible and intelligible, even when heard in a car on a motorway.

While broadcasters and makers of CDs usually have no control over the environment where the listener will hear their recording. There are exceptions. Examples include *Son et Lumière* and tourist guides on cassette or CD heard on headphones in Museums, monuments and historic buildings. At places like these a wide dynamic range can be part of the entertainment.

8.3 Panning

There is a school of thought that says that all speech should be in mono in the centre and only music and effects heard in stereo. This may seem a little severe until you realize that all modern cinema films are balanced this way, as is much of the stereo sound on TV programmes.

In practice, speech inserts into programmes are in mono because they are easier to edit, if the image is not constantly moving around. A common convention is to put the presenter into the centre and then put contributors in panned 6 dB left or right. Panning more than this becomes gimmicky and potentially reduces compatibility if heard in mono; a contributor panned too far in either direction will sound quieter than the central presenter. Another convention is to aim for a symmetrical balance with interviews panned equally left and right, or discussions panned symmetrically across the sound stage. (Studio interviews, using the presenter, should be recorded contributor panned hard left and the presenter panned hard right, using separate microphones. This gives total flexibility of positioning in production.)

8.4 Cross fading

Cross fades have several functions:

- equal power cross fades to hide the join between separate, but similar, sections of audio;
- long passage of time/substantial change of location where a dip in level, combined with a change of sound, indicates a passage of time. This is effectively a modern contracted form of the old fade-out, pause and fade-in convention;
- simple change of sound without level dip. This indicates a short passage of time and a small change of location (say, to get from one room to another).

8.5 Stereo/binaural

Conventionally, audio is heard on loudspeakers. Normally there are two of them but, for surround sound systems like Dolby stereo, there will be more. These surround systems either use special coder/decoders to play phase tricks with two-channel sound, or they use multiple channels. DVD has standardized on a 5.1 channel surround system. The '.1' is an engineer's jokey way of saying that the sixth channel is low bandwidth, suitable only for conveying low frequency audio (earthquakes, spaceships, etc.).

Binaural sound has been around since the beginning of stereo when Clément Adèr relayed the Paris Opera to an exhibition in 1881. Visitors to the Paris Exhibition listened using two earpieces fed separately from different microphones at the theatre.

Techniques for recording binaural sound usually involve some form of dummy head. This can range from an actual model of a human head, with microphones in its ears, to something more stylized but just as effective; a Perspex disc approximately the size of a head, with small omni-directional microphones mounted either side about 9 inches (22 cm) apart. (The relative transparency of the Perspex disc makes it acceptable for slinging above orchestras at public concerts. A disembodied head might disturb the audience). There are also microphones designed to fit in a person's ears and thus a real head can be used. This does have the disadvantage that it is very difficult for the person wearing the microphones not to turn their head towards different sounds. This produces a very disturbing sound image for the listener.

Binaural seeks to emulate how we actually hear. The resulting sound is surprisingly good heard on loudspeakers and can be dramatically effective heard on headphones (open headphones work well than enclosed ones). When it works for you, a totally three-dimensional sound is heard. However, it is said that only 60 per cent of the population can get the best out of the system. A large proportion of the rest get a 3D effect but not one emulating real life. A common problem is everything seeming to be coming from behind the listener.

It works using phase and frequency response variations caused by the obstruction represented by the dummy head. This gives sounds outside the listener's head. For example, a mono voice panned fully left, heard on headphones, will seem to be coming from the left ear itself. A binaural sound coming from the left can seem to be metres further out. The phase accuracy and excellent frequency response of digital recording and editing make this an ideal medium for binaural sound.

Binaural recording can make an exciting alternative for productions aimed at headphone listeners, such as a headphone tour around a historic building. Although the phase accuracy of compact cassette is not that good, the binaural effect survives, provided everything is kept digital up to the master that the cassettes are copied from. Portable CD or Minidisc players preserve the directional effects because of their inherent phase accuracy. (Small timing errors between tracks are often expressed as phase differences. Thus, a difference of 1/48 000th of a second, is equivalent to a quarter of a wavelength at 1 kHz.

Because of the mathematical nature of waves, a whole cycle of audio is said to have gone through 360°. Halfway through a cycle is 180°. Here complete cancellation will occur if the two sources are mixed at equal level. Cassette machines have a degree of jitter where the track relationships vary; how much is dependent on how good the tape transport is.)

Mono sound, added to the mix, remains firmly 'between the ears' and this can be used to distinguish narration from other voices. There have been a number of dramas where characters, mute from a stroke, have had 'think speech' coming from within the listener's head, while all the 'real world' action takes place outside of the head.

8.6 Surround sound

There are various systems purporting to give surround sound to the listener. They succeed to greater or lesser degrees. In the 1970s there was a brief burst of popularity of quadraphonic sound. This played little regard to the actual theory of how we hear and suffered from the fact there were only two-channel delivery systems available. This meant that some form of compatible encoding matrix system had to be used. In reality there were many systems that led to the public being confused. The need to have four full-sized loudspeakers in the room, also conflicted with many people's furniture (Figure 8.1).

As Quad began to fail commercially there was a separate development in the cinema. Dolby Laboratories developed a way of not only dramatically improving the sound quality on ordinary optical film sound tracks but also were able to shoehorn in stereo and surround at the same time. They modified one of the existing quadraphonic matrix systems and reallocated the sound channels. Quadraphonic audio had used the seemingly obvious method of having front left and right speakers combined with back left and right speakers. Instead, Dolby Laboratories used the encoded channels for a centre speaker and

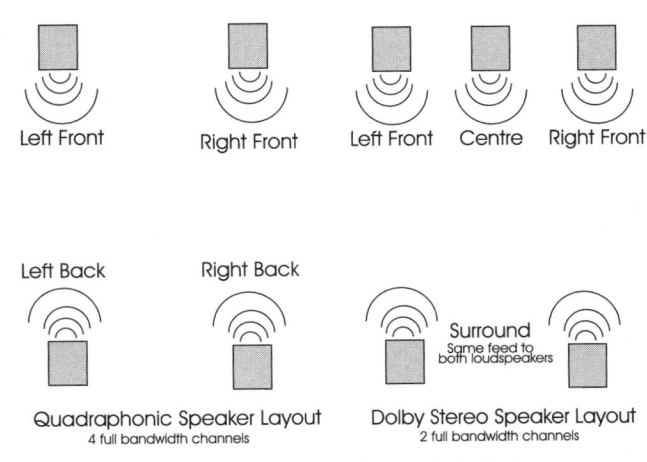

Figure 8.1 Quadraphonic speaker layout

Figure 8.2 Dolby stereo speaker layout

the other channel for a diffuse rear feed that was fed with surround information to give depth to the sound. The centre speaker is there to harden the centre of the sound of the centre sound image where the dialog is put. This means that someone at the side of the audience does not feel that the dialog is 'over there'. This still allows directional effect and even voices to come from either side or behind you. It is not good at making sounds come from beside you.

Dolby stereo educated audiences enjoyed the surround effects on commercial films. At the same time video recorders were becoming ubiquitous. They increasingly had stereo playback facilities when plugged into the home viewer's Hi-Fi. The sound track that was available already had the surround information encoded for the cinema and so surround sound or 'Home Theatre' decoders became increasingly popular (Figure 8.2).

You need a Dolby surround encoder properly to obtain surround. However, without this you can still get impressive results by listening to the output of your stereo audio editor while adding extra effects like thunder, rain or reverberation in mono out-of-phase. This is 'heard' by the Dolby Surround decoder as surround information. Because the system cannot reproduce every positional effect, even professionals using fully featured surround encoders will listen as they mix to loudspeakers fed by a surround encoder in tandem with a decoder so as to have some idea of the end effect (Figure 8.3).

Film sound tracks progressed and new method of encoding sound became available. At the same time DVD arrived. This has the capability of handling much more data than a CD. There is room not only for pictures but also for extra audio tracks. The standard that has been agreed upon is based on the Dolby stereo concept but with discrete channels to six loudspeakers. Before the centre channel and a single surround channel were derived from a two-channel feed. Now the centre channel has it own feed and there are two rear surround feeds. This gives much stronger directional feed although there are still positions that cannot be reproduced. Again you should listen through a proper playback system to judge what you are achieving. Additionally a sixth low bandwidth channel is used directly to feed a subwoofer instead of the feed being derived from the others. This is intended for low frequency effects like explosions, earthquakes and thunder.

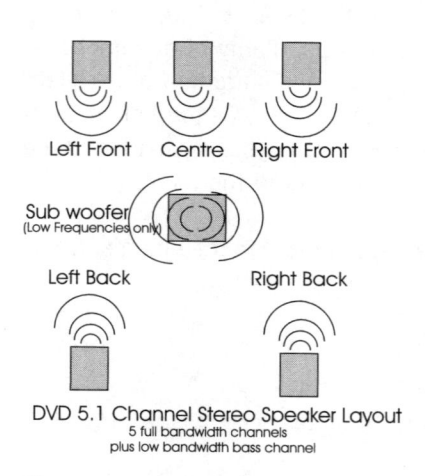

Figure 8.3 DVD 5.1 channel stereo speaker layout

When presenting the finished material on multi-track tape the assignment supported by the European Broadcasting Union (EBU) and Society of Motion Picture and Television Engineers (SMPTE) is the following:

Ch 1   Left
Ch 2   Right
Ch 3   Centre
Ch 4   Low frequency effects

Ch 5   Left Surround (or Mono Surround 3 dB down)
Ch 6   Right Surround (or Mono Surround 3 dB down)

Tracks 7 and 8 can be used for left and right of a stereo mix

For classical music, a different system called Ambisonics can be used. This uses a coincident of microphones and a control box effectively produces three figure-of-eight microphones pointing forward, sideways and upwards along with an omnidirectional output.

In practice, the upward output is often omitted as most playback systems cannot make use of height information as this needs a ceiling loudspeaker. But full scale 'Periphonics' can produce spectacular results.

8.7 Multi-layer mixing

Dramatized features sometimes need to create a sound for which there is no recording, such as the Biblical story of the Fall of Jericho. Here, building up the sound will require a great number of tracks. A large army can be simulated by sound FX recordings of suitably selected crowds, added on top of each other. If necessary, the same crowd can have different sections mixed with itself to make it sound larger. Various screams and shouts have to be added and panned appropriately, plus the trumpets that bring down the walls.

Before the arrival of digital multitrack editors this sort of thing was difficult to build up, because of the limited number of play-in machines, and often required multi-generation dubs, not to mention enormous tape loops precariously run round the room. Audition can loop wave files so even short sections can be extended in definitely.

9

Audio design

9.1 Dangers

Audio editors come with a multitude of 'toys' that you may never need – until that day when they rescue you from certain disaster!

Remember that, just because a toy is available, you do not HAVE to use it. There is no substitute for clear, interesting, relevant speech or good music, well performed. With documentary, have faith in the power of the spoken word. One of the more depressing of my activities as a studio manager was working with producers who did not. They would ruin exciting speech by blurring it with reverberation or obscuring it with sound effects.

All the facilities available allow you, very easily, to ruin recordings. Audio and music technology is full of controls that, set in one direction, do nothing and, set in the other, make everything sound dreadful. Somewhere in between, if you are lucky, is a setting where something magical happens and a real improvement is obtained. When trying to find this setting, always refresh your ears regularly by checking the sound of the recording with nothing done to it.

9.2 Normalization

The first of these toys is normalization, which is so massively useful that you are likely to use it constantly.

One of the very first things that is done when making a recording is to 'take level'. The recording machine is adjusted to give an adequate recording level with 'a little in hand' to guard against unexpected increases in volume. However, that little bit in hand will vary, as will the volume of the speaker or performer.

Most recordists know that people, in general, are 6 dB louder on a take than when they give level – except those occasions when they are quieter! The result is that, even for those most meticulous with levels, different inserts into a programme will vary in level and loudness. While level and loudness are connected they are not the same thing.

The three illustrations (Figures 9.1–9.3) show the same piece of audio but at different loudness. Figure 9.1 illustrates audio which is at a lower level (6 dB) than the other two. Figures 9.2 and 9.3 are the same level but Figure 9.3 sounds much louder. When we talk about level we actually mean the 'peak' level – we talk about audio 'peaking' on the meter. On average speech or music most of the time the instantaneous level is much less.

Audio design

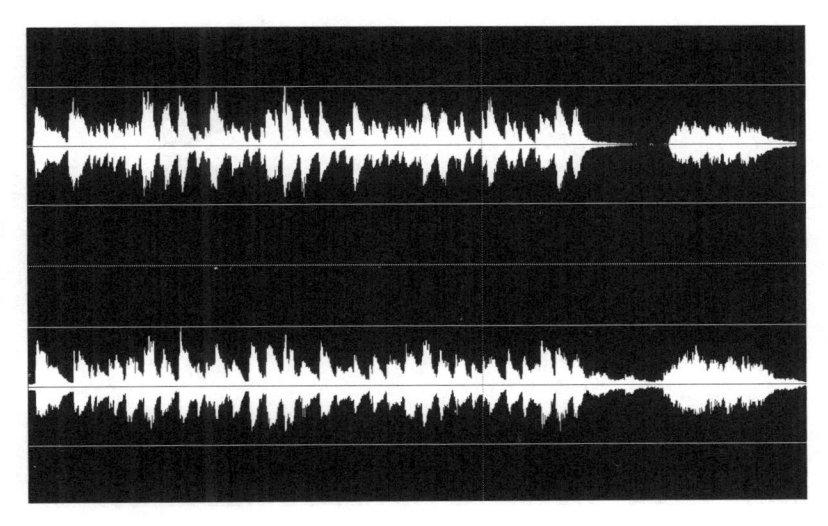

Figure 9.1 Audio at −6 dB

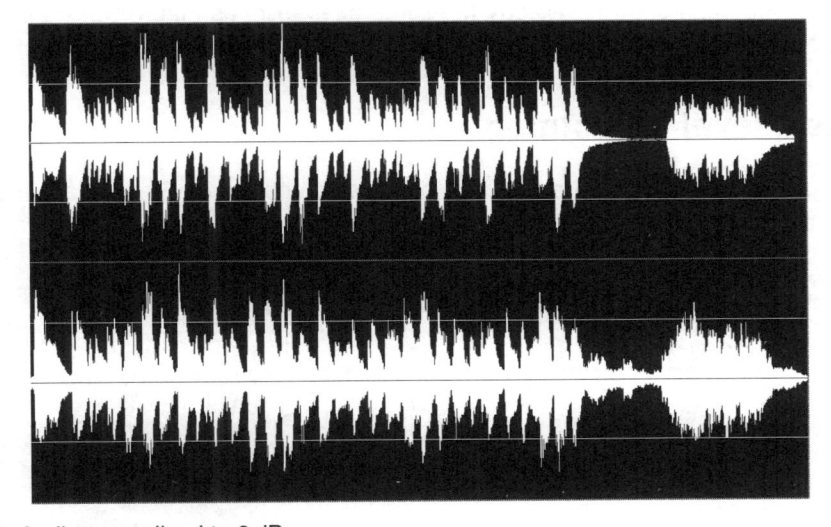

Figure 9.2 Audio normalized to 0 dB

Audio, within the computer, is stored as a series of numbers so it is very easy for the computer to multiply those numbers, by a factor, so that the highest number in your audio signal is set to the highest value that can be stored. If this is done to all your audio inserts when you start, this will ensure that your levels are consistent.

However, their relative loudness may not be the same. It all depends on what proportion of your recording is at the higher levels. As a generalization, while the peak level is represented by the highest points reached on the display, the loudness is equivalent to the area of the audio display.

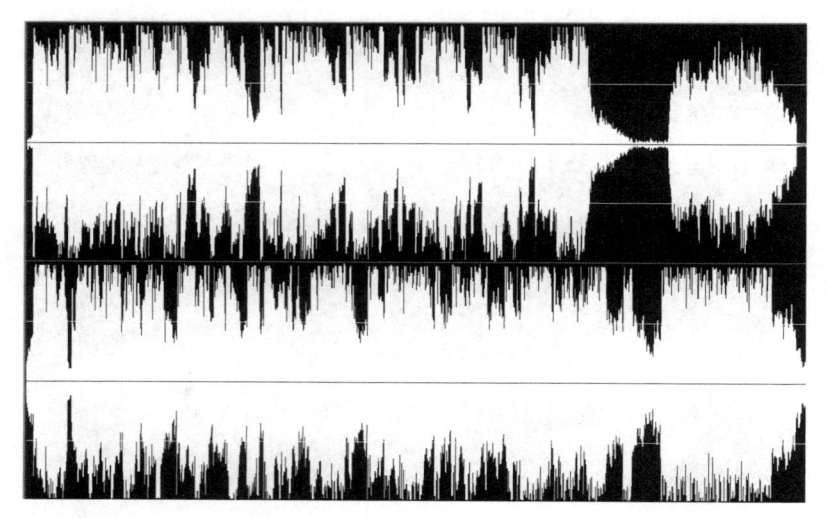

Figure 9.3 Audio compressed to 0 dB.

Figure 9.3, although showing exactly the same peak level as Figure 9.2, is much louder because it has been compressed.

9.3 Basic level adjustment

Effects/Amplitude/Amplify (Figure 9.4) can be used to change the level of a file or a selection by sliding or typing the figure in.

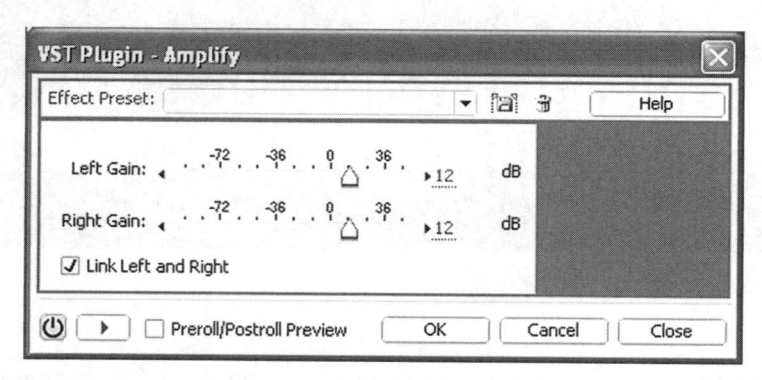

Figure 9.4 Effects/Amplitude/Amplify

Effects/Amplitude/Amplify/Fade (Figure 9.5) does the same thing but can transition from one value to another over the length of the file or selection.

The Effects/Amplitude/Binaural Auto-Panner tweaks the phase of the two channels and rotates sound spatially from left to right in a seemingly circular pattern. This effect delays either the left or right channel so sounds reach each ear at different times, tricking the brain into thinking sounds are coming from either side (Figure 9.6).

Audio design

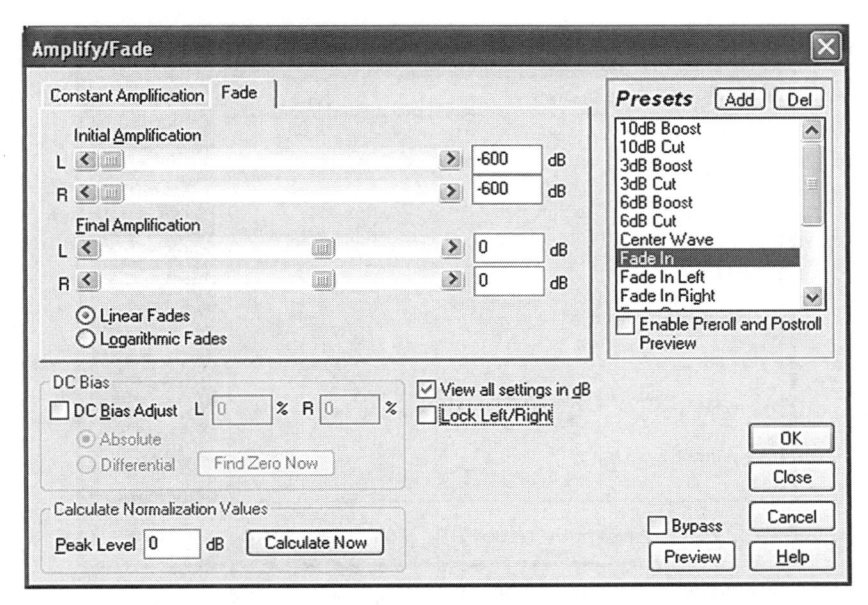

Figure 9.5 Effects/Amplitude/Amplify/Fade

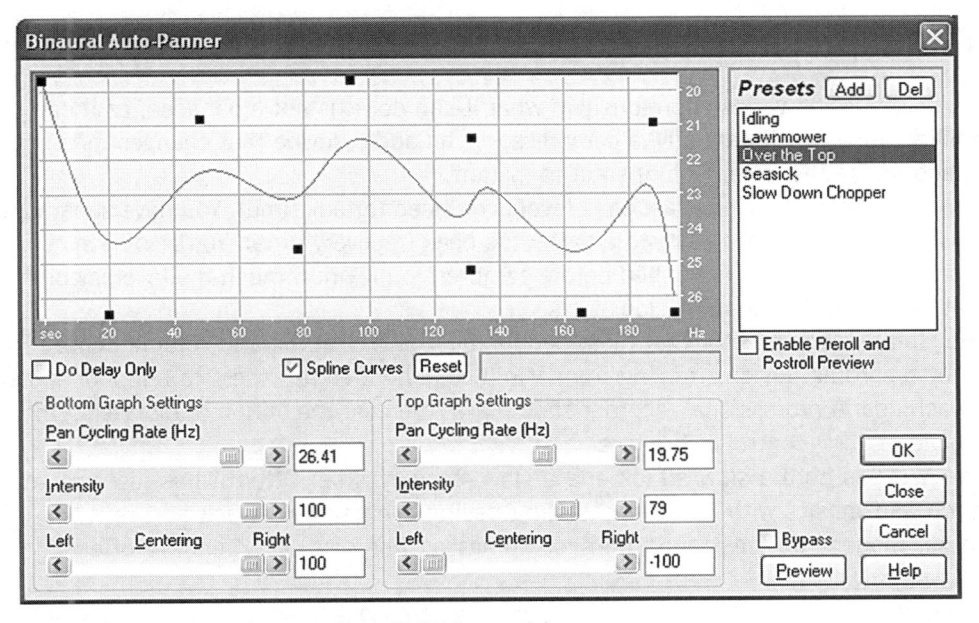

Figure 9.6 Binaural Auto-Panner effect (Edit View only)

Effects/Amplitude/Amplify/Channel mixer (Figure 9.7) facilitates cross mixing between the two stereo channels. The invert options allow you to use phase tricks including generating and decoding MS (Middle–Side) stereo (see Section 9.12 'Spatial effects').

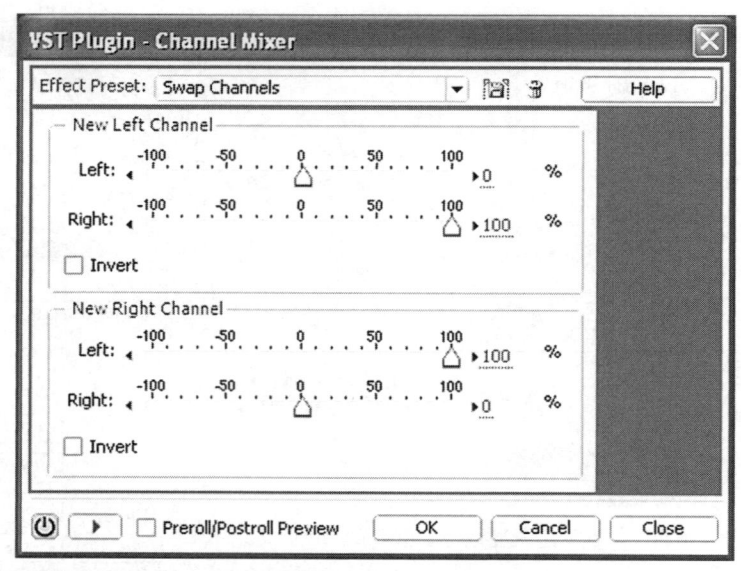

Figure 9.7 Effects/Amplitude/Amplify/Channel mixer

9.4 Compression/limiting

Audio compression varies enormously in what it can do. It is not a panacea, as it can cause audibility problems. Essentially, a compressor is an audio device that changes its amplification, and hence the level, from instant to instant.

You have control over how fast peak level is reduced (attack time). You have separate control over how fast the level is restored after the peak (recovery time). You also have control of how much amplification is applied before compression, by how much the level is controlled, and at what level the compression starts.

The simplest compressor that most people meet is the 'automatic volume control' (AVC) used by recording devices ranging from Minidisc and video recorders to telephone answering machines. A professional recorder should give the user the option of bypassing the AVC, to enable manual control of the level.

The AVC has both a slow attack and a slow recovery time. Often it has just two settings (speech and music) with the music setting having even slower attack and recovery times. Even so, modern devices found on such things as Minidisc recorders are often relatively unobtrusive.

The normal advice, with analog recording, is not to use AVC where the resulting recording is going to be edited. This is because the background noise goes up and down at the same time and any sudden change at an edit will be very obvious. With digital recording and editing the arguments are more evenly balanced, especially when digital editing is being used.

Analog recorders overload 'gracefully' and the increase in distortion is relatively acceptable, for short durations. Digital recorders are not so forgiving. They record audio as a string of numbers and have a maximum value of number that they can record. This corresponds to peak level on the meter.

The CD standard, also used by Minidisc and DAT, is 16 bit, which allows 65 536 discrete levels occupying the numbers +32 768 through 0 to −32 767. Any attempt to record audio of a higher level than this will cause a string of numbers set at the same peak value. When played back, this produces very nasty clicks, thumps or grating distortion.

When recording items like vox pops, by the very nature of an in-the-street interview, there is likely to be a very high background noise. This has always been a classic case where switching off any AVC was regarded as essential, because of the resulting traffic noise bumps on edits. However, the likelihood of short- or medium-term overload is quite considerable. On the other hand, sudden short, very high-level sounds such as exhaust backfires and gunshots can 'duck' the AVC to inaudibility for several seconds.

Digital editors make matching levels at edits the work of seconds. The danger of overload is such that this correction may be preferable to risking losing a recording owing to digital overload. In reality, it is down to the user to decide how good their recording machine's AVC is compared with how well it copes with overloads. They may prefer deliberately to under-record, giving 'headroom'.

Most portable recorders have some form of overload limiting which may allow you to return to base with 'usable' recordings, but it is better to return with good recordings. In the end, a saving grace is that it is the nature of these environments that the background noise is quite high and will drown any extra noise from deliberately recording with a high headroom.

Compression may make the audio louder, but it also brings up the background noise rather more than the foreground. This can mean that noise that was not a problem, becomes one. It will also bring up the reflections that are the natural reverberation of a room. This can change an open, but OK, piece of actuality into something that sounds as if the microphone had been lying on the floor pointing the wrong way.

If you are creating material for broadcast use you should also be aware that virtually all radio stations compress their output at the feed to the transmitter. While the transmitter processors used are quite sophisticated, they will be adding compression to any you have used. This can mean that something that sounds only just acceptable on your office headphones may sound like an acoustic slum on air.

There are standard options on any compressor that are emulated in any compression software. The compressor first amplifies the sound and then dynamically reduces the level (compression software is often found in the menu labelled as 'Dynamics'). Compressors can be thought of as 'electronic faders' that are controlled by the level of the audio at their input. With no input or very quiet inputs they have a fixed amplification (gain), usually controlled by separate input and output gain controls. Once the input level reaches a threshold level, their amplification reduces as the input is further increased. The output level still increases, but only as a set fraction of the increase in input level. This fraction is called the compression ratio (Figure 9.8).

A compression ratio of 2 : 1 means that once the audio is above the threshold level then the output increases by 1 dB for every 2 dB increase in input level.

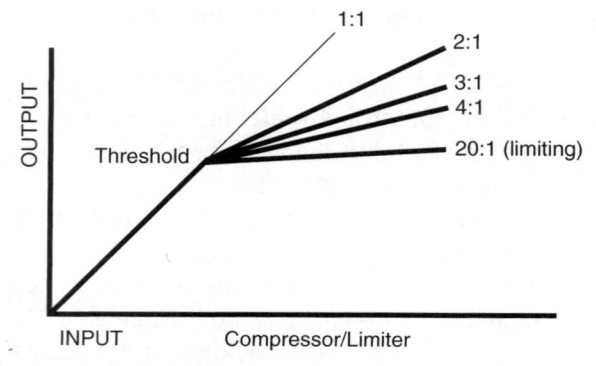

Figure 9.8 Compression ratios

A ratio of 3 : 1 means that once the audio is above the threshold level then the output increases by 1 dB for every 3 dB increase in input.

Compression ratios of 10 : 1, or greater, give very little change of output for large changes of input. These settings are described as limiting settings and the device is said to be acting as a limiter.

Hard limiting

Hard limiting can be useful to get better level out of a recording that is very 'peaky'. Some people have quiet voices with some syllables being unexpectedly loud. Hard limiting can chop these peaks off with little audible effect on the final recording, except that it is now louder. Adobe Audition 2.0 has a separate effects transform for this.

Threshold

The Threshold control adjusts at what level the compression or limiting starts to take place. By the very nature of limiting, the threshold should be set near to the peak level required. With compression, it should be set lower; 8 dB below peak is a good starting place.

Attack

Control for adjusting how quickly the gain is reduced when a high-level signal is encountered. If this is set to be very short then the natural character of sounds is softened and all the 'edge' is taken out of them. It can be counterproductive as this tends to make things sound quieter. Set too slow an attack time and the compression is ineffective. A good place to start when rehearsing the effect of compression is 50 milliseconds. With musical instruments much of their character is perceived through their starting transients. Too much compression with a poor choice of attack time can make them sound wrong. On the other hand, a compressor set with a relatively slow attack time can actually improve the sound of a poor bass drum by giving it an artificial attack that the original soggy sound lacks.

Release/recovery

These are alternative names for the control that adjusts how fast the gain is restored once the signal is reduced. Set too short and the gain recovers between syllables giving an audible 'pumping' sound which is usually not wanted. This makes speech very breathy. A good starting point is 500 milliseconds to 1 second.

Input

Adjusts the amplification before the input to the compressor/limiter part of the circuit. With software graphical display controls this may, at first, appear not to be present but is represented by the slope of the initial part of the graph.

Output

Adjusts the amplification after the compressor/limiter part of the circuit. Because it is after compression has taken place it also appears to change the threshold level on the output. Software compressors often have the option automatically to compensate for any output level reduction due to the dynamic gain reduction, by resetting the output gain.

In/out, bypass

A quick way of taking a device out of the circuit for comparison purposes.

Link/stereo

Two channels are linked together for stereo so that they always have the same gain despite any differences in levels between the left and right channels. If this is not done then you may get violent image swinging on the centre sounds such as vocals.

9.5 Expanders and gates

Adobe Audition 2.0's dynamics can also act as expanders and gates. These work in a very similar way to compressor/limiters but have the reverse effect. With high-level inputs they have a fixed maximum gain. Once the input level decreases to a threshold level then their gain decreases as the input is further decreased. When used as expanders the output levels still decrease but only as a set fraction of the decrease in input level. This fraction is called the expansion ratio (Figure 9.9).

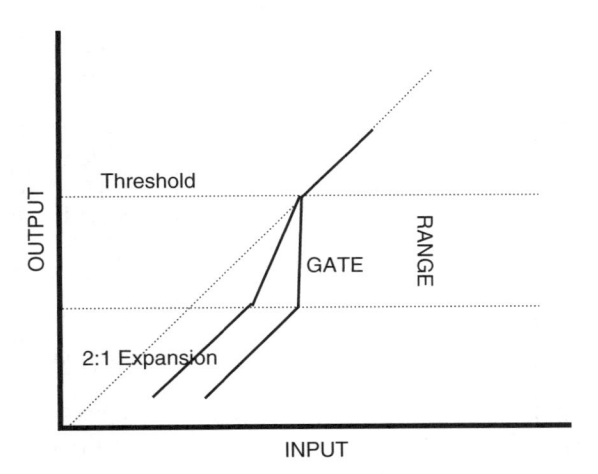

Figure 9.9 Expander and gate actions

A expansion ratio of 2:1 means that once the audio is below the threshold level then the output decreases by 2 dB for every 1 dB decrease in input level. A ratio of 3:1 means that the output decreases by 3 dB for every 1 dB decrease in input. Expansion ratios of 10:1 or greater give a very large change of output for little changes of input and the device is said to be acting as a gate.

Unlike compressors, which were originally developed for engineering purposes, the expander is an artistic device and must be adjusted by ear. Changing its various parameters can dramatically change the sound of the sources. Its obvious use is for reducing the amount of audible spill. This is usually restricted by the fact that too much gating or expansion will change the sound of the primary instrument. However, this very fact has led to gates and expanders being used deliberately to modify the sounds of instruments – particularly drums. Their use is almost entirely with multitrack music; bass drums are almost invariably gated.

Expanders and gates are bad news for speech. They change the background noise. The ear tends to latch on to this rather than listening to what is said. While I can imagine circumstances where they might help a speech recording, I have not yet met any.

Dynamics transform (Figures 9.10–9.13)

Adobe Audition 2.0's Dynamics transform is a very flexible implementation of dynamics control. It provides you with alternative ways of inputting what you want, along with a profusion of presets. A tour through these will educate your ears as to what can be done. There are four tabbed pages within the dialog box; 'Graphic', 'Traditional', 'Attack/Release' and 'Band Limiting'. The first two tabs give you alternative ways of entering what you want. All four tabs allow you access to the presets.

With the graphical tab (Figure 9.10) you can 'draw' the dynamics that you want. Click on the line to give you a 'handle' and then move it where you want. It is usually quicker to start

Figure 9.10 Graphic

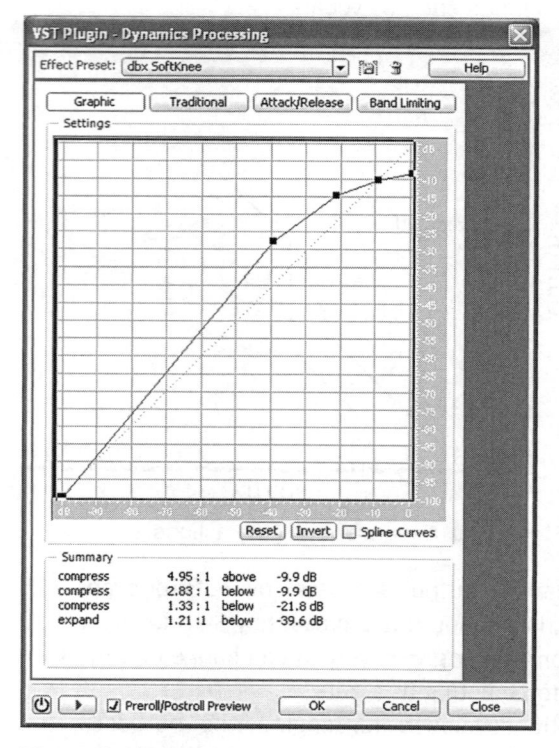

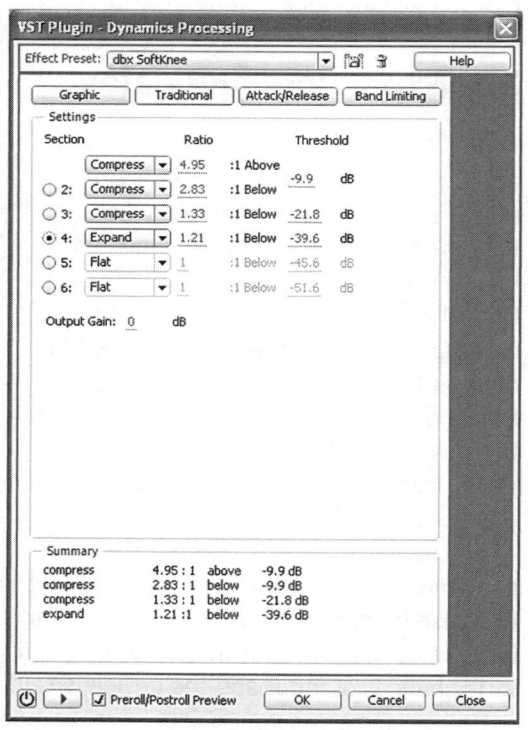

Figure 9.11 Traditional

Audio design

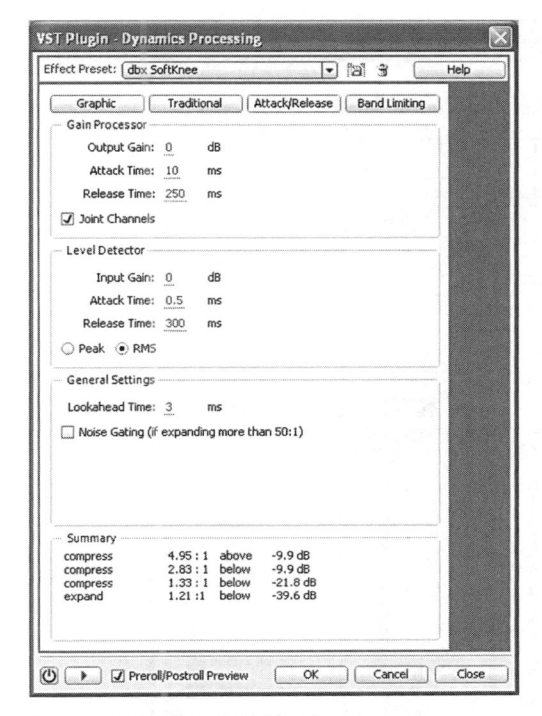

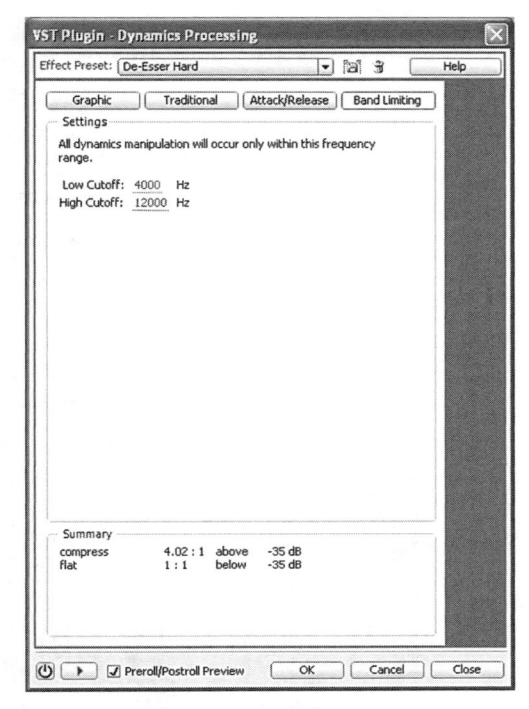

Figure 9.12 Attack/release

Figure 9.13 Band limiting

from a preset and adjust the settings if required. Illustrated is a 'soft knee' compression of the sort associated with the *dbx* company.

Figure 9.11 shows the same setting in the 'Traditional' tabbed page, with each section labelled as text. 'Output compensation allows you to boost the level after compression, if this is needed.

Figure 9.12 is where you can setting the attack and release times. How you set these times will substantially affect the sound of an instrument as it will modify its start and finish transients. A very quick recovery time with compression will bring up background noise and increase the apparent amount of reverberation.

Figure 9.13, 'Band Limiting' allows you to restrict the frequency range that limiting occurs. The setting shown is one that is used in one of the 'De-essing' presets. De-essers do what you would think they reduce the level of sibilants, 'S' sounds in a vocalist's voice. You should apply this after any full frequency range compression as otherwise the compression can restore the sibilance!

9.6 Multiband compressor

Adobe Audition 2.0 has a full scale multiband limiter of the sort often used to make CDs and broadcasts sound louder. They split the audio into bands and separately compress each band.

The one supplied with Audition has four bands. (Figures 9.14 and 9.15)

The controls can be a bit overwhelming at first so begin familiarizing yourself by trying the various presets. You can hear what is being done by soloing each band with the 'S' button or you can

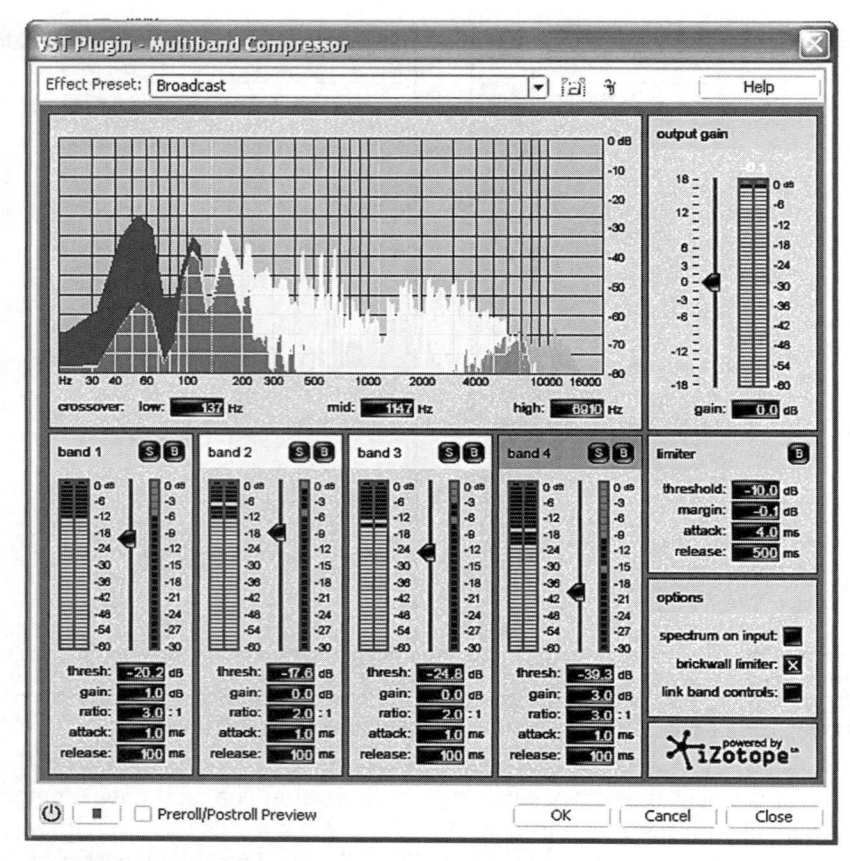

Figure 9.14 Multiband compressor

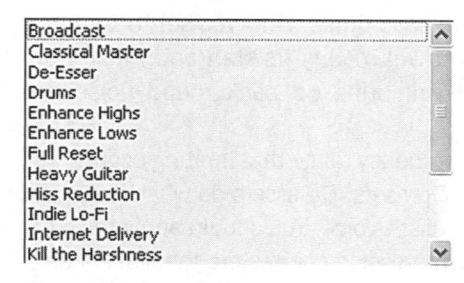

Figure 9.15

bypass it altogether with the 'B' button. You can apply the multiband compression in real time during a mix – conveniently it can be inserted into the main bus – or you can apply it to the mixdown afterwards in the edit view when the controls at the bottom will appear.

This is probably the prettiest of the Adobe Audition 2.0 displays and there is certainly a lot going on. The panel is dominated by the spectrum display at the top with the output level and gain control at the top. Three vertical white lines indicate the crossover between bands. They can be slid with the mouse or the numbers typed or dragged beneath. You might expect the lines to be in the centre of the spectrum for each band but this is not apparent because the frequency display is logarithmic. Each band has its own control panel with an output level meter, a threshold control and a gain reduction meter which operates downwards. There are numerical boxes for threshold (working with the slider) gain, compression ratio, attack and release times.

Audio design

Beware of being seduced by multiband processing; it makes for a forward impressive sound and make for extra clarity in difficult listening conditions such as in an automobile. However, it is surprising how often something that impresses for 5 minutes becomes tiring and headache-inducing in 30 minutes.

Tube-modelled compressor

The 'tube-modelled' compressor tries to simulate the old valve (vacuum tube) devices. It is bought in from the same company that provides the multiband compressor and has an identical control panel to a single band of its brother. It is actually configured as a VST plug in and can be found using the Effects/VST menu selection (Figure 9.16).

Normalize

It make sense to normalize files before working on them. Normalization is explained in Chapter 4.

Pan/Expand

This allows you to change the width and pan settings of a file. Its basically an MS control which can pan the M signal and change the level of the S signal (see Section 9.12 'Spatial Effects') (Figure 9.17).

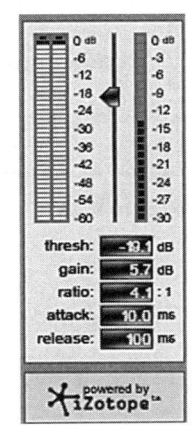

Figure 9.16 Normalize

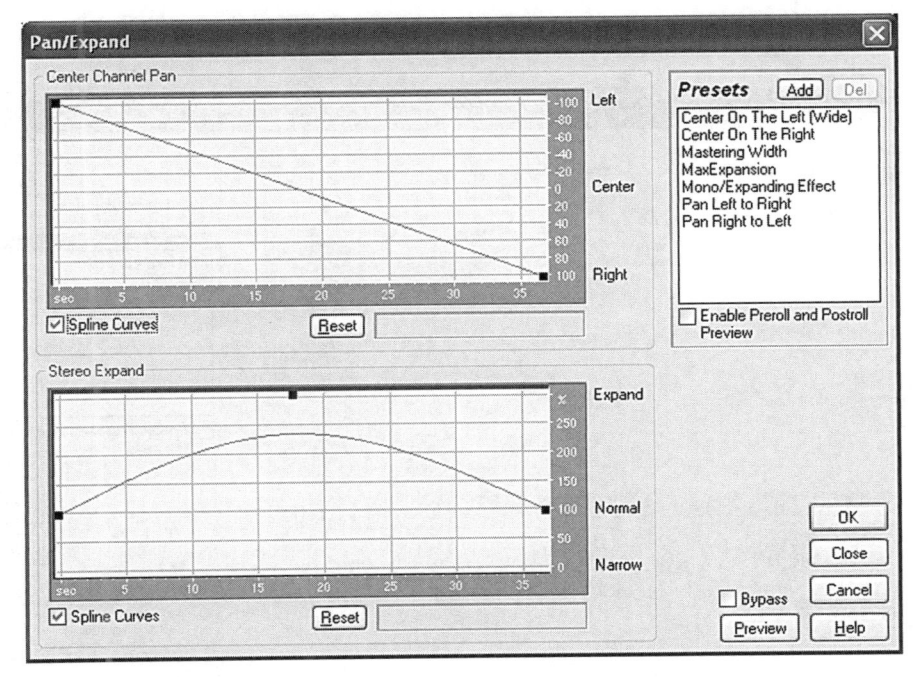

Figure 9.17 Pan/Expand

The stereo expander option (Figure 9.18) is a non-dynamic version which applies a constant change throughout a files or selection.

The stereo field rotate has two options, process and non-process. The VST plug-in version can be used as a real-time effect in the multitrack view and be controlled by an automation lane. The process version can be varied over the length of a file in Edit view and then fixed by saving the file (Figure 9.19).

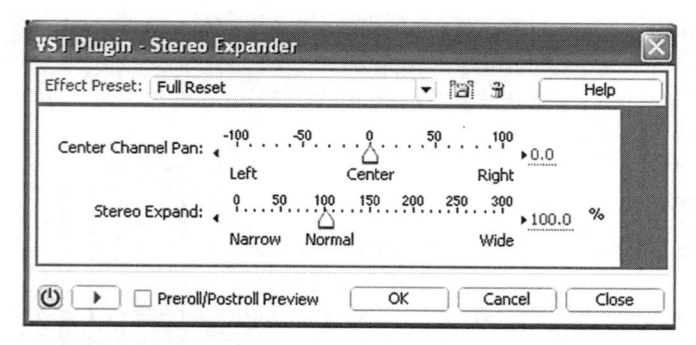

Figure 9.18 The stereo expander option

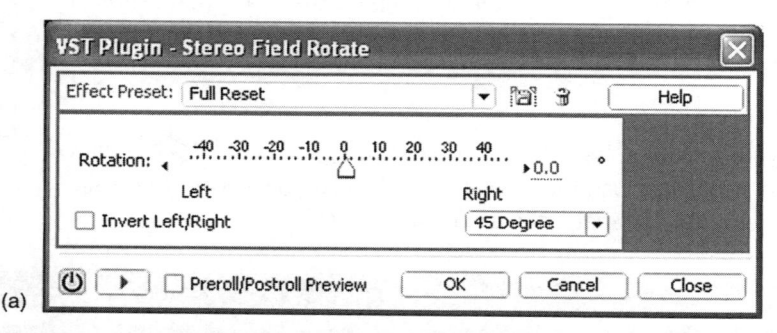

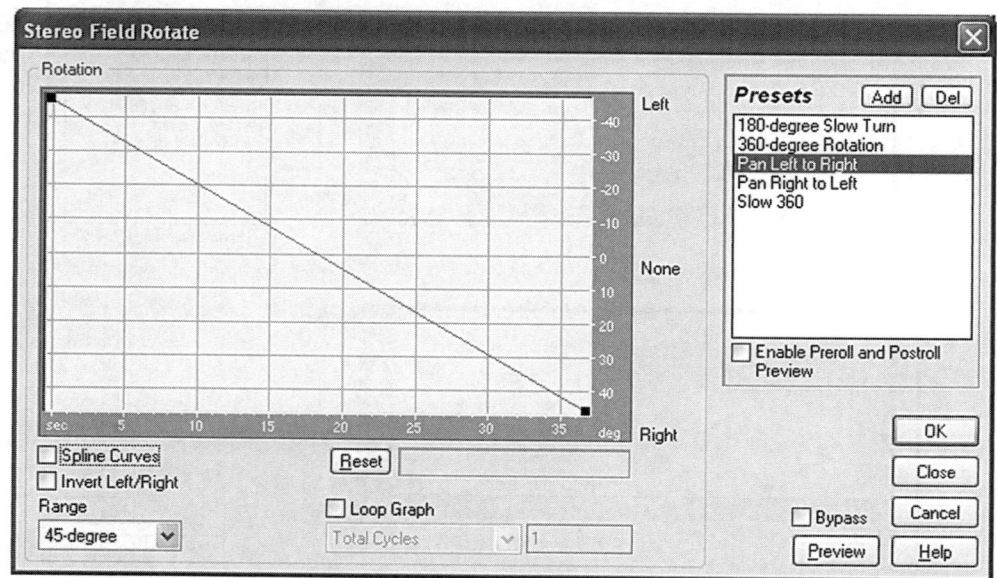

Figure 9.19 The stereo field rotate

Audio design

9.7 Reverberation and echo

Reverberation (often abbreviated to 'reverb') and echo are similar effects.

Echo is a simpler form where each reflection can be distinctly heard (Echo ... Echo ... Echo ... Echo). Reverberation has so many different reflections that it is heard as a continuous sound. For reasons more owing to tradition than logic, broadcasters often refer to reverberation as 'echo' and use the term 'flutter echo' for echo itself.

Reverberation time

The amount of reverberation is usually quoted in seconds of reverberation time. This is defined as the time it takes for a pulse of sound (such as a gun shot) to reduce (or decay) by 60 dB. This ties in well with how the ear perceives the length of the reverberation.

Diffusion

Simulates natural absorption so that high frequencies are reduced (attenuated) as the reverb decays. Faster absorption times simulate rooms that are occupied and have furniture and carpeting, such as night clubs and theatres. Slower times (especially over 1000 milliseconds) simulate rooms that are emptier, such as auditoriums, where higher frequency reflections are more prevalent. In acoustic environments, higher frequencies tend to be absorbed faster than lower frequencies.

Perception

Adds subtle qualities to the environment by changing the characteristics of the reflections that occur within a room. Lower values create smoother reverb without as many distinct echoes. Higher values simulate larger rooms, cause more variation in the reverb amplitudes, and add spaciousness by creating distinct reflections over time (Figure 9.21).

Devices

Reverb is the first and oldest of studio special effects. Originally it was created by feeding a loudspeaker in a bare room and picking up the sound with a microphone (Figure 9.20). Although this worked, there was no way of changing the style of reverberation. It was also prone to extraneous noises ranging from traffic, hammering, underground trains or even telephones ringing inside the room.

The next development was the 'echo plate'. A large sheet of metal (the plate) about 2 m × 1 m was suspended in a box. There was a device to vibrate the plate with audio and two pickups (left and right for stereo) to receive the reverberated sound.

This worked on a similar principle to a theatre 'thunder sheet'. The trick was to prevent it sounding like a thunder sheet! This was achieved by surrounding it with damping sheets. By moving these dampers away from the sheet, you would increase the reverberation time (usually

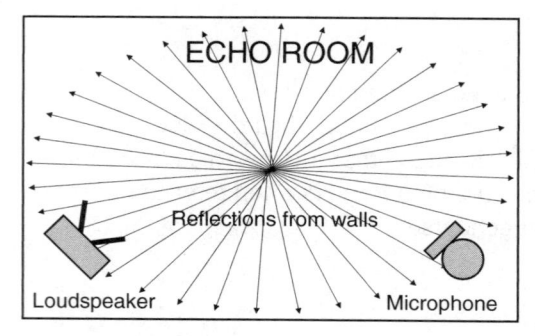

Figure 9.20 'Echo' Room set-up

at the cost of a more metallic sound). Often, a motor was fitted allowing remote control. In a large building with many studios, a handful of the devices could be shared using a central switching system.

Devices were also made using metal springs. Their quality was often not good but they took up little space and were 'good enough' for many purposes like reverb on disc jockeys on Hallowe'en.

Digital

These days digital reverberation devices reign supreme. The software used on these machines is equally able to be written for use on a PC. Like the separate devices it emulates, the PC will usually offer a number of different 'programs' with different acoustics (Figure 9.22). This will often include emulation of the old plates and room, but also include concert hall acoustics, along with special effects that could not exist in 'real life'.

Reverberation is a complex area as it triggers various complex subconscious cues in the brain. There are two main pieces of information we get from how 'echoey' a recording of

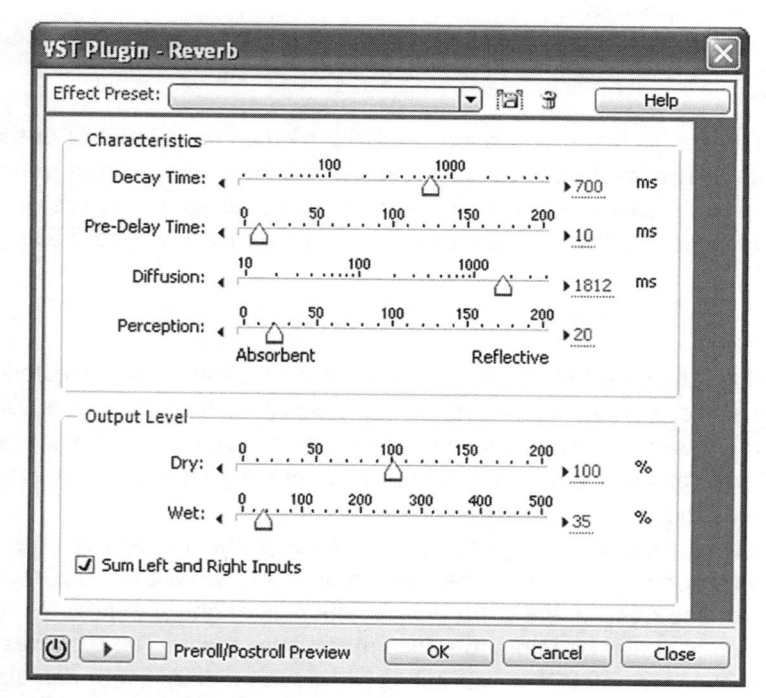

Figure 9.21 Reverb dialog giving the simplest of options

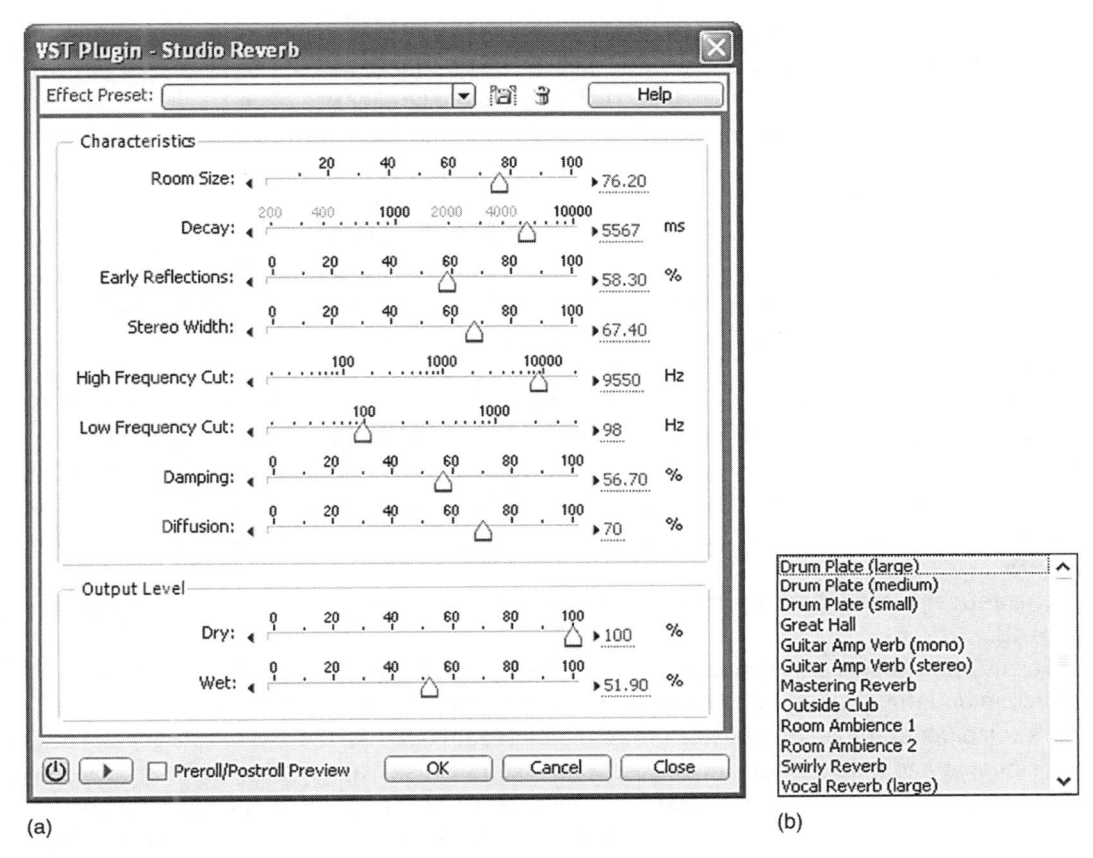

Figure 9.22 Studio Reverb dialog from Adobe Audition 2.0 showing presets.

someone speaking, or talking, is. They both interact, but need separately to be controlled. These are the size of the room, and how far away from the microphone the voice is. Move the voice further from the mic and the more echoey the recording becomes. If the microphone is held and the person walks into a larger room, the more echoey it becomes. However, we are also adept at distinguishing between different types of room. We can hear the difference between an underground car park and a concert hall, a hallway from a living room. How do we do this?

Let us imagine we are recording in the centre of a large, open acoustic, suspended in the air, as well as centred over the floor. Now make a sharp impulsive noise like a single hand clap, while making a recording. When we examine the waveform of the recording we will see that there is a moment of silence between the hand clap and the start of the reverberation. This is because reverberation is caused by sound reflecting off surfaces such as walls, floors and ceilings. Sound does not travel instantaneously. Its actual speed varies with the temperature and humidity of the air (one of the reasons, the acoustic of a hall can change so dramatically once the audience arrives, even if the seats have been heavily upholstered to emulate the absorption of a person sitting in them). Sound travels at roughly 330 metres per second, or about 1 foot per second. This means that if the nearest reflecting surface is 50 feet away then you will not hear that first reflection for 100/1000th of a second (100 milliseconds). It is the size of this delay that is our major cue for perceiving how large a room a recording was made in. An unfurnished living room may be as reverberant as a cathedral but it will have short delay on the reflections from the walls (10–20 milliseconds) and so can never be confused with the church.

Returning safely to floor level, we can make another recording, with someone speaking and moving towards, and then away from, the microphone. The microphone picks up two elements of the voice (as opposed to the ambient background noise from the environment). These elements are the sound of the voice received directly and the sound received indirectly via reflections.

As the voice moves nearer to, and further from, the microphone, the amount of indirect sound does not really change in any practical way. However, the voice's direct level does change. It is very easy to imagine that it is the reverberation that is changing. This is because a recordist will be setting level on the voice. As it gets closer so the louder it will become. The recordist compensates by turning down the recording level and so apparently reduces the Reverb.

Theoretically, every time the distance from the microphone is doubled, the level will drop by 6 dB.

This is often known as the inverse square law. While this is true for an infinitely small sound source in an infinitely large volume of air, this is not totally true for real sound sources. A person's voice, for example, comes not only from their mouth but from their whole chest area. Only the sibilants and mouth noises (like loose, false teeth and saliva) are small so they disappear much faster than the rest of the voice (one reason why mic placement can be so critical).

As a generalization, if the delays on the reflections remain the same, then the ear interprets relative changes in Reverb level as movement of the sound source relative to the microphone. If the delays change then the ear interprets this as the mic moving with the voice into a different acoustic.

The reason why reflections die away is that only a fraction of the energy is returned each time. Hard stone surfaces reflect well; soft furnishings and carpets do not. Such things as good quality wallpaper on a hard surface will absorb high frequencies but reflect low frequencies as well as does the bare hard surface.

Another reverberation parameter is its smoothness. Lots of flat surfaces give a lumpy, hard quality of reverberation. A hall full of curved surfaces, heavily decorated with carvings, will give a much smoother sound. This does much to explain why many nineteenth-century concert halls sound so much better than their mid-twentieth-century replacements, built when clean, flat surfaces were fashionable. So much more is now known about the design of halls that there is little excuse for getting it wrong in the twenty-first century.

When creating an artificial acoustic, all these mistakes are yours to make! As always with audio processing, it is always very easy to make a nasty 'science fiction' sound, but so much more difficult to create something magical and beautiful. The effects transforms within all good digital audio editors are provided with presets. It saves time to use them albeit modifying them slightly. When you find a setting that you really like, you can save it as a new preset.

Audio design

Summarizing, a simple minimalist external reverberation device would include options such as:

1 *Program*: Setting the basic simulation; reverb-plate, garage, concert hall, etc.
2 *Predelay*: This controls how much the sound is delayed before the simulated reflections are heard. This normally varies from zero to 200 milliseconds. As sound travels at approximately 1 foot per millisecond (1/1000th of a second), a predelay of 200 milliseconds (1/5th second) will emulate a large room with the walls 100 feet (30 metre) away (200 feet there and back).
3 *Decay*: This controls the reverberation time. This is the apparent 'hardness' or 'softness' of a room. A room with bare shiny walls will have a long reverberation time. A room with carpets, curtains and furniture will have a short Reverb time.
4 *High Frequency rolloff*: This is an extra parameter that shortens the Reverb time for the high frequencies only, making the 'room' seem more absorbent and well furnished.
5 *Input and output*: Adjusts the input or output level.
6 *Mix*: Controls a mix between reverberation and the direct sound you are starting with. Sometimes, there are separate controls labelled 'dry' and 'wet'. As a dry acoustic is one without much reverberation, so pure reverberation is thought of as the wet signal.

Let us now take a closer look at how Adobe Audition 2.0 implements this. Reverberation and echo comes under the submenu of delay effects and the choice is potentially overwhelming! (See Figure 9.23.)

Chorus (Figure 9.24)

This is an effect used a great deal in pop music. By producing randomly delayed electronic copies of a track, it can make one singer sound like several or many. It can also be used to thicken musical tracks and other sounds. The random delays inherently cause a random pitch change. This can sound very nasty especially on spill behind a vocal.

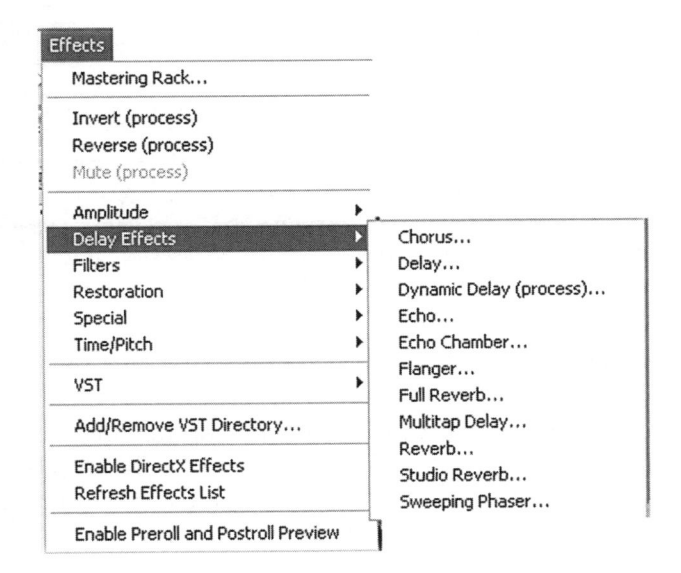

Figure 9.23 Delay Effects submenu.

Outside of music, chorus is not particularly useful except that the various options provided by chorus can produce interesting stylized acoustic effects, not dissimilar to adding echo. The variable delay pitch variation can help produce effects suitable for science fiction.

103

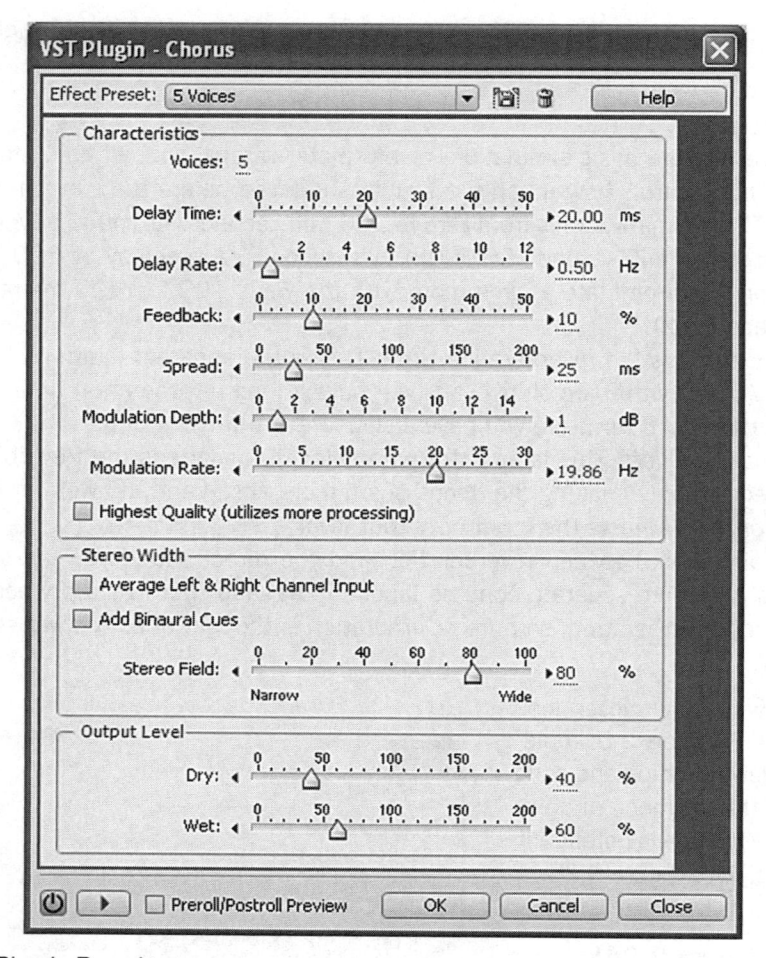

Figure 9.24 Plug-in Reverb

As well as the pitch changing caused by the variable delay, the vibrato settings introduce random amplitude changes.

Rather than me trying to describe all the effects, try changing them yourself using the demo version of Adobe Audition 2.0 on the CD supplied with this book. Using a recording of a single voice, give yourself a tour of the preset effects using the preview button. There will be sounds there you will find useful one day.

Delay (Figure 9.25)

This effects transform is the equivalent of tape flutter echo where the output of the playback head was fed back to the record input. While not the most subtle effect, it does have its uses. Emulating Railway station announcements and 1960s Rock 'n' Roll come to mind. Again,

Audio design

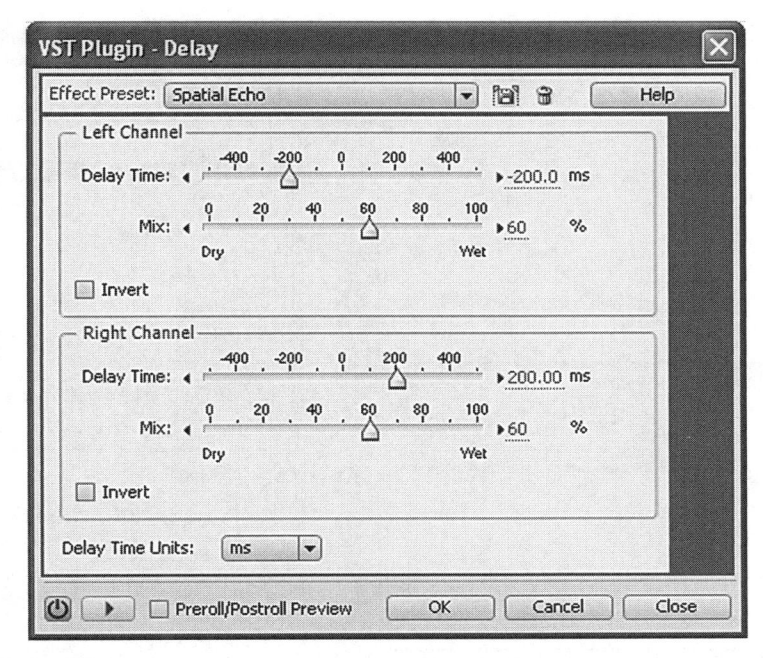

Figure 9.25 Delay effect

give yourself a tour around the presets. Used as a real-time effect in the multitrack view Delay's parameters can be controlled by envelopes. With the parameter box displayed in the effects rack you can see the sliders moving in response to the envelope settings as you play though the session.

Dynamic Delay (Figure 9.26)

The Dynamic delay effects transform allows you to flange the sound in a graphically controlled way. By varying the delay between the original and a copy you get a comb filter effect; the 'psychedelic' sound associated with 1960s pop music. This is a 'Process' effect that has to be applied to a file and cannot be used in real time.

In those days the only practicable way of apply variable delays was to feed the music to be treated to two identical tape recorders and combine the outputs of the two playback heads. Because of small mechanical imperfections, the two delays introduced by the replay heads were very slightly different. They would vary slightly as the tension varied slightly and changed the stretch of the tape. This became known as *phasing*. It causes a notch in the frequency response giving a drainpipe type sound as shown in Figure 9.27–9.29. This is because the two signals cancel where half the recorded wavelength of the audio becomes equal to the delay. Depending on the delays involved, not only will the frequency cancelled-out change but there may well be a multiplicity of cancellations producing what is known as a *comb filter* effect.

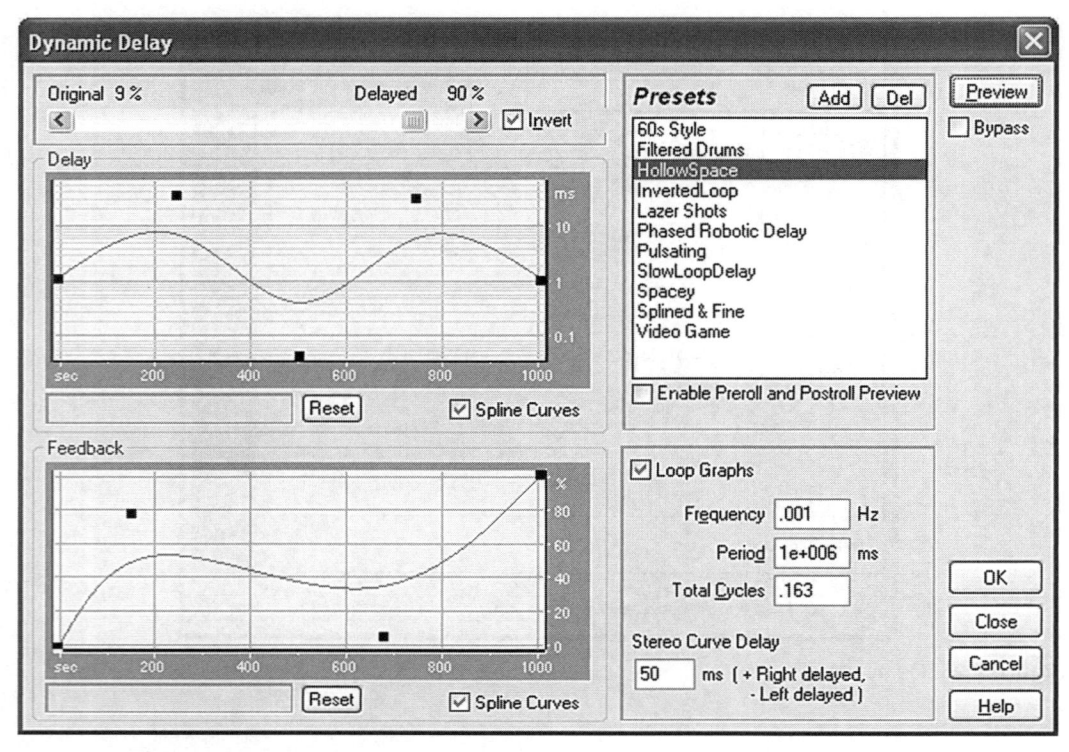

Figure 9.26 Dynamic Delay

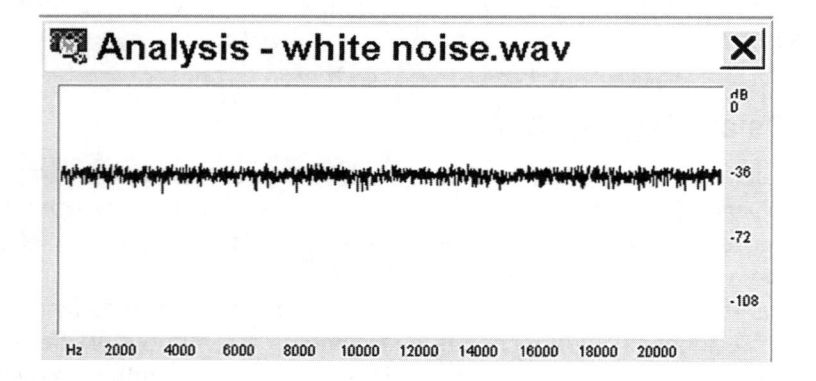

Figure 9.27

According to legend it was John Lennon of *The Beatles* who found that the delay difference could be modified subtly and controllably by putting a finger on the flange of the feed spool of one of the tape recorders. The effect became known as *flanging*. It gives the classic skying sound of the 1960s. This is a comb filter effect moving up and down the sound spectrum (Figure 9.30).

Audio design

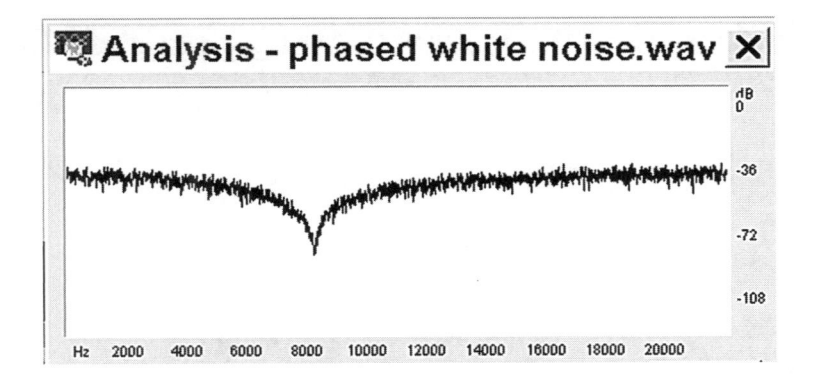

Figure 9.28

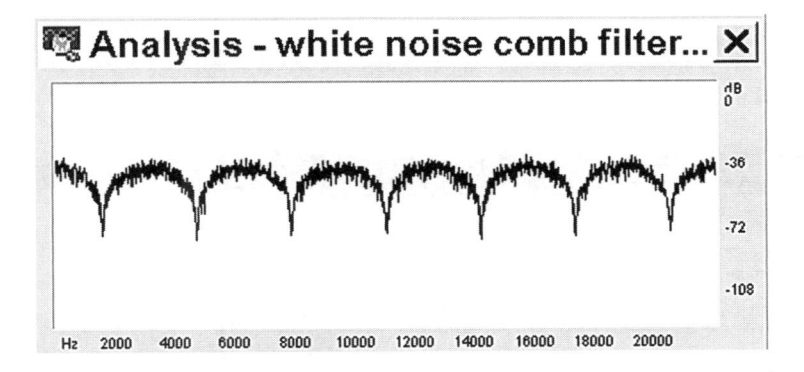

Figure 9.29

Echo (Figure 9.31)

This is very much like the delay effects transform with the additional option of adding equalization (EQ) to the delayed echo. Again some useful presets are supplied.

Echo chamber (Figure 9.32)

This approaches designing a reverberant effect in a totally different way.

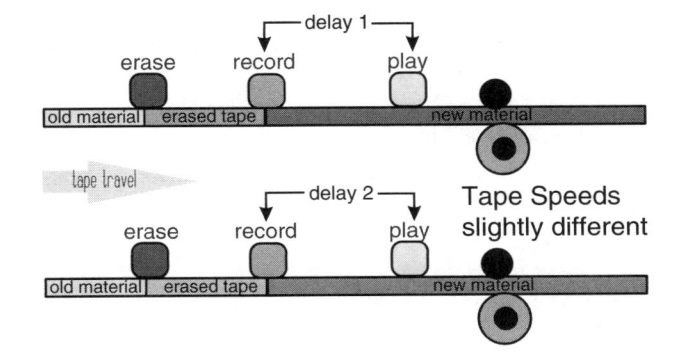

Figure 9.30 Original tape phasing

You are invited to enter the dimensions of the room you wish to simulate and where the microphone is placed. You can also set the absorption (damping factor) of the floor, ceiling and each wall. Some useful dramatic environments are supplied as presets.

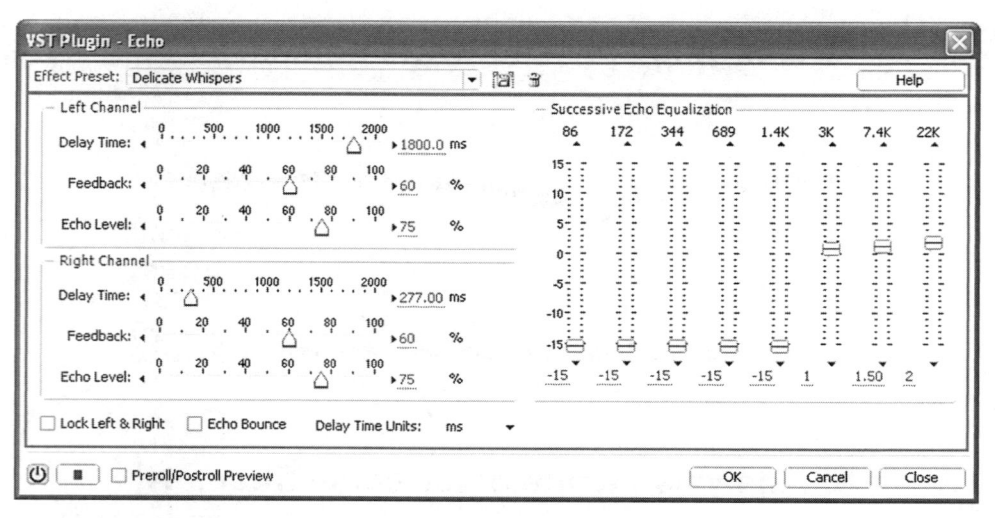

Figure 9.31 Echo

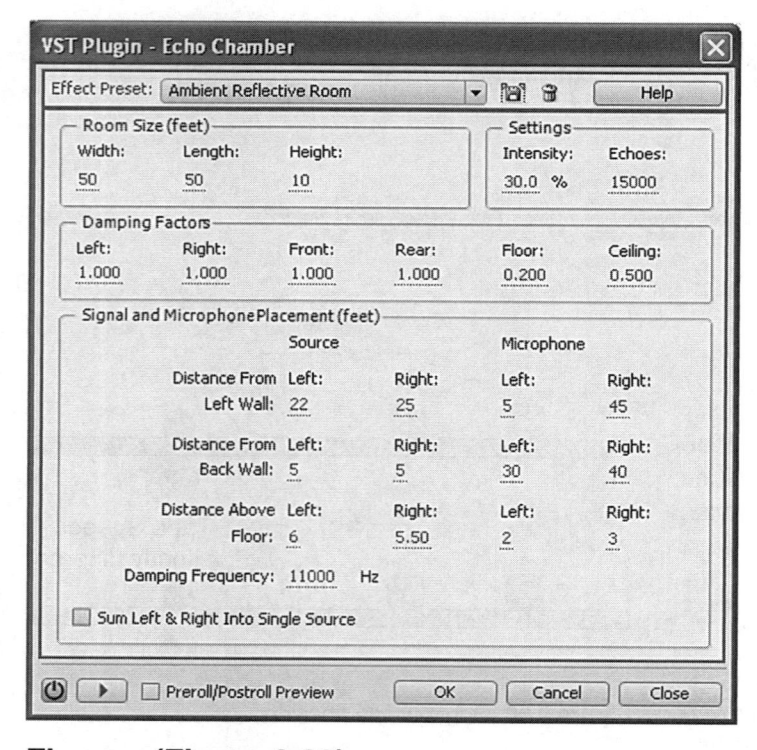

Figure 9.32 Echo chamber

Flanger (Figure 9.33)

As already described in the section above on Dynamic delay. Flanging was so named because the cancelled frequencies could be changed dynamically by pressing a finger on the flange of the feed spool on one of the tape machines. The resulting 'skying' effect is irretrievably

Audio design

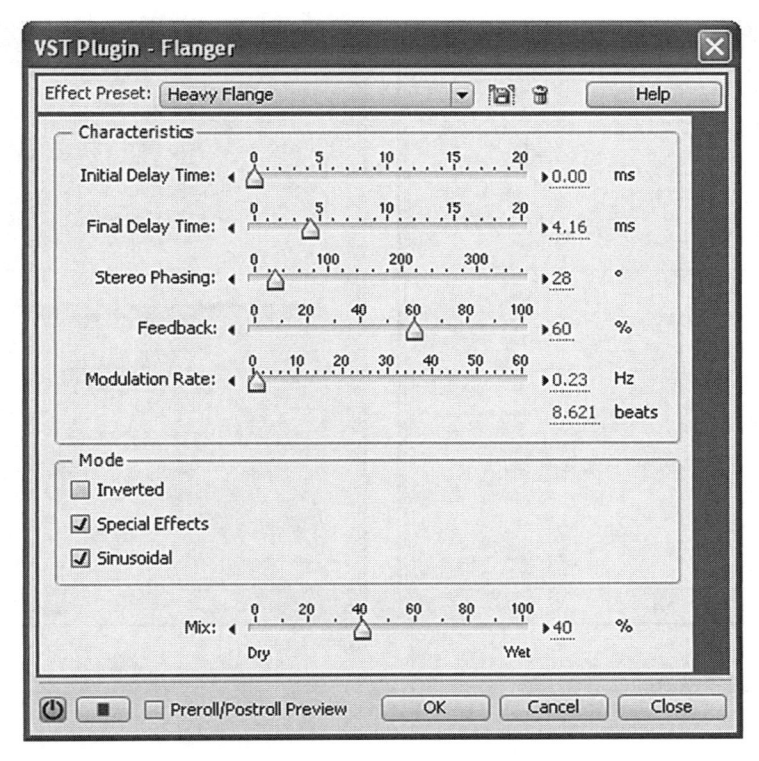

Figure 9.33 Flanger

associated with late 1960s pop music. Both phasing and flanging are available and have their own useful presets.

Full Reverb (Figure 9.34)

This is the all-singing, all-dancing Reverb option. Because of its complication, I do not recommend using this unless you have a specific need and have the time and knowledge to navigate around the options. However, as before, there are some presets that you may find useful.

There are two pages for settings (the vital ones controlling the level of the original signal (dry), early reflections and the Reverb (wet) are repeated on each page). The first page, 'general', contains the basic level controls. Section 1 has the reverb settings; Section 2 sets the early reflections which are so critical in our 'picture' of the simulated room, effectively controlling the room size.

The other page, 'coloration', gives you the ability to equalize the Reverb within the algorithm. You can move the three bands, low, mid and high, around the frequency band. The amount of coloration is adjusted by the three sliders.

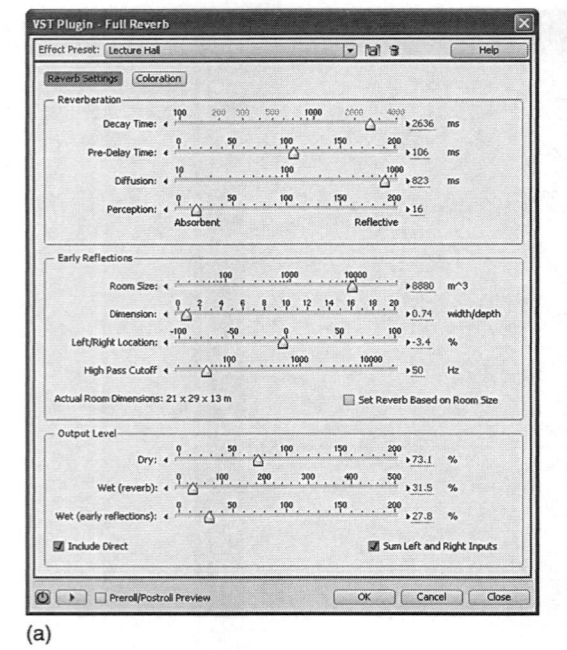

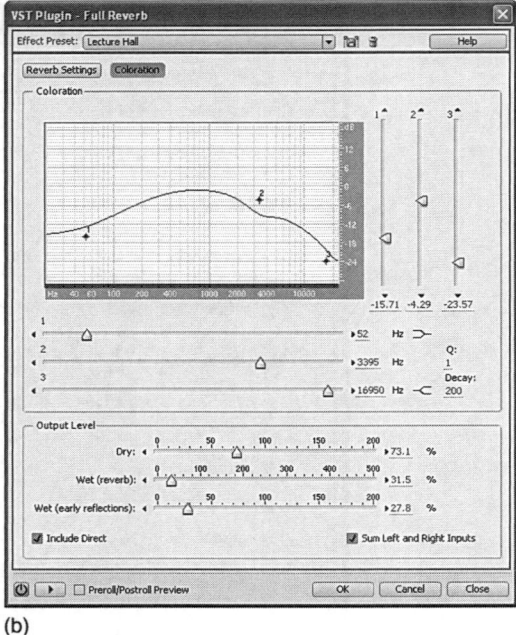

Figure 9.34 Full Reverb

You can separately control the wet/dry mix of the general reverb and the short reflections in the third panel which appears on both pages. As with the other effects transforms, the easiest approach is to use the presets and then modify as required.

Multitap delay (Figures 9.35 and 9.36)

This simulates the old tape loop delay systems. The effects produced are relatively primitive but suit certain types of music or stylized effects.

Reverb (Figure 9.37)

This is the most useful general purpose reverberation effects transform. It is flexible, but still simple to operate.

Figure 9.35 Multitap tape delay

Audio design

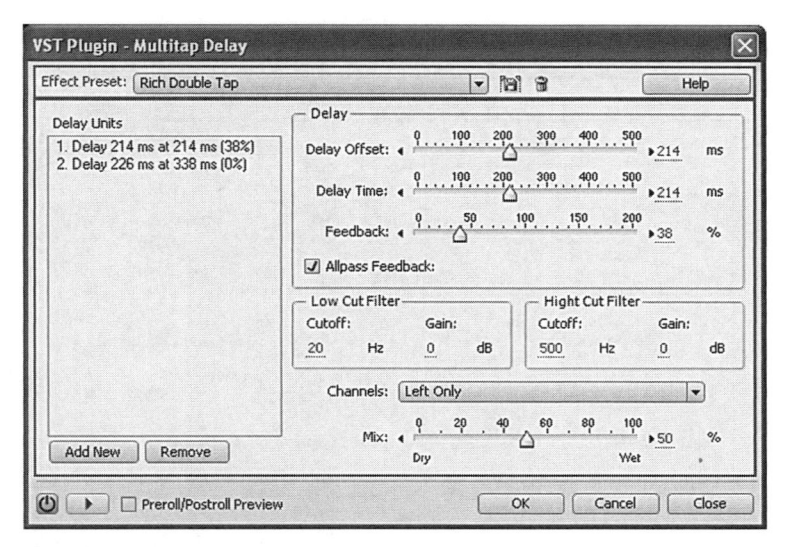

Figure 9.36 Multitap Delay

As well as a sensible range of presets, the Reverb effects transform has the basic logical controls. This is the effects transform to use if you 'just want a bit of reverb' and you are not too critical as to subtlety.

Studio Reverb (Figure 9.38)

Yet another reverb option. This has a useful reverberation stereo width option. We are so used to hearing music in stereo these days that it is very easy to forget that music that is broadcast will also be heard in mono by

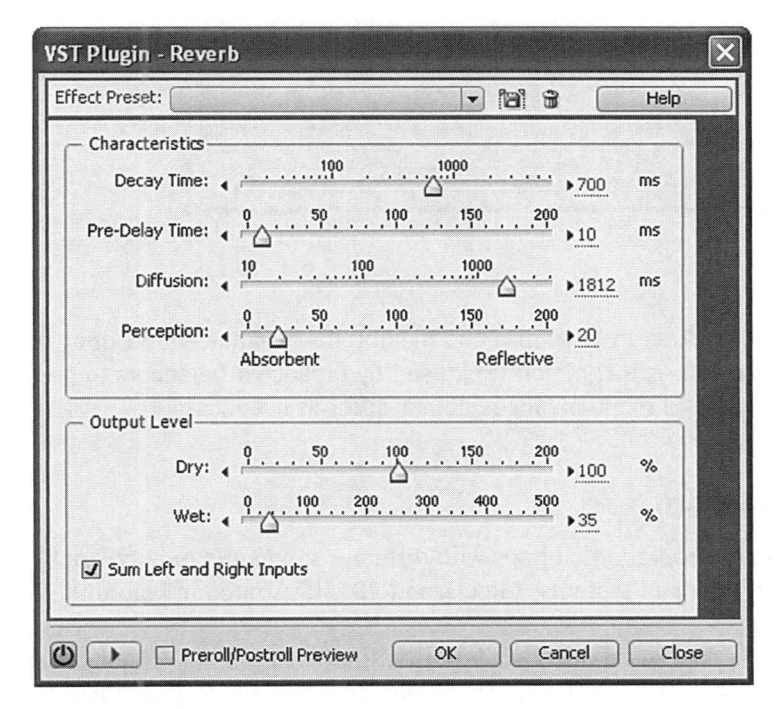

Figure 9.37 Reverb

many listeners. This is obtained by summing the left and right signals. Out-of-phase material is lost. The width of stereo reverb is caused by out-of-phase audio. This means that if the

111

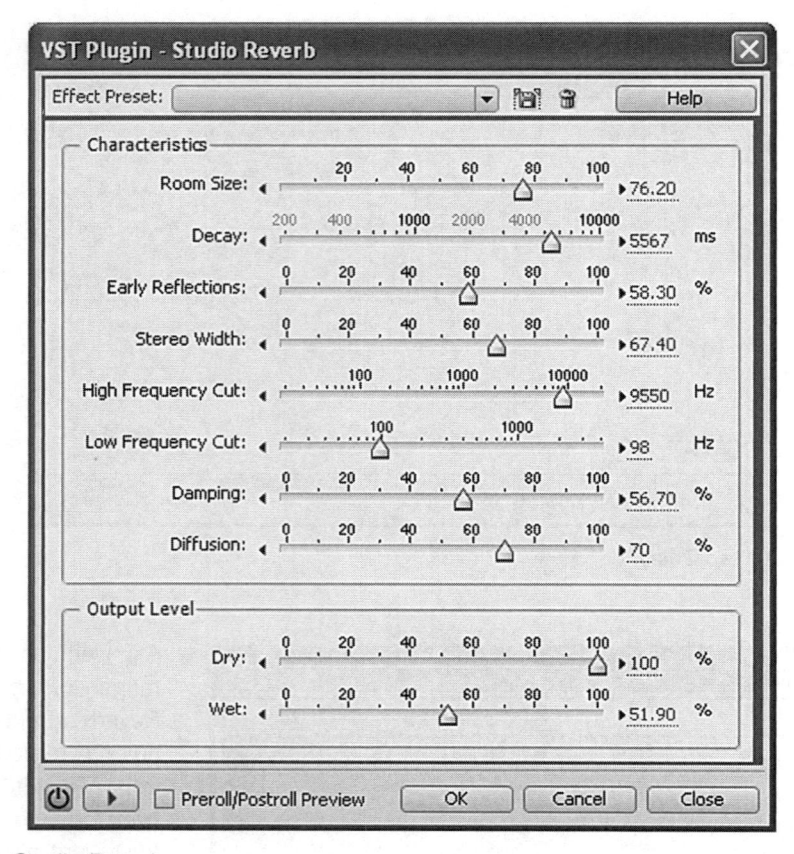

Figure 9.38 Studio Reverb

reverb is very wide then it will reduce in level in mono making the balance sound drier in mono. Slightly narrowing the reverb is a common trick used by broadcast balancers to preserve compatibility so the stereo and mono balances do not differ in liveliness.

Plug-in Reverb (Figure 9.39)

You can also buy or download reverberation plug-ins with different sound and quality. Adobe Audition 2.0 can handle both the main systems; DirectX and VST. Illustrated in Figure 9.39 is the free 'Freeverb VST' plug-in.

Many of the DirectX plug-ins can be downloaded from the Internet in demo versions like the Arboretum Hyperprism plug-in (Figure 9.40). These demos allow you to hear them to check that they do what you want. It's worth saying not all are better, or even, as good as those already built-in to Adobe Audition 2.0. The demos are invariably limited in some way. Some only allow a certain number of uses, a limit to the number of days they work or they put random bleeps into their output so that they cannot be used in a practical way.

Audio design

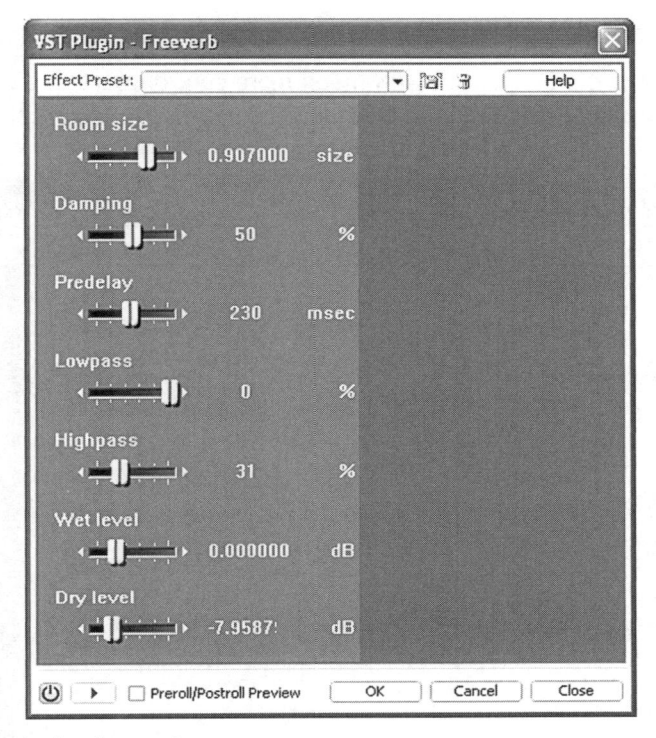

Figure 9.39 Freeverb plug-in reverb

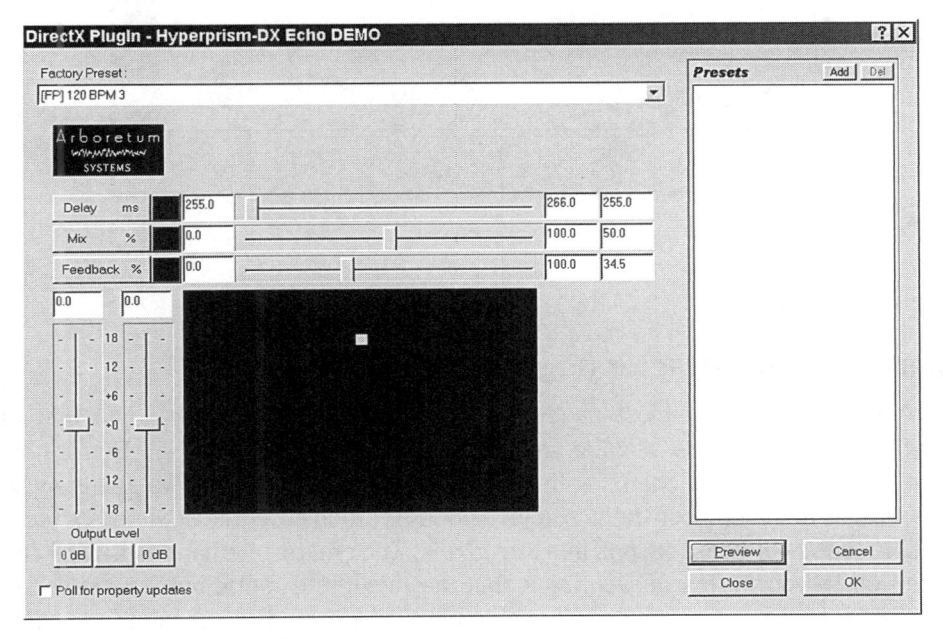

Figure 9.40 Hyperprism echo demo

113

Phaser (Figure 9.41)

This is very similar to the flanger but with much more control over the effects available. Again, start by using the presets and play with the settings to modify the sounds.

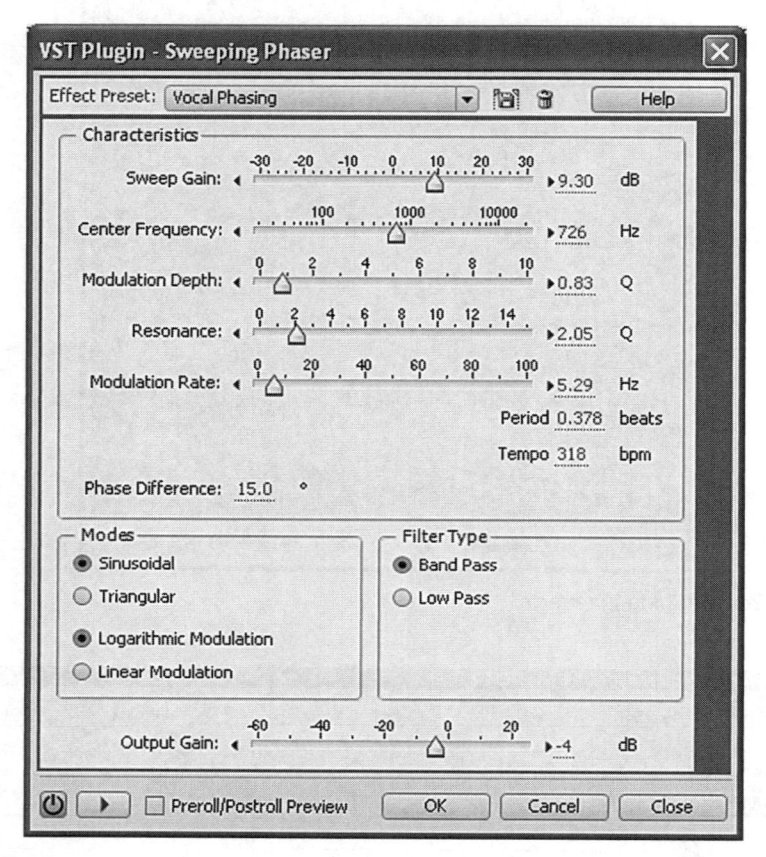

Figure 9.41 Phaser

9.8 Filters

Center Channel Extractor (Figure 9.42)

Adobe Audition 2.0 has almost an embarrassment of EQ and filters. The first on the Effects/filters sub menu is a peculiar one and is an attempt at 'unmixing'. This is not really possible as it would involve changing the laws of physics but, in audio, we are rarely dealing with perfection but the art of the possible. Adobe Audition's Center Channel Extractor is a tool to help reduce or boost sounds in a stereo mix. The greatest demand is to alter the relative levels of the vocals. The convention is that they are in the centre but Audition's tool also has the ability to tweak for offset audio using either the drop down at the top of the dialog or sliding the values underneath. A need for what this filter supplies indicates a degree of desperation. However, it can get you out of a jam. While the output may often be a bit flakey, it is surprising what you can get away with if it becomes a new element of a mix.

Equalization, or EQ, as it is usually referred to, has two functions. The first is to enhance a good recording and the second, to rescue a bad one.

The best results are always obtained by first choosing the best microphone and placing it in the best place. This is not always possible and so, in the real world, EQ has to come to the rescue.

The most common need for speech EQ is to improve its intelligibility. A presence boost of 3-6 dB around 2.8 kHz usually makes all the difference. Here the simple graphic equalizer effects transform will be fine.

Another common problem is strong sibilance where 'S's are emphasized; reducing the high frequency content can be advantageous (another option is to use the

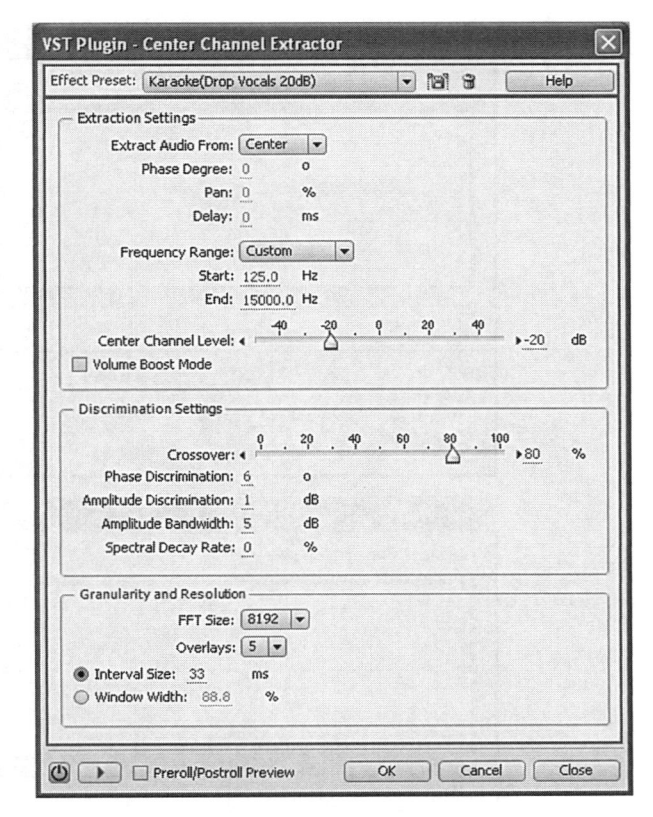

Figure 9.42 Center Channel Extractor

De-essing preset in the Amplitude/dynamics preset. This has the effect of reducing the high frequencies only when they are at high level).

Bass rumbles are usually best taken out by high-pass filters. This naming convention can be confusing, at first, as it has a negative logic. A filter that passes the high frequencies is not passing the low frequencies! Therefore a high-pass filter provides a bass cut, and a low-pass filter provides a high frequency cut. This sort of filter works without changing the tonal quality of the recording.

As always, it is best to avoid the problem in the first place. If your recording environment is known to have problems with bass rumbles, like the Central London radio studios which have underground trains running just below them, then use a microphone with a less good bass response. A moving coil mic will not reproduce the low bass well, unlike an electrostatic capacitor microphone.

Dynamic EQ (Figure 9.43)

This effect allows you to vary the effect over the selection. There are three graphs selected by the tabs at the top. The first tab allows you to vary the operating frequency with time. The

PC Audio Editing with Adobe Audition 2.0

(a)

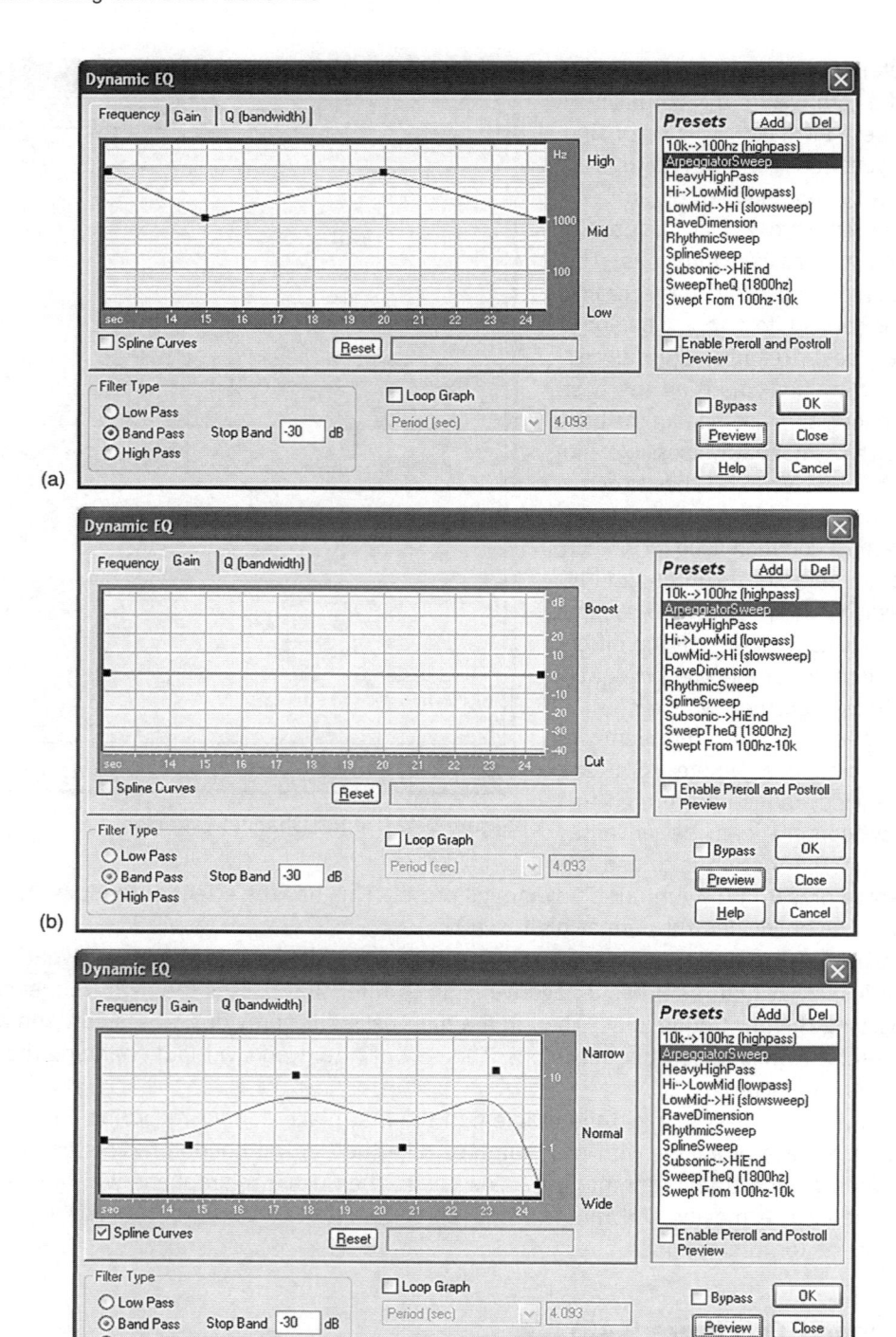

(b)

(c)

Figure 9.43 Dynamic EQ

Audio design

second, the gain and the third the 'Q' or bandwidth. Three type of filter are provided low, band and high pass. The first and last terms can be confusing. A 'low-pass' filter removes high frequencies (and passes low frequencies). Similarly a 'high-pass' filter does the opposite; it passes high frequencies and therefore filter out low frequencies. A band-pass filter passes frequencies within its range while filtering those outside. It can be thought of as a 'middle-pass' filter although it can also be used at the extremes of the frequency range. This is a process effect and cannot be used as a real-time multitrack effect.

FFT filter (Figure 9.44)

FFT stands for Fast Fourier Effects transform. It is nothing less than the Techie's dream of designing a totally bespoke filter. Great fun if you know what you are doing. Totally confusing if you do not! Possibly best avoided if you did not start your career as a technician.

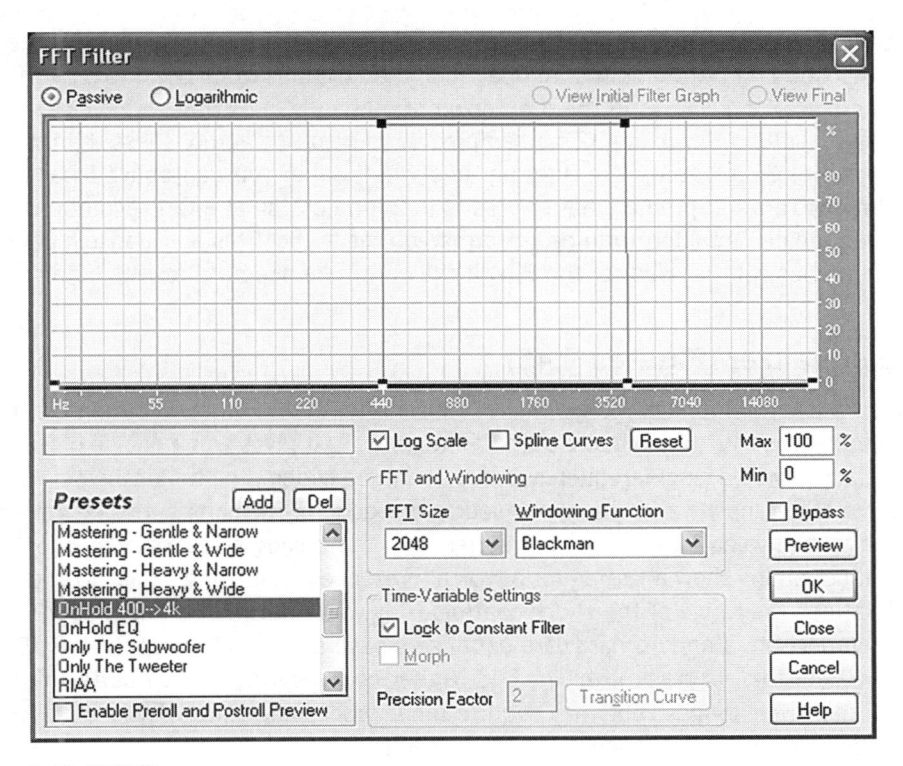

Figure 9.44 FFT filter

Designing your own filter does not need any mathematical ability but you have to have some idea what frequencies you want to filter. You can literally design a filter to remove frequencies that have broken through on to your recording at a gig. You do this by drawing the frequency response that you want. The line on the graph is modified just like the envelopes on the multitrack mixer. Click on the line to create a 'handle' and then use the mouse to move it. If the spline curves box is ticked then a smoother curve is achieved with the handles acting more as though they are attracting the line toward themselves rather than actually being 'fence posts' on the line.

The windowing functions are different ways of calculating the filters. There is a trade-off between accuracy in generating the curve and artefacts that can spoil the sound. Adobe recommend the Hamming and Blackman functions as giving good overall results.

You can also set up two different filter settings and automatically get a transition between them. This will either be a cross fade between the two settings or if the 'morph' box is ticked the settings will gradually change. This means that if you have a transition between a lot of bass boost and a lot of top boost, morphing will cause the boost, as Syntrillium used to describe it, to 'ooze' along the frequencies between one and the other. With the box unticked the frequencies between are not affected. The precision factor controls how large the steps are in the transition a higher number means smaller steps but more processing time.

The FFT size controls the precision of processing higher number giving better resolution but slower processing time. Setting to a low value like 512 gives speed when previewing and listening to how your do-it-yourself filter performs. When you actually want to run the filter set to a higher value for better results. Adobe recommend values from 1024 to 8196 for normal use. The power of 2 values in the drop down are the only ones that will work.

For the best results, filter using 32-bit samples. If your source audio is less than this then convert the file to 32 bit to do the filtering, and when done, you can convert back to the lower resolution. This will produce better results than processing at lower resolutions, especially if more than one transform will be performed on the audio. This is because higher resolution mathematics will be used for the 32-bit file and more accurate results obtained.

Graphic equalizer (Figures 9.45)

This effects transform emulates the graphic equalizer often found on Hi-Fi amplifiers. Rather than just offering the five or six sliders of the average Hi-Fi, it gives you a choice of 10, 20 or 30 sliders. Each covers a single frequency band. The more sliders, the better the resolution. It represents a more friendly way of designing your own frequency response. It works rather better than analog equivalents as it does not suffer from the analog system's component tolerances. The illustrations show a setting for a typical presence boost using 10 or 30 sliders.

'Range' defines the range of the slider controls. Any value between 4 and 180 dB can be used. (By comparison, standard hardware equalizers have a range of about 30–48 dB.)

'Accuracy' Sets the accuracy level for EQ. Higher accuracy levels give better frequency response in the lower ranges, but they require more processing time. If you equalize only higher frequencies, you can use lower accuracy levels.

Graphic phase shifter

The Graphic Phase Shifter effect lets you adjust the phase of a waveform by adding control points to a graph. This is something I have never wanted to do and is a candidate for the least-likely-to-be-used prize. However Adobe suggest that 'you can create simulated stereo by creating a zigzag pattern that gets more extreme at the high end on one channel. Put two

Audio design

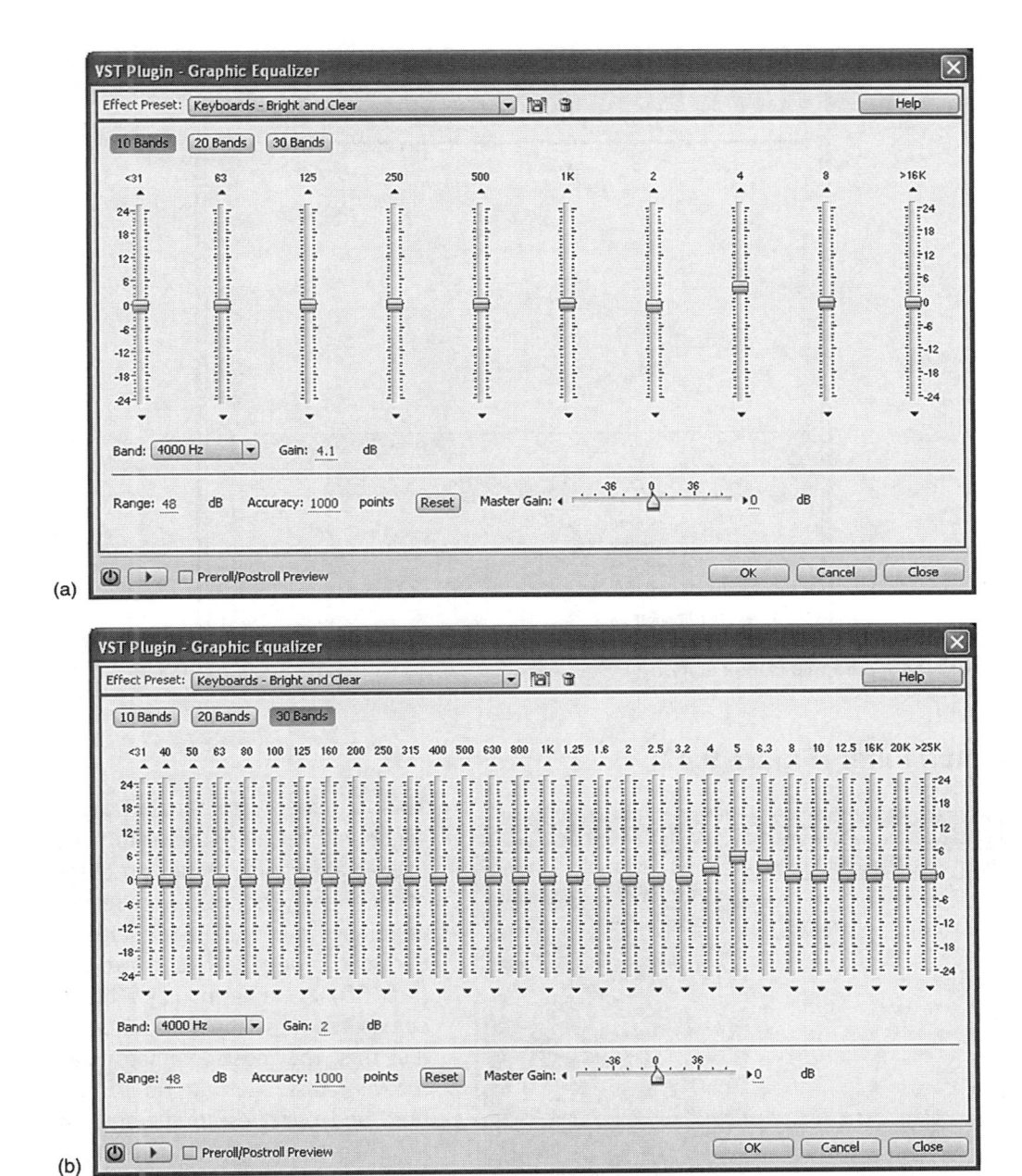

Figure 9.45 (a) Graphic Equalizer with 10 sliders (b) Graphic Equalizer with 30 sliders

channels together that have been processed in this manner (using a different zigzag pattern for each) and the stereo simulation is more dramatic. (The effect is the same as making a single channel twice as "zigzaggy".)' Beware with this method of faking stereo that it is unlikely to be compatible if heard in mono (see 'Faking Stereo', Page 137) (Figure 9.46).

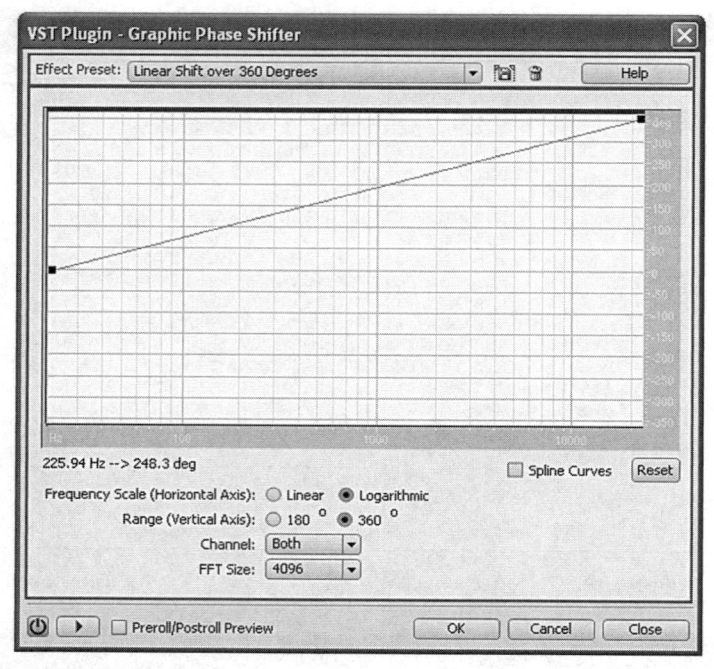

Figure 9.46 Graphic Phase shifter

Notch Filter (Figure 9.47)

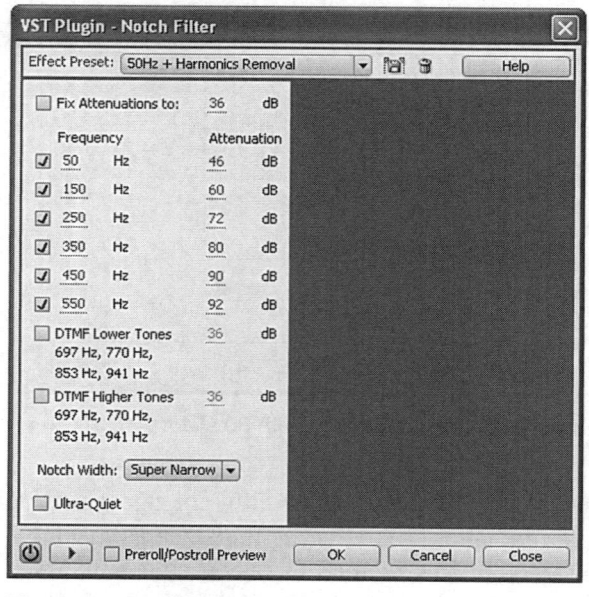

Figure 9.47 Notch filter

This is a specialist effects transform which you will hope never to have to use. It allows you to set up a series of notches in the frequency response to remove discrete frequencies. The Parametric equalizer is shown set to filter 50 Hz and harmonics. This filter comes ready set-up for undertaking just this, the most common disaster recovery task: getting rid of mains hum and buzz resulting from poorly installed equipment (theatre lighting rigs are a common source of buzzes).

In Europe, the frequency of the mains power supply is 50 Hz. In the USA it is 60 Hz. Pure 50 Hz or 60 Hz is a deep bass note. However, mains hum is rarely pure. It comes with harmonics; multiples of the original frequency; 100 Hz, 150 Hz, 200 Hz, etc. (or their equivalents for a 60 Hz original). By 'notching' them out, the hope is that they can be removed without affecting the programme material too much. Except in severe cases this works surprisingly well.

The ultra-quiet option throws more processing for an even lower noise figure but Adobe say that you need top quality monitoring to hear the difference, making the processing overhead rarely worthwhile.

The other preset options are to filter DTMF (Dual Tone Multi-Frequency) tones. These are the tones used to 'dial' telephone numbers. Some commercial radio stations use these to control remote equipment. Apparently they can end up getting mixed with programme material and have to be unscrambled from it. You can do this yourself by using 'Generate/DTMF tones in Adobe Audition 2.0.

Parametric equalizer (Figure 9.48)

This is another Techie-orientated effects transform. While there is a graphical display like the FFT filter, here the control is applied by putting in figures or by moving sliders. The sliders each side of the graph control the low and high frequencies while the horizontal slider underneath sweep the centre frequencies of five 'middle' controls. The vertical sliders to the right control the amount of boost or cut for each filter. Most mixers which have middle control

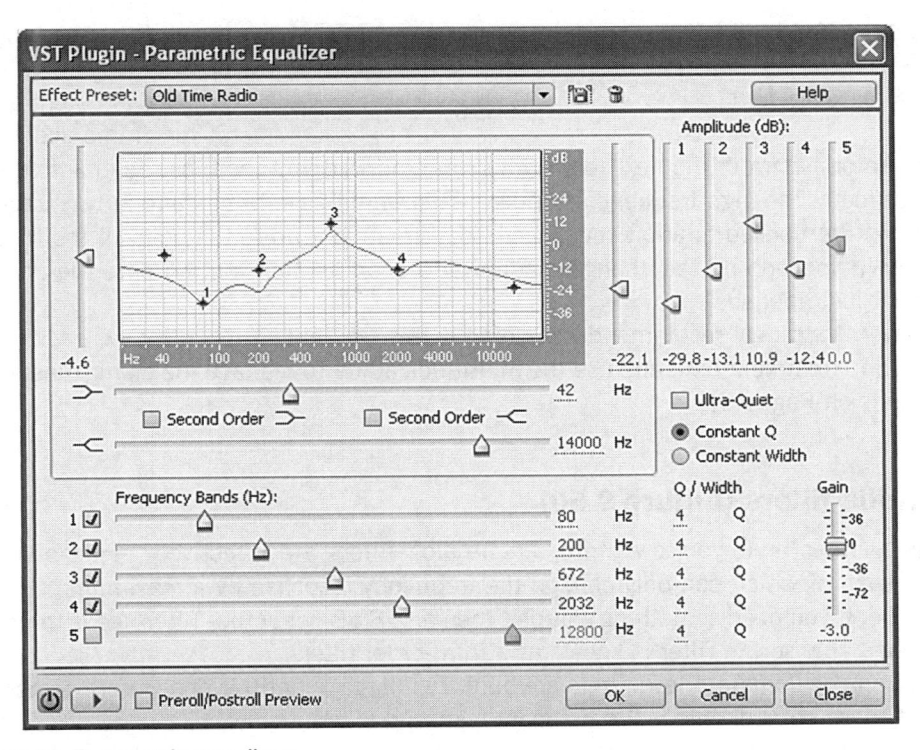

Figure 9.48 Parametric equalizer

usually only have two knobs the '±' control and the frequency sweep. More expensive mixers also have a 'Q' knob. This controls the width of the cut or boost. A very high number makes narrow notches a low number makes the boost or cut very wide. Technically 'Q' is the ratio of width to centre frequency. Adobe Audition 2.0's Parametric equalizer also has the option of constant bandwidth as measured in Hertz wherever you are in the frequency range. Like the notch filter it also has an 'ultra-quiet' option.

Quick filter (Figure 9.49)

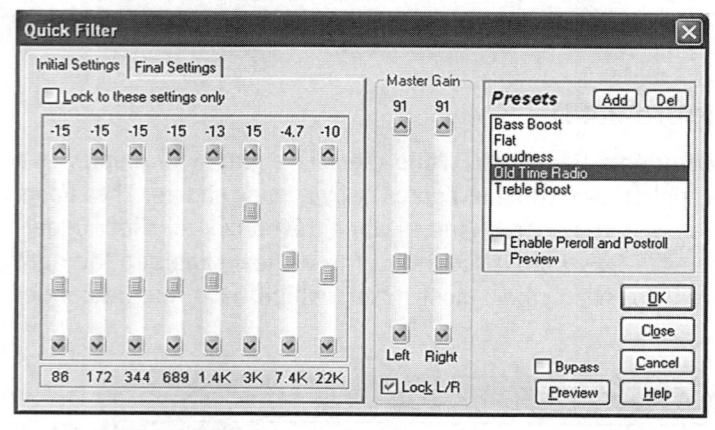

Figure 9.49 Quick filter

In many ways this is a little like another graphic equalizer except that the way it does things is slightly different. Its major difference is that it is two equalizers and implements a transition between the two settings. This can be useful to match different takes where not enough care has been taken to get them consistent in the first place.

It is also potentially useful for drama scene movements. An orchestra in the next room will sound muffled and then, as the characters move into that room, the high frequencies appear. This would have to be done in two stages with the Quick filter locked to apply the 'other room' treatment up to the point of the transition. It is then unlocked and the change made over a short section, butted onto the end of the treatment just applied.

However, I strongly recommend not doing it this way but to make a new track with the 'other room' treatment on it and use the multitrack editor to manage the change instead (see Transitions on Page 76).

Scientific filters (Figure 9.50)

This is another build-your-own filter kit. Analog filters are effectively made of building blocks, each of which can only change the frequency response by a maximum of 6 dB per octave. So, if you need something sharper, then an 18 dB per octave filter needs three building blocks. This sort of filter is known as a third-order filter.

In the analog domain, these filters are built from large quantities of separate components. Each of these components will have manufacturing tolerances of 10 per cent, 5 per cent, 2 per cent or 1 per cent, depending on how much the designer is prepared to pay. This means

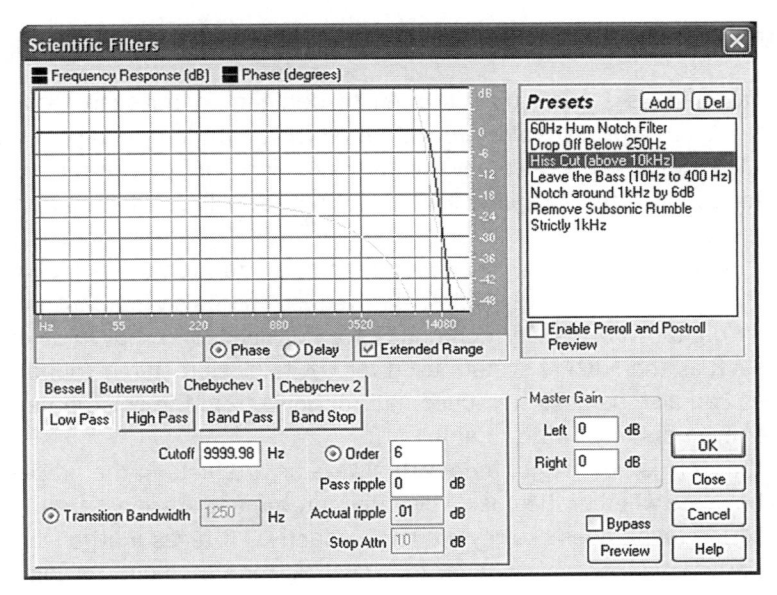

Figure 9.50 Scientific filters

that anything more than a third- or fourth-order filter is either horrendously expensive or impossible to make, because of the component tolerances blurring everything.

In the digital domain, these components exist as mathematical constructs. They have no manufacturing errors and therefore are exactly the right value. This means that a digital filter can have as many orders as you like, limited only by the resolution of the precision of the internal mathematics used.

Even so you cannot get something for nothing. Heavy filtering will do things to the audio other than change its frequency response. Differential delays will be introduced along with relative phase changes. Some frequencies will tend to ring. Leaving the jargon aside, what this means is that you may achieve the filtering effect that you want, but with the penalty of the result sounding foul.

The Scientific Filter Effects transform offers a set of standard filter options from the engineering canon. The graph shows not only the frequency response, but also the phase or delay penalties. In the end your ears will have to be the judge. If you have a tame audio engineer to hand – assuming that you are not one yourself – then you may persuade them to design you some useful filters optimized for your requirements and saved as presets.

A neat trick to eliminate phase shifts can be to create a batch file which runs the filter you want, then reverses the files and runs the filter again – effectively backwards. It then reverses the file back to normal. The idea is that the shifts cancel themselves out. It also means that you can use a 9th order filter twice rather than an 18th order once.

Plug-ins

Again VST and DirectX plug-ins can be downloaded often with specialist applications.

9.9 Restoration

'Digitally Remastered' is a label that is often to be seen on reissued recordings. Adobe Audition 2.0 has a suite of restoration tools that allow you to do this to your own collection of recordings. The filters and EQ we have already examined are part of the armoury but the core is noise reduction software applied to noise from recordings ranging from tape hiss to clicks and plops.

In the world of analog recording, noise reduction refers to the various techniques, mainly associated with the name of Dr Ray Dolby, that reduce the noise inherent in the recording medium. In our digital world, such techniques are not needed except in some data reduction systems such as the NICAM system used for UK television stereo sound. Noise reduction techniques can also be used to reduce natural noise from the environment, such as air conditioning or even distant traffic rumble.

Be warned that this is not magic; there will always be artefacts in the process. A decision always has to be made whether there is an overall improvement. Be very careful when you first use noise reduction techniques as they are very seductive. It takes a little while to sensitize yourself to the artefacts and it is very easy to 'overcook' the treatment. As you would expect, the more noise you try to remove, the more you are prone to nasties in the background.

The most common noise reduction side-effect is a reduction of the apparent reverberation time as the bottom of Reverb tails are lost with the noise. In critical cases, it can be best to add, well chosen, artificial Reverb at very low level, to fill in these tails. Getting a good match between the original and the artificial requires good ears.

The basic method of noise reduction is very simple. You find a short section (about 1 second) of audio that is pure noise, in a pause, or the lead into the programme material. This is analysed and used by the program to build up a filter and level guide. Individual bands of frequencies are analysed in the rest of the recording and, when any frequency falls below its threshold in the, what is sometimes known as fingerprint, then this is reduced in level by the set amount. The quieter the noise floor the more effective this is.

However, trying to reduce high wide band hiss levels can leave you with audio suitable only for a science fiction effect. The availability of NR should not be used as an excuse for sloppiness in acquisition of recordings. Don't neglect obtaining the very best quality original recording in the first place.

The classic example of this is old 78 rpm recordings. These used a much wider groove. Theoretically, you could use an LP gramophone at 33 rpm to dub a 78 and then speed correct, and noise reduce, in the computer. This would sound foul. Played with the correct equipment, with purpose-designed cartridges and correct sized stylus at the correct playing weight, the quality from a 78 can be astonishing. Remember that most of them were 'direct-cut'; they were created direct from the output of the live performance mixer. There was no intervening tape stage. Old Mono LPs had wider grooves than stereo records and a wider stylus will produce better outcomes. A turntable equipped with an elliptical stylus will produce the best results from both mono and stereo LPs.

The most effective noise reduction can be achieved just by cleaning the record. Even playing the record through, before dubbing, can clean out a lot of dirt from the groove.

A really bad pressing, with very heavy crackles, will sometimes benefit from wet playing. Literally, pour a layer of distilled water over the playing surface and play it while still wet. This can improve less noisy LPs as well, but once an LP has been wet played, it usually sounds worse once dry again and needs always to be wet played. If it is not your record, then the owner may have an opinion on this!

The subject of gramophone records reminds us of the other major form of noise reduction, namely de-clicking. There are a wide number of different specialist programs that do this, as well as general noise reduction. They vary in how good they are, and how fast they are. If you are likely to be doing a lot of noise reduction then it may be worth your while to invest in one of these specialist programs. Some of these plug-ins can cost more than the program that they plug into!

Click/Pop eliminator (Figure 9.51)

While Adobe Audition 2.0's de-clicker is not as fast as some, it is very effective and has improved with each revision of the program. You can correct an entire selection or instantly remove a single click if one is highlighted at a high zoom level. You can use the Spectral View feature with the spectral resolution set to 256 bands and a Window Width of 40 per cent to see the clicks in a waveform. You can access these settings in the Display tab of the Preferences dialog box. Clicks will ordinarily be visible as bright vertical bars that go all the way from the

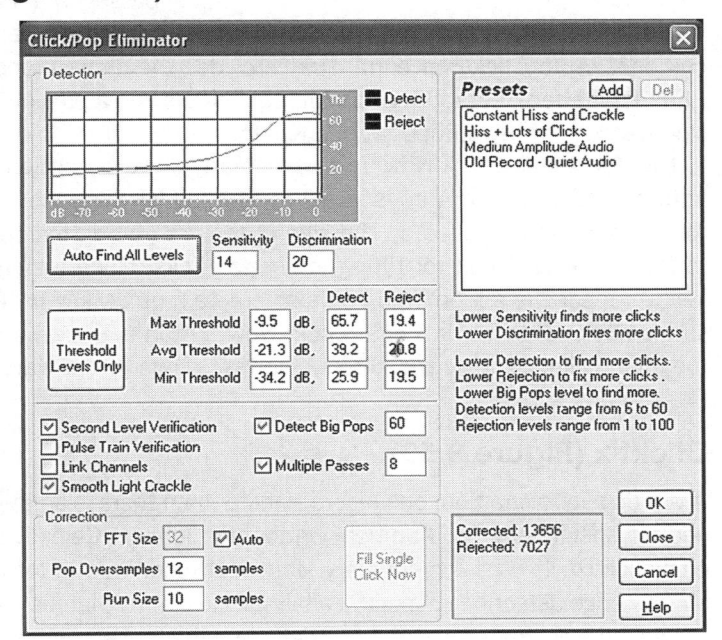

Figure 9.51 Click/pop eliminator

top to the bottom of the display. Unfortunately Auditions click eliminators do not have a 'Keep Clicks' option which can be very helpful to enable you to increase the threshold until you start to hear distorted programme material and then back off a bit. To satisfy yourself that only clicks rather than music were removed you can save a copy of the original file somewhere, then Mix Paste it (overlap it) over the corrected audio with a setting of 100 per cent and Invert enabled (Figure 9.52).

It may take a little trial and error to find the right settings, but the results are well worth it – much better than searching for and replacing each click individually. The parameters that make the most difference in determining how many clicks are repaired are the Detection and

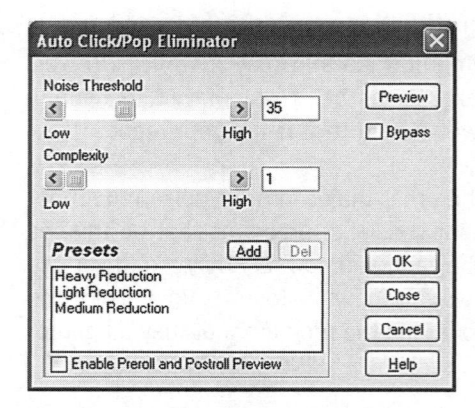

Figure 9.52 Auto Click eliminator

Rejection thresholds (the latter of which requires Second Level Verification). Making adjustments to these will have the greatest effects; you might try settings from 10 for a lot of correction, 50 for very little correction on the detection threshold, or 5 to 40 on the rejection threshold.

'Second level verification' slows the process but enables Adobe Audition 2.0 to distinguish clicks from sharp starting transients in the music. The next parameter that affects the output most is the Run Size. A setting of about 25 is best for high-quality work. If you have the time, running at least three passes will improve the output even more. Each successive pass will be faster than the previous one. 'Auto Click/Pop eliminator' provides the same processing quality as the Click/Pop Eliminator effect, but it offers simplified controls and a preview not available with the main version because of its multipass options, etc.

The 'Fill Single Click Now' is a particularly powerful option for clicks and plops that resist automatic deletion. If Auto is selected next to FFT Size, then an appropriate FFT size is used for the restoration based on the size of the area being restored. Otherwise, settings of 128 to 256 work very well for filling in single clicks. Once a single click is filled, press the F3 key to repeat the action. You can also create a quick key in the Favorites menu for filling in single click . If the 'Fill Single Click' Now button is unavailable, the selected audio range is too long. Click Cancel, and select a shorter range in the waveform display.

Clickfix (Figure 9.53)

If you do a lot of gramophone record transfer then there is an inexpensive commercial de-clicker add-on specifically for Audition written by Jeffery Klein available from http://www.jdklein.com/clickfix/. It uses a proprietary statistical technique to locate clicks and pops. It does not employ edge detection or spectral analysis. He claims that his special technique results in *much faster*, more accurate click and pop detection than other methods. It is certainly very fast. It does have a 'Keep only clicks' option so you can preview the de-clicking and adjust the threshold just short of the breakthrough of programme material. The auto setting covers nearly all my needs for most records.

Clip Restoration (Figure 9.54)

If your audio is overloaded, you 'run out of numbers' and the effect is to clip the peaks, giving them flat tops. You'll hear this as distortion which can be very unpleasant. With this transform, Adobe Audition 2.0 has a fist at restoring the audio. It will never be perfect as data is missing. It does it in two stages. The first is to attenuate the file 'to make room' for the restored peaks. It then goes through the file trying to work out what those peaks might have been. How

Audio design

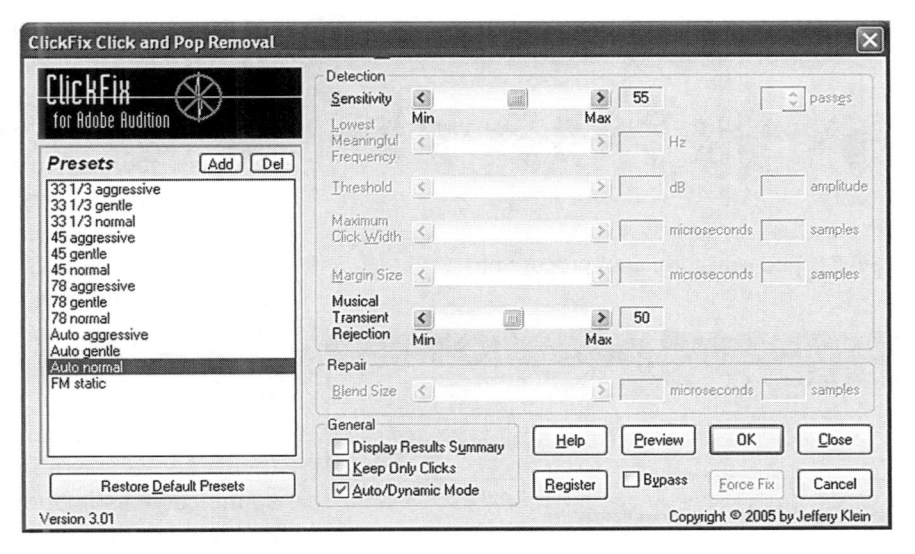

Figure 9.53 Clickfix

successful this is depends a great deal on the programme material and the extent of clipping but it can make the difference between a usable file and one that has to be junked. (Figures 9.54–9.56)

'Input Attenuation' Specifies the amount of amplification that occurs before processing.

'Overhead' Specifies the percentage of variation in clipped regions. A value of 0 per cent detects clipping only in perfectly horizontal lines at maximum amplitude. A value of 1 per cent detects clipping beginning at 1 per cent below maximum amplitude. (A value of 1 per cent detects almost all clipping and leads to a more thorough repair.)

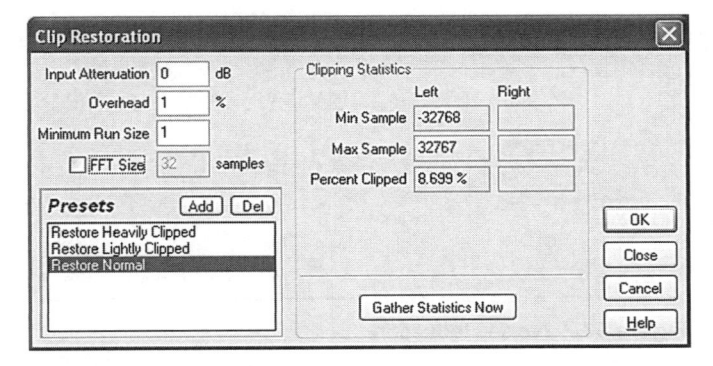

Figure 9.54 Clip restoration

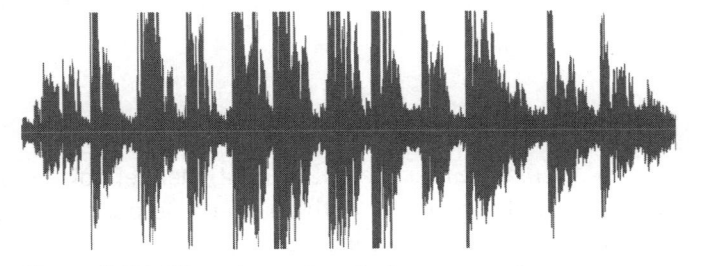

Figure 9.55 Clipped waveform before processing

'Minimum Run Size' Specifies the length of the shortest run of clipped samples to repair. A value of 1 repairs all samples that seem to be clipped, while a value of 2 repairs a clipped sample only if it's followed or preceded by another clipped sample.

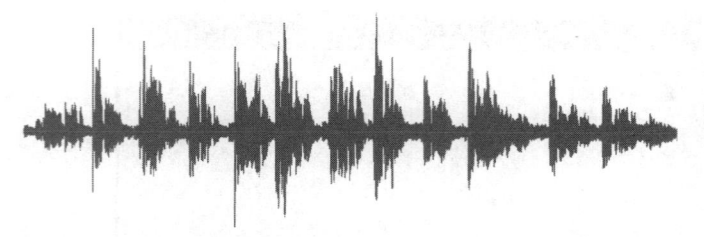

Figure 9.56 Clipped waveform after processing

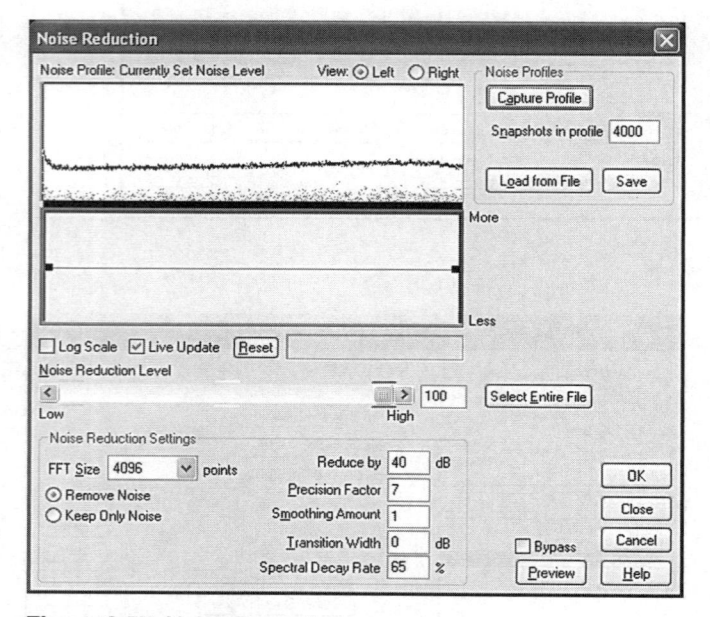

Figure 9.57 Noise Reduction

'FFT Size' Sets an FFT Size, measured in samples, if audio is severely clipped (e.g., because of too much bass). In this case, you want to estimate the higher frequency signals in the clipped areas. Using the FFT Size option in other situations might help with some types of clipping. (Try a setting of 40 for normal clipped audio.) In general, however, leave FFT Size unselected as then Adobe Audition uses spline curve estimation.

'Clipping Statistics' Shows the minimum and maximum sample values found in the current selected range, as well as the per cent of samples clipped based on that data.

'Gather Statistics Now' Updates the Clipping Statistics values for the current selection or file.

To retain amplitude when restoring clipped audio, work at 32-bit resolution for more precise editing. Then apply the Clip Restoration effect with no attenuation, followed by the Hard Limiting effect with a Boost value of 0 and a Limit value of −0.2dB (Figure 9.57).

Noise Reduction options

'View' displays either the left or right channel noise profile. The amount of noise reduction is always the same for both channels. To perform separate levels of reduction on each channel, edit the channels individually.

Noise Profile graph represents, in yellow, the amount of noise reduction that will occur at any particular frequency. Adjust the graph by moving the Noise Reduction Level slider.

'Capture Profile' extracts a noise profile from a selected range, indicating only background noise. Adobe Audition gathers statistical information about the background noise so it can remove it from the remainder of the waveform.

If the selected range is too short, Capture Profile is disabled. Reduce the FFT Size or select a longer range of noise. If you can't find a longer range, copy and paste the currently selected range to create one. (You can later remove the pasted noise by using the Edit > Delete Selection command.)

'Snapshots In Profile' determines how many snapshots of noise to include in the captured profile. A value of 4000 is optimal for producing accurate data.

Very small values greatly affect the quality of the various noise reduction levels. With more samples, a noise reduction level of 100 will likely cut out more noise, but also cut out more original signal. However, a low noise reduction level with more samples will also cut out more noise, but likely will not disrupt the intended signal.

'Load From File' opens any noise profile previously saved from Adobe Audition in FFT format. However, you can apply noise profiles only to identical sample types. (e.g., you can't apply a 22 kHz, mono, 16-bit profile to 44 kHz, stereo, 8-bit samples.)

'Save' saves the noise profile as an .FFT file, which contains information about sample type, FFT size and three sets of FFT coefficients: one for the lowest amount of noise found, one for the highest amount and one for the power average.

'Select Entire File' lets you apply a previously captured noise reduction profile to the entire file.

Reduction graph sets the amount of noise reduction at certain frequency ranges. For example, if you need noise reduction only in the higher frequencies, adjust the chart to give less noise reduction in the low frequencies, or alternatively, more reduction in the higher frequencies.

The graph depicts frequency along the x-axis (horizontal) and the amount of noise reduction along the y-axis (vertical). If the graph is flattened (click Reset), then the amount of noise reduction used is based on the noise profile exactly. The readout below the graph displays the frequency and adjustment percentage at the position of the cursor.

Select Log Scale to divide the graph evenly into 10 octaves.

Deselect Log Scale to divide the graph linearly, with each 1000 kHz (for example) taking up the same amount of horizontal width.

'Live Update' enables the Noise Profile graph to be redrawn as you move control points on the Reduction graph.

'Noise Reduction Level' adjusts the amount of noise reduction to be applied to the waveform or selection. Alternatively, enter the desired amount in the text box to the right of the slider.

Sometimes the remaining audio will have a flanged or phase-like quality. For better results, undo the effect and try a lower setting.

'FFT Size' determines how many individual frequency bands are analysed. This option causes the most drastic changes in quality. The noise in each frequency band is treated separately, so the more bands you have, the finer frequency detail you get in removing noise. For example, if there's a 120 Hz hum, but not many frequency bands, frequencies from 80 Hz on up to 160 Hz may be affected. With more bands, less spacing between them occurs, so the actual noise is detected and removed more precisely. However, with too many bands, time slurring occurs, making the resulting sound reverberant or echo-like (with pre- and post-echoes). So the trade-off is frequency resolution versus time resolution, with lower FFT sizes giving better time resolution and higher FFT sizes giving better frequency resolution. Good settings for FFT Size range from 4096 to 12000.

'Remove Noise, Keep Only Noise' removes noise or removes all audio except for noise.

'Reduce By' determines the level of noise reduction. Values between 6 and 30 dB work well. To reduce bubbly background effects, enter lower values.

'Precision Factor' affects distortions in amplitude. Values of 5 and up work best, and odd numbers are best for symmetric properties. With values of 3 or less, the FFT is performed in giant blocks and a drop or spike in volume can occur at the intervals between blocks. Values beyond 10 cause no noticeable change in quality, but they increase the processing time.

'Smoothing Amount' takes into account the standard deviation, or variance, of the noise signal at each band. Bands that vary greatly when analysed (such as white noise) will be smoothed differently than constant bands (like a 60 cycle hum). In general, increasing the smoothing amount (up to 2 or so) reduces burbly background artefacts at the expense of raising the overall background broadband noise level.

'Transition Width' determines the range between what is noise and what remains. For example, a transition width of zero applies a sharp, noise gate-type curve to each frequency band. If the audio in the band is just above the threshold, it remains; if it's just below, it's truncated to silence. Conversely, you can specify a range over which the audio fades to silence based upon the input level. For example, if the transition width is 10 dB, and the cut-off point (scanned noise level for the particular band) is −60 dB, then audio at −60 dB stays the

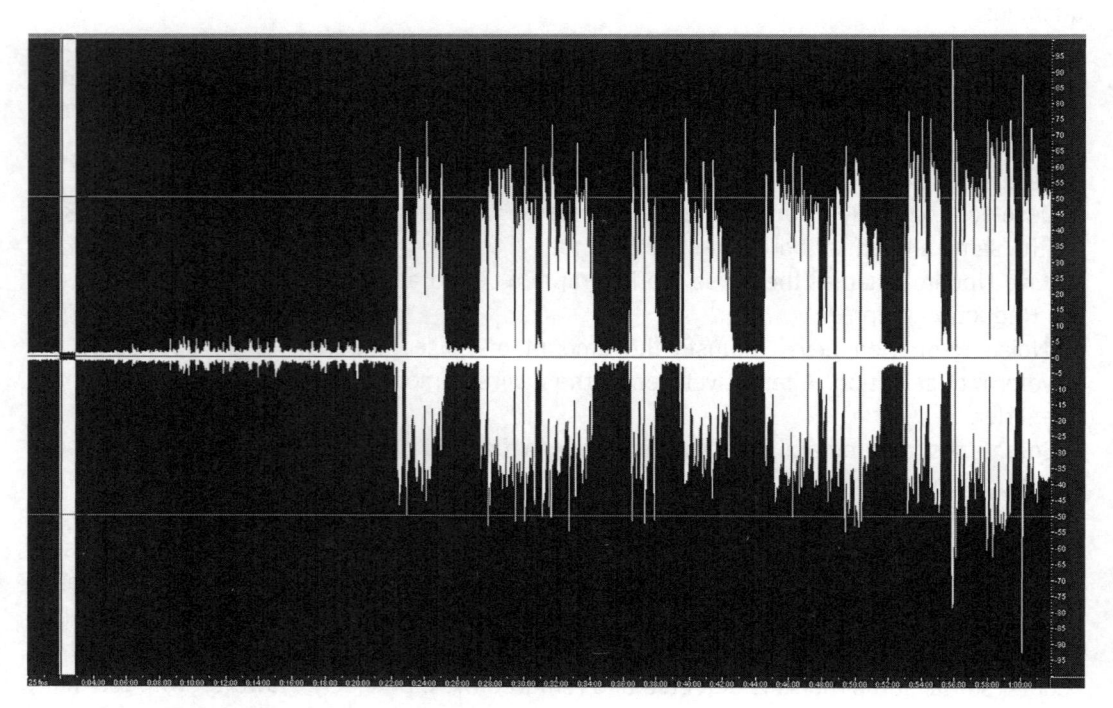

Figure 9.58 Selecting noise at start of track

same, audio at −62 dB is reduced (to about −64 dB) and so on, and audio at −70 dB is removed entirely. Again, if the width is zero, then audio just below −60 dB is removed entirely, while audio just above it remains untouched. Negative widths go above the cut-off point, so in the preceding example, a width of −10 dB creates a range from −60 to −50 dB.

'Spectral Decay Rate' specifies the percentage of frequencies processed when audio falls below the noise floor. Fine-tuning this percentage allows greater noise reduction with fewer artefacts. Values of 40 to 75 per cent work best. Below those values, bubbly-sounding artefacts are often heard; above those values, excessive noise typically remains.

Figure 9.58 shows the beginning of the vocal track from a cassette-based four-track machine with its dbx noise reduction switched off. Looking closely at the track you can see that some of the noise is from tape hiss and hum, but there is also spill from the vocalist's headphones. I have selected a couple of seconds where there is no spill, only machine noise. If we take a noise print we can see that it has picked up the hum on the output of the four-track machine and raises the threshold at the hum frequencies (Figure 9.59).

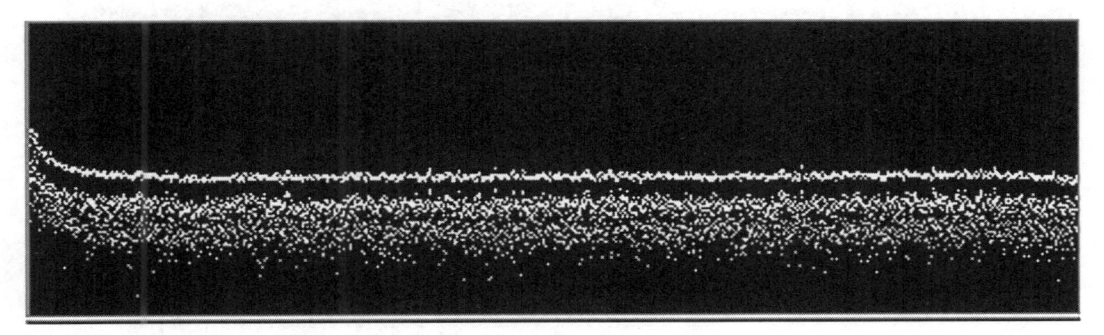

Figure 9.59 Noise profile showing hum

Having taken the noise print, we return to Adobe Audition 2.0 and select the whole wave and initiate the noise reduction process. In Figure 9.60, you can clearly see that it has removed the hiss and hum but left the headphone spill. It will depend on the material whether this amount of noise reduction sounds acceptable. The spill can be removed selectively using the Effects/mute menu option.

9.10 Special effects

Convolution (Figure 9.61)

Convolution is the effect of multiplying every sample in one wave or impulse by the samples that are contained within another waveform. In a sense, this feature uses one waveform to 'model' the sound of another waveform. The result can be that of filtering, echoing, phase shifting or any combination of these effects. That is, any filtered version of a waveform can be echoed at any delay, any number of times. For example, 'convolving' someone saying

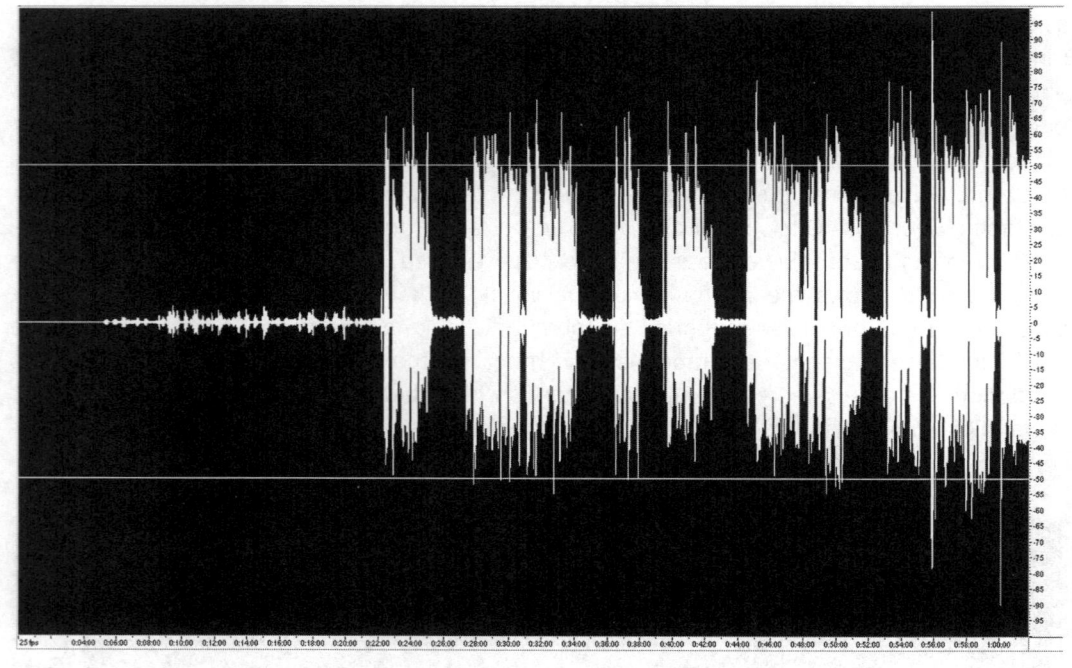

Figure 9.60 Hiss and hum removed but headphone spill still visible

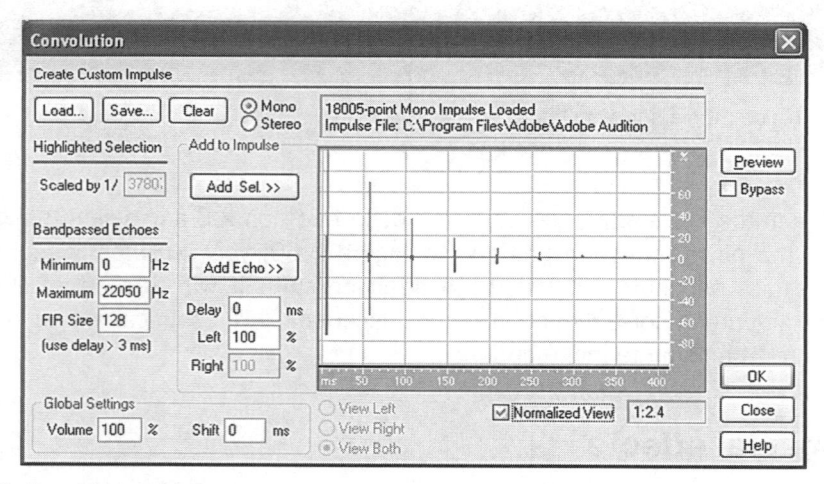

Figure 9.61 Convolution Dialog

'Hey' with a drum track (short full spectrum sounds such as snares work best) will result in the drums saying 'Hey' each time they are hit. You can build impulses from scratch by specifying how to filter the audio and the delay at which it should be echoed, or by copying audio directly from a waveform.

To get a feel for Convolution, load up and play with some of the sample Impulse files (.IMP) that were installed with Adobe Audition 2.0. You can find them in the /IMPS directory inside of the directory where you have installed Adobe Audition 2.0.

With the proper impulses, any reverberant space can be simulated. For example, if you have an impulse of your favourite cathedral, and convolute it with any mono audio (left and right channels the same) then the result would sound as if that audio were played in that cathedral. You can generate an impulse like this by going to the cathedral in question, standing in the spot where you would like the audio to appear it is coming from, and generating a loud impulsive noise, like a 'snap' or loud 'click'. You can make a stereo recording of this 'click' from any location within the cathedral. If you used this recording as an impulse, then convolution with it will sound as if the listener were in the exact position of the recording equipment, and the audio being convoluted were at the location of the 'click'. Most of Adobe Audition 2.0's Reverb transforms are based on convolution algorithms.

Another interesting use for convolution is to generate an infinite sustained sound of anything. For example, one singing 'aaaaaah' for 1 second could be turned into thousands of people singing 'aaaaaah' for any length of time by using some dynamically expanded white noise (which sounds a lot like radio static).

To send any portion of unprocessed 'dry' signal back out, simply add a full spectrum echo at 0 millisecond. The Left and Right volume percentages will be the resulting volume of the dry signal in the left and right channels.

Distortion (Figure 9.62)

We spend most of our time trying to avoid distortion but there are times when we want it. This can be for dramatic purposes or for things like a 'fuzz' guitar. This transform does this by 'remapping' samples to different values.

9.11 Time/pitch

Adobe Audition 2.0 has five effects transforms to do with pitch changing and time stretching.

The first creates Doppler shifts (Figure 9.63). Doppler shifts are the pitch changes you hear when a noise making object passes you. As it approaches the pitch is increased and then once it passes the pitch is reduced compared with what its pitch is when stationary. Perhaps, the most common time we hear this is when an emergency services

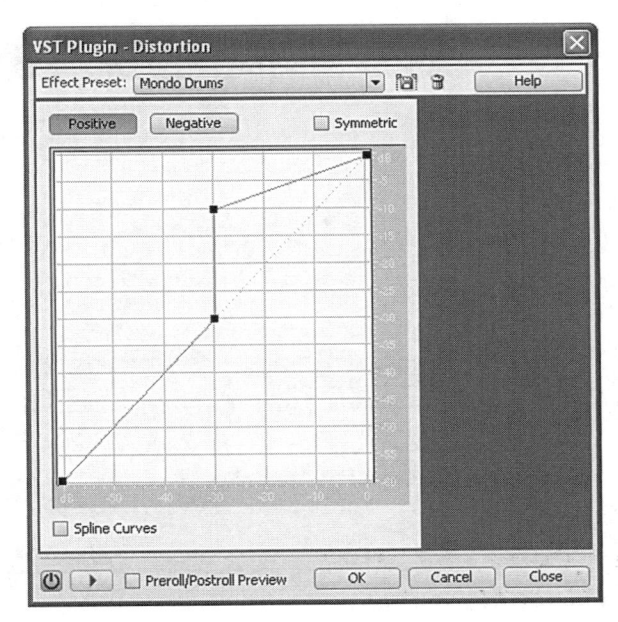

Figure 9.62 Distortion

vehicle speed by sounding its siren. This transform allows you to take a static recording of, say, a siren and process it so that it sounds as though it is moving. The volume can be handled as well if you want. The transform can handle circular movement like a merry-go-round as well.

The second time/pitch transform is the pitch bender (Figure 9.64). This is the direct equivalent of varying the speed of a recording tape. It can handle a fixed change, or make a transition from one 'speed' to another. You can draw the curve as to how it does this. As such it forms the ideal basis to repair a recording made on an analog machine with failing batteries.

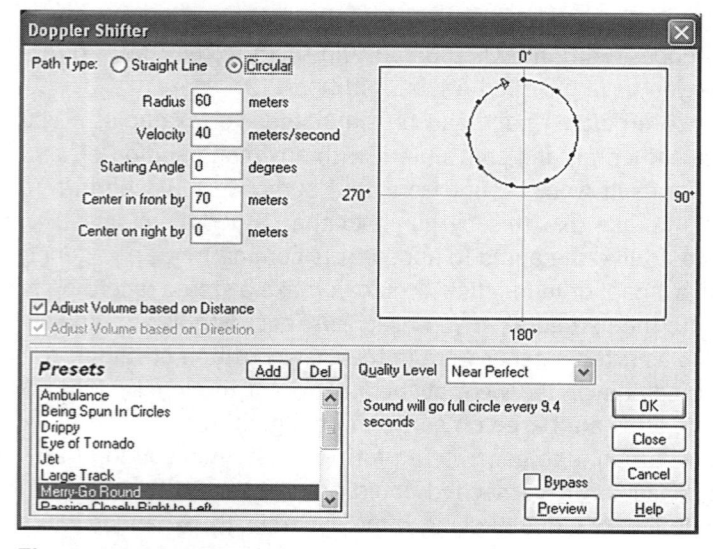

Figure 9.63 Doppler shifter

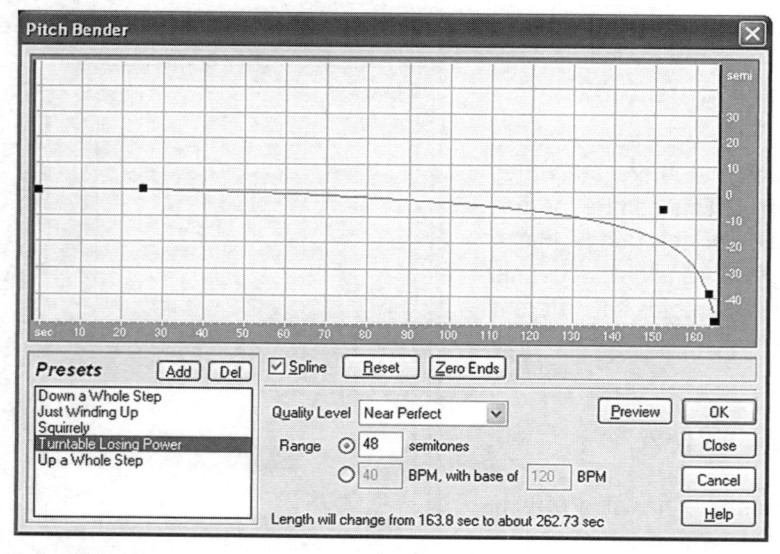

Figure 9.64 Pitch Bender

What happens when the tape speed varies on an analog tape is that, as the speed decreases, so the pitch decreases. If the tape is running at half speed, everything takes twice as long and the pitch is 1 octave lower. However, with computer manipulation this relationship no longer need be maintained.

Adobe Audition 2.0's next effects transform is Pitch correction (Figure 9.65). This can be a boon if your vocal talent is not that talented.

The Time/Pitch > Pitch Correction effect provides two ways to adjust the pitch adjustments of vocals or solo instruments. Automatic mode analyses the audio content and automatically corrects the pitch based on the key you define, without making you analyse each note. Manual mode creates a pitch profile that you can adjust note-by-note. You can even overcorrect vocals to create robotic-sounding effects.

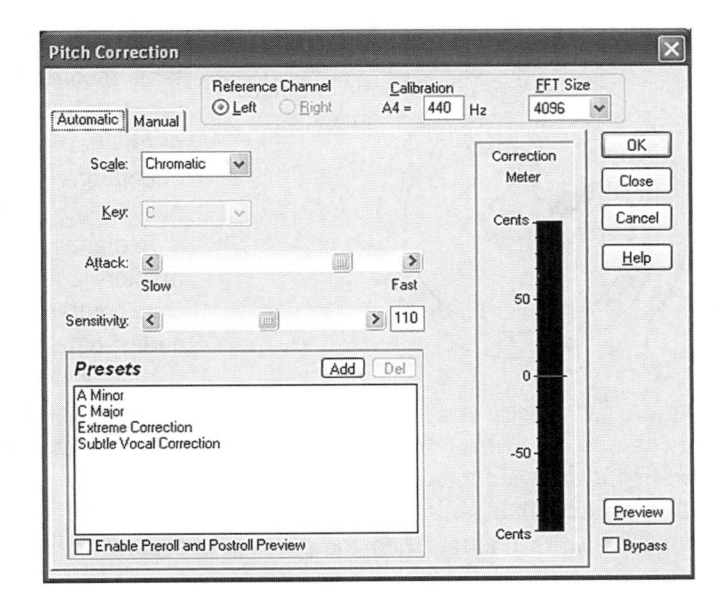

Figure 9.65 Pitch correction

The Pitch Correction effect detects the pitch of the source audio and measures the periodic cycle of the waveform to determine its pitch. The effect is most effective with audio that contains a periodic signal (i.e. audio with one note at a time, such as saxophone, violin or vocal parts). Non-periodic audio, or any audio with a high noise floor, can disrupt the effect's ability to detect the incoming pitch, resulting in incomplete pitch correction (Figure 9.66).

The Pitch shifter allows you to change the overall speed of a recording without changing the pitch or, alternatively, change the pitch without changing the speed.

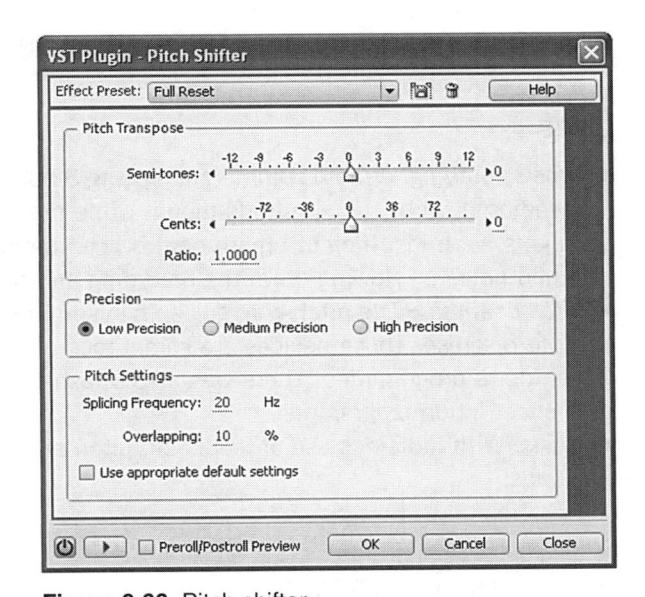

Figure 9.66 Pitch shifter

This is not done by magic, but by manipulation of the audio. Indeed there were analog devices that did the same, just not very well. Even in the digital domain you are less likely to get good results with extreme settings.

Pitch manipulation is a specialized form of delay and works because, as in the chorus effect, changing a delay while listening to a sound changes its pitch while the delay is being changed. As soon as the delay stops being changed, the original pitch is restored whatever

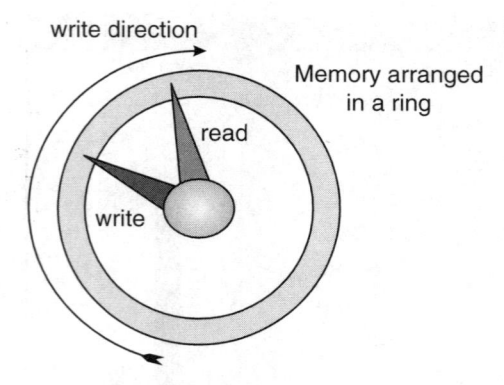

Figure 9.67 Digital delay and pitch change

the final delay. Anyone who has varispeeded a tape recorder while it was recording and listened to its output will know the temporary pitch change effect lasts only while the speed is changing. It returns to normal the moment a new speed is stabilized.

A digital delay circuit can be thought of as memory arranged in a ring, as in Figure 9.67. One pointer rotates round the ring, writing the audio data. A second pointer rotates round the same ring, reading the data. The separation between the two represents the delay between the input and output. In a simple delay, this separation is adjusted to obtain the required delay.

In a pitch changer, the pointers 'rotate' at different speeds. If the read pointer is faster than the write pointer, then the pitch is raised. If the read pointer travels slower than the write pointer then the pitch is lowered.

Quite obviously, there is a problem each time the pointers 'cross over' when a 'glitch' occurs. The skill of the software writer is to write processing software that disguises this. The software either creates, or loses, data to sustain the differently pitched audio output.

Uses

- *Time stretching:* A radio commercial lasting 31 seconds can be reduced to 30 seconds by speeding it up by the required amount while maintaining the original pitch.
- Singers, with insufficient breath control, can be made to appear to be singing long, sustained notes by slowing down the recording and pitch changing to the original.
- *Pitch changing:* The pitch change can be useful for correcting singers who are out of tune.
- *Voice disguise:* These devices are sometimes used to disguise voices in radio and television news programmes. There was also a fashion for them to be used for alien voices in science fiction programmes.
- *Music:* With judicious use of feedback, pitch changing can be used to thicken and enrich sounds.

9.12 Spatial effects

By spatial effects, we usually mean effects where the sound appears to be coming from somewhere other than between the speakers. A general atmosphere seems to fill the room, or a spaceship flies overhead, from behind you.

Techniques like Dolby stereo use extra loudspeakers, and special processing, to get their directional effect. DVD recordings can use five channels of audio to move sounds, but simple stereo can be beefed up, with a little simple manipulation.

Normally we think of stereo consisting of left and right channels. The relative volume of these channels controls where a sound comes from. A more sophisticated way of thinking about stereo is as an MS signal, MS standing for 'middle' and 'side' NOT mono and stereo. The middle signal consists of a simple mix of the left and right signals in phase. The side signal is again a simple mix of the left and right signals with the right channel phase reversed (inverted).

The middle signal is what is used as the mono signal in broadcasting. If you feed your stereo channel with a simple mono signal panned to the centre – say a presenter's voice – then there will be no side signal; the left and right channels are identical and, because the right-hand signal is inverted, they cancel themselves out.

As the mono signal is panned to one side the side signal increases until the sound is panned fully to one side. At this point, the middle and side signals are equal in level. Once you are thinking in this way, you can consider what happens when you increase the side signal so that it is larger than the middle signal. Doing this can pan the sound image outside the speakers. How effective this will be depends on the programme material. As the out-of-phase level is increased, the image will soon collapse and appear to come from inside the head.

With a really complex stereo signal, tweaking the side level can give dramatic dimensional effects. It is also a means of creating a fake stereo signal from a mono original.

M & S

An alternative way of dealing with stereo is as an MS signal, the letters standing for Middle and Side. They can be converted back and forward. Technically

$$(L + R) \div 2 = M$$
$$(L - R) \div 2 = S$$

equally

$$(M + S) \div 2 = L$$
$$(M - S) \div 2 = R$$

What the '$\div 2$' means is a matter of controversy. The long-standing BBC Radio convention is that it represents 3 dB (half power) of attenuation. An alternative view is that it should represent 6 dB (half voltage) of attenuation. The are known colloquially as 'M3' and 'M6'. M6 is becoming popular as it can be louder than M3.

Faking stereo

You sometimes need to fake a stereo sound from a mono source. This can range from a unique background sound to a period train, only ever recorded in mono, passing across the sound stage between the speakers.

There are several techniques open to the balancer. With continuous sounds such as applause, trains, atmospheres, etc., it is often possible to get a spread effect by playing two copies of the same (or different) sources. One is panned left and the other panned right with, perhaps, a third in the centre. If they get into close synchronization then compatibility problems may arise with phasing on a mono radio receiver. In Adobe Audition 2.0, this is easily done by using duplicate copies on different tracks and sliding the tracks relative to each other to get the right effect.

A better way, which gives no compatibility problems, is to play one source in mono and the other phase-reversed so that it only contributes to the stereo signal. This bonus sound is not heard by the mono listener as it cancels out.

Again duplicate tracks are used except that one (not both) of the duplicate tracks is converted to a stereo wave file and one channel (say the right) is phase-reversed (inverted). For a discrete sound, like an aircraft taking off or a train passing, sliding this track to delay the sound by 50 to 100 milliseconds can produce good results, especially if the volume envelope on this difference channel is manipulated.

If you want a directional sound, like a train passing between the speakers, you need to do this in two stages. First generate a 'spread' train and mix this down to create a new stereo wave file. This can then be panned using the pan envelope control.

With atmospheres, very long delays can be used – try 30 seconds (if the FX is long enough). Changing the level of the out-of-phase signal in sympathy with the sound can improve realism.

Thunder claps can be particularly effective, if you run different mono claps on the in-phase and out-of-phase tracks. It can be difficult to find genuine stereo thunder that rolls around the room as effectively.

The same techniques can be used to beef up stereo sounds which do not come up to your expectations. Provided the sound is OK for any mono listener – if the recording is to be broadcast – then anything you add out of phase to your mix will be a bonus to the stereo sound without affecting the mono. If you can guarantee your recording will never be heard in mono then compatibility is not a problem. Remember that, in addition to mono radios, there are mono cassette players.

9.13 Basic effects

There are three basic effects at the top of the Effect menu drop down. Although they are very simple they are extremely useful.

Invert

The Invert menu option is a fairly specialized effects transform. All it does is to invert the waveform so that positive becomes negative and negative becomes positive.

In music balancing, this is the equivalent of the phase reverse switch on a mixing console. You can find, when you come to a mix down, that a microphone was out of phase with others used at the same time.

Microphones should be in phase. This means that their output voltages are going in the same direction; an impulse that makes the diaphragm move inwards on the microphones produces the same direction voltage. All the microphones should produce a positive voltage, or all the microphones should produce a negative voltage. It doesn't matter which, as long as they are the same. If they are out of phase, they tend to cancel, and the mix can sound a bit like a drainpipe.

The left and right channels of stereo should be in phase too as, otherwise, a central image seems to be coming from inside the head, rather than from between the speakers. Worse, if the recording is broadcast, people listening on mono receivers will not hear anything in the centre as it will cancel out. See description of Adobe Audition 2.0's Phase Analysis on Page 164 for more information.

Reverse

Reverse does just literally that; it reverses the sound so that the beginning becomes the end. Its main uses are for producing reversed echo and alien voices. It can also be used in batch files to reduce phase shift errors using filters by running the filter, reversing the audio, running the filter again so that any phase timing errors are reversed and then reverse the audio back to normal again.

Reversed echo is often used for magical effects: witches and aliens. It sometimes has a role in music, although it is an entirely unnatural effect.

A multistage process is used. First the programme material is reversed. Then it has a reverberation effects transform applied to it. That transformed recording is then reversed again. This restores the original speech, etc. to the right way round, but with the Reverb now reversed. What was a Reverb die-away now becomes a build up. Satisfyingly weird effects can be created this way. However, the relative balance is critical so, again, it is best if the reversed Reverb is created as a 'wet-only' track, so that it can be mixed with the original while you can actually hear the intended effect.

If you have a performer who is a good mimic, you can create a really good alien voice by getting them to record the lines and then reversing them. The performer now imitates the sound of the reversed speech – this takes some practice. He, or she, is then recorded imitating the backwards speech and then that recording itself is reversed. All being well, an inhuman voice will be heard speaking very strangely but intelligibly. This sort of trick has to be done in relatively short takes to work well.

Mute

This mutes the selected section of the wave file. Its main purpose is for music balancing, as you can use it to remove spill, coughs and sneezes from, say, a vocal track. This means that you do not have to set the volume envelope for each vocal entry and exit, but can guarantee absolute digital silence in breaks in the performance.

Don't confuse this with Generate/silence. The difference between the two is that this transform mutes a section of existing audio, making no change to its length. Generate/silence inserts a given number of seconds of silence at the cursor so that inserting 10 seconds of silence will make the file 10 seconds longer.

Generate silence, DTMF, noise and tones are discussed in Chapter 11 'Mastering'.

9.14 Multitrack effects

The Multitrack view of Adobe Audition 2.0 has three effects transforms of its own. They are here because they generate a new track or tracks in the multitrack view.

The envelope follower and the vocoder combine, non-destructively, two different audio tracks to create a third. It can be very frustrating to try to use these as the menu options are grayed out unless the conditions are right. To use Adobe Audition 2.0's vocoder or envelope follower, you have to both select two tracks *and* highlight the section to be processed. The easiest way to do this is to have the control and process tracks adjacent to each other and to highlight both with one action. Place the mouse cursor over the first track, press and hold down the left mouse button at the beginning of, or just before, the wanted section and drag the cursor down to the next track and along to the end of the section required. You can do this starting from the end and moving to the beginning if you prefer. This is often the best way if you are starting at the very beginning (0:00.00). Both tracks will now be highlighted and the Effects/vocoder and Effects/Envelope follower menu options will be enabled (this happens only if two tracks are selected).

If the two tracks are not adjacent, you can still do the same, except that the tracks in between will also be enabled. You need individually to deselect these by holding the Control key down and single clicking on each.

Selecting the section and then enabling two tracks by holding the control key down and single clicking on them will also work.

Envelope follower (Figure 9.68)

The Envelope follower varies the output level of one waveform, based on the input level of the other. The amplitude map, or envelope, of one waveform (the analysis wave) is applied to the material of a second waveform (the process wave), which results in the second waveform taking on the amplitude characteristics of the first waveform. This lets you, for example, have a bass guitar line which only sounds when a drum is being hit. In this example you would have the drum waveform as the analysis wave, and the bass guitar waveform as the process wave.

The dialog has a go at guessing which way round the tracks should be but often gets this wrong. You can switch them by clicking on the drop-downs at the top of the dialog. Once you have found a setting that works you can save it as a preset.

When you click OK the normal progress dialog will appear and a new track will appear. Normally the next available one will be used but you can change it with the 'Output to' drop down.

Frequency Band Splitter

The frequency band splitter takes the output of a single track and creates up to 9 new tracks each with a different frequency band of audio. This presents intriguing possibilities as you

Audio design

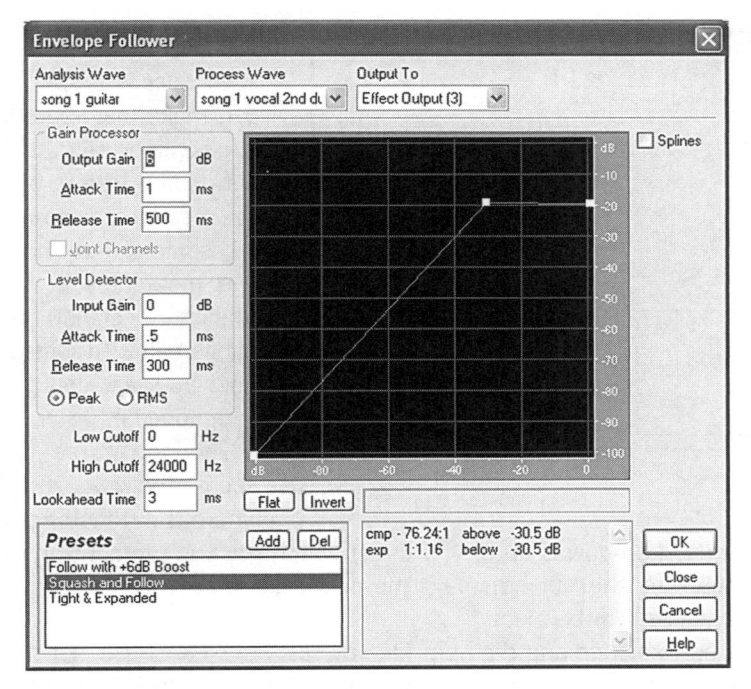

Figure 9.68 Envelope follower

can separately treat these bands with any of the processing available to Adobe Audition 2.0. At the simplest, you could boost the level of a cymbal to change the sound or you could even compress it. By putting compression on each track you have the basis of a multiband compressor for mastering. This is one of those tools which, while appearing simple opens up a host of possibilities.

You can have up to 9 bands and you can change the crossover frequencies by typing them in the left hand boxes. The number of tracks produced is selected by the buttons at the side, These are not toggles but clicking one will produce that number of bands (Figure 9.69).

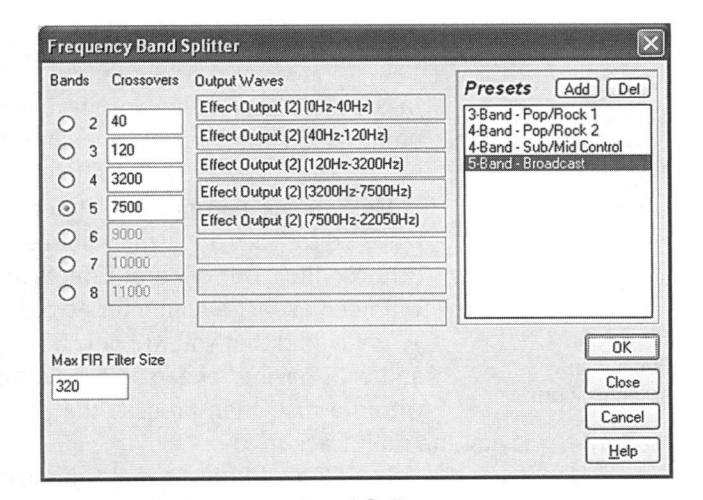

Figure 9.69 Frequency Band Splitter

141

Vocoder (Figure 9.70)

A vocoder is a special effect which can make inanimate objects appear to speak or sing. If you want the leaves of a tree to talk or sing, then the vocoder may do what you want. It takes one sound (the process signal) and modulates it with the second (the control signal). Modulate a telephone bell with someone saying 'ring ring' and you could have a talking telephone bell.

The top three boxes tell you about the two input tracks and where the transform will be placed. The left and centre drop-downs allow you to set which track is the control wave and which is the process file. The third box displays the next track empty at the selected area where the resulting transform will be placed.

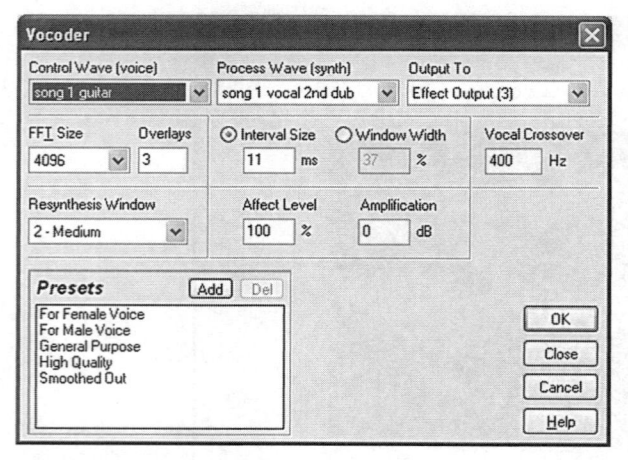

Figure 9.70 Vocoder

Some presets are provided and the help file advises 'For ease of use, set Window Width to about 90 per cent, use 3 or 4 overlays, set Resynthesis Window to 1 or 2, and choose an FFT size between 2048 and 6400.' A fair amount of fiddling must be expected to get good results.

9.15 Real-time controls and effects

In the multitrack editor you have various controls in the left hand column. Which ones you see are controlled by the icons at the top (Figure 9.71 – see Chapter 7 'Multitrack').

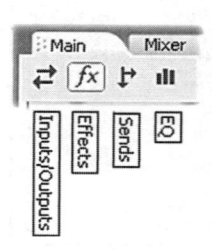

Figure 9.71 Facility view selector icons

Most of the non-process effects that were described earlier in the Wave Editor view are also available as real-time effects within the multitrack editor. But beware! Effects are very processor intensive and you will need as fast a computer as possible. If you have a slower machine you may find that you will only be able to use a small number at a time and will, having rehearsed the setting to get them how you like them within a mix, then process the actual wave file to reduce processing load when playing the multitrack mix.

In an ideal world all your FX should be in real time just like on an ordinary sound mixer. This is because everything interacts. A vocal line that sounded fine may need hardening if you have made the backing music brighter.

On the other hand the ability to pre-record your effects does allow you to use the slow computer you have now rather than having to save up for something faster before you can do anything.

Audio design

A useful compromise, especially for Reverb is not to alter the original wave file but to generate a new wave file which is just the Reverb for a track. Controlling the level of this additional track requires much less processing power.

Adobe Audition 2.0 has a neat halfway house. You can lock or 'freeze' the effects once you have a setting that you like. There will then be a pause while the progress bar shows it processing and saving the FX. It can be left untouched and the processing overhead will be small but a click of a button will restore the track to real-time effects (Figure 9.72).

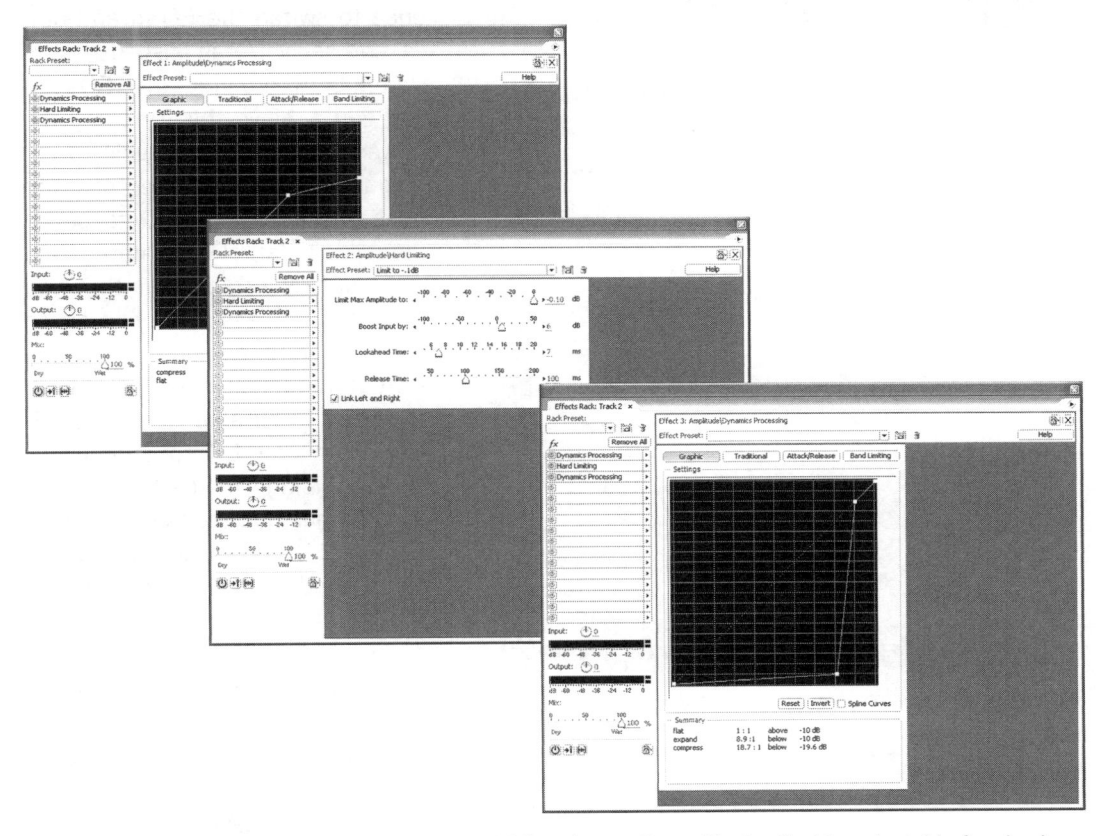

Figure 9.72 Single channel Effects rack containing three effects. Each effect is selectable for viewing. Shown is a vocal track with soft compression, followed by hard limiting and then gating to remove spill.

Track Equalizer

Each track comes with its own real-time parametric equalizer. If you have clicked the EQ icon at the top of the column you see the basic settings in numerical form (Figure 9.73).

Clicking the EQ button will reveal a graphical display so that you literally can see, as well as hear, what you are doing. (Figure 9.74)

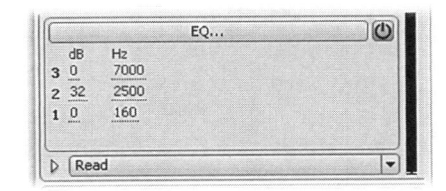

Figure 9.73 Numerical track equalizer view

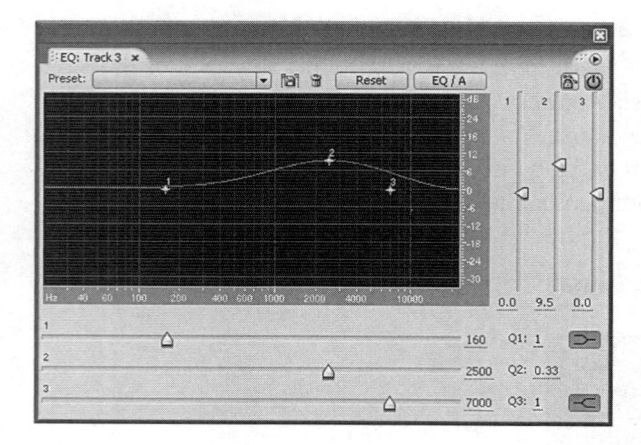

Figure 9.74

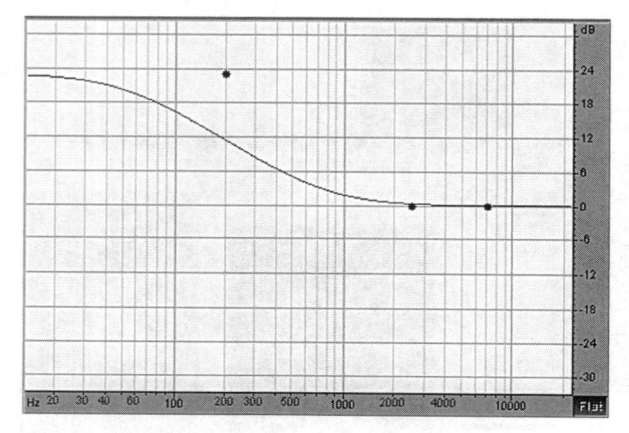

Figure 9.75 Low frequency shelf setting

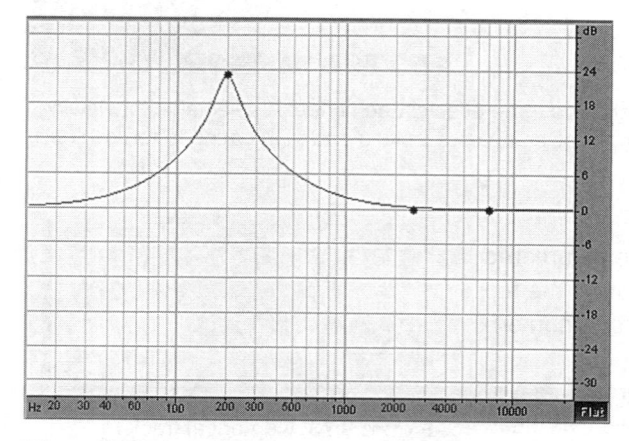

Figure 9.76 Low frequency Band setting

In fact there are two alternative equalizers, 'A' and 'B' which can be switched between easily by single clicking the 'EQ' button. It will toggle between showing 'EQ/A' or 'EQ/B'. At its simplest, this allows you to use 'EQ/B' as a flat reference to switch the EQ in and out. Of course, you can also use it to switch between alternative EQs but you can achieve this by using the power button which toggles from green and white. Sometimes you need the alternative EQ to be an 'equalized flat'. For example, you may already established that you need a high-pass filter setting on the low control but now want to try a bit of middle boost on top of that. You used to be able to copy a setting from one EQ to another by double-clicking in the 'EQ' button but this has not been continued in version 2.0. However presets can be created in one and used in another.

The low and high controls can be switched between shelf and Band EQ. (Figures 9.75 and 9.76)

Shelf EQ operates like an ordinary bass or treble control on a Hi-Fi. Unlike an ordinary Hi-Fi, you can alter where the 'turnover' from flat to sloping takes place. The boost out cut levels out after a couple of octaves so as not to excessively boost extreme high or low frequencies.

Clicking the button toggles the filter into a band type equalizer that operates like a middle control. Thus you can have three middle

Audio design

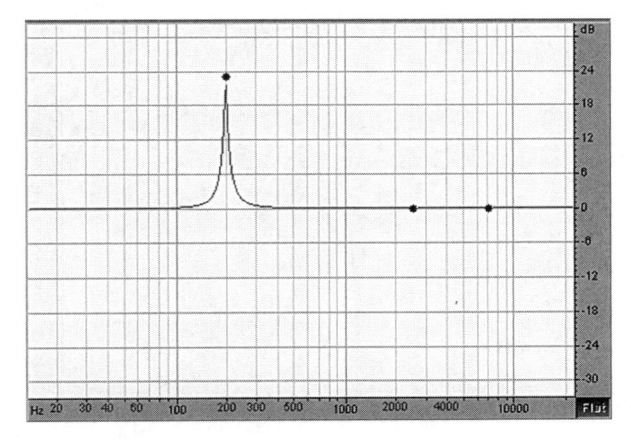

Figure 9.77 Middle control high Q setting

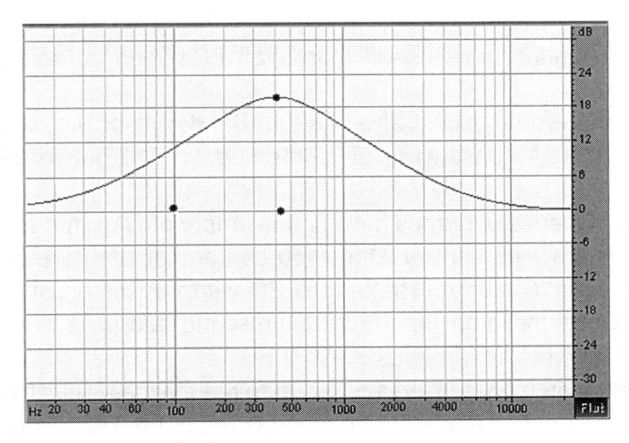

Figure 9.78 Middle control low Q setting

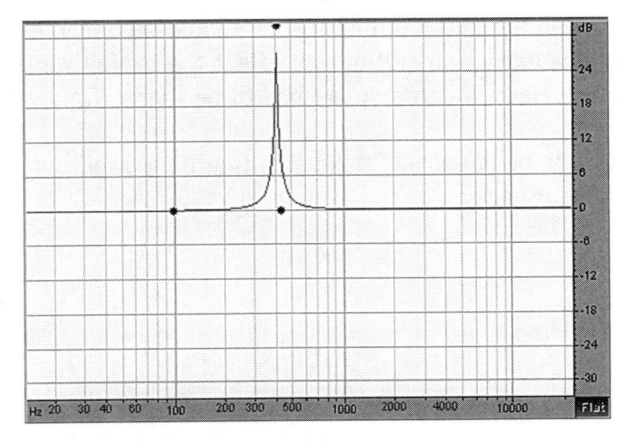

Figure 9.79 Maximum boost to isolate unwanted frequency.

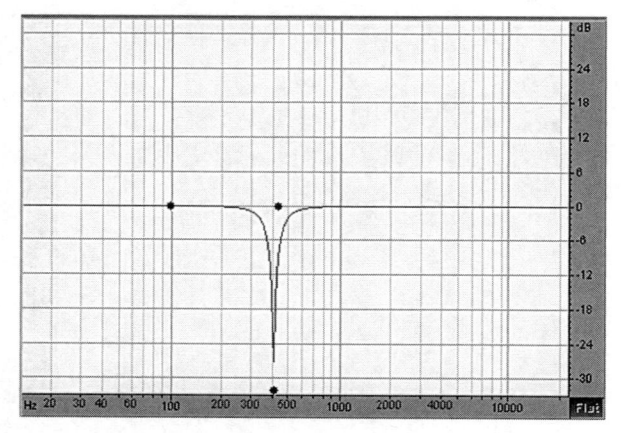

Figure 9.80 Maximum cut to remove unwanted frequency

controls which can be much more useful than shelf EQ. Shelf is indicated by the Q value turning black.

The middle is, as you would expect always set to Band. Any of the bands can be adjusted for Q. Any of the controls can always be controlled for the active frequency they start operating.

Semi-professional mixers sometimes have just a simple plus or minus control for the middle. Better ones have a sweep control where you can adjust the frequency of boost or cut. The very best also have a 'Q' control that adjust the width of the boost or cut. The very narrow setting can be useful for removing interference sound that have been picked up ranging from buzzes to lighting whistles (Figures 9.77–9.80).

(But don't forget the notch filter is already set up to reduce mains electricity related buzzes).

A good technique to remove a single frequency is to set the band control to maximum Q, the narrowest setting, at maximum boost. move the frequency control until the unwanted sound is found and has become much louder. Now move the level control from maximum boost to maximum cut and the unwanted frequency should be tamed or eliminated. In practice, you may find that a maximum setting will affect the overall wanted sound as well as removing the unwanted. Here, a compromise has to be made. Only your ears can tell you how much.

If you are not using a track equalizer, make sure that its 'power' button is off so as not to use unnecessary processor power.

10
Reviewing material

10.1 Check there are no missed edits

In news and current affairs broadcasting, many compromises have to be made. Elsewhere, there is less excuse. Traditionally, with quarter inch tape, you would spool back from the end, with the tape against the heads, controlling the speed so that you could hear the speech rhythms. A missed edit, a gap or talkback, would show as a break in the rhythm. Any missed edits that were found could be razor bladed in seconds.

With material prepared on an audio editor, life is not so simple. If the final item is transferred to CDR or DVD there is no way of listening to the recording while this is done. Transferring to DAT involves a real-time copy which can be a good opportunity to review the item.

The snag is that if a missed edit is found then the copy has to be restarted from the beginning. Ironically, this can lead to a programme being prepared entirely using digital technology only to be transferred to analog quarter inch tape so that any missed edits can be edited the old fashion way, with a razor blade, rather than having to start the transfer from the beginning.

10.2 Assessing levels

Very rapidly, the eye becomes used to assessing levels from the appearance of the waveform display. However, there is no substitute for listening to the item in one go, on good equipment.

You should also make a habit of listening to your material in the environment that it will be used. If you are making a promotional cassette to be heard in a car, then listen to your tape in a car, preferably a noisy one so that you can check that levels don't drop to inaudibility.

If the item is to be used on the telephone, as an information line recording, then this presents a virtually unique balancing challenge. This is genuine monaural sound: one eared. Check your balance with headphones worn so that only one ear is used. If you have facilities to feed the audio down a phone line, then do so. Remember, even if you have been asked to provide the material as analog audio, it will almost certainly be reduced to, at best, 8-bit digital audio on the play-out device. Lower bit rates are often used, with the resulting quality being much below what the telephone system (which is an 8-bit system) is capable of.

10.3 Listening on full quality speakers

This will reveal every deficiency in your recording. It becomes a production decision as to what the listener will hear the recording on. It is as important to listen on poor speakers. Music studios always have a set of small speakers to check balances on low bandwidth speakers. A good balance travels well. If your item sounds fine on good speakers, and on a portable cassette player, then it is likely to be successful.

The 'grot' speaker also reveals whether effects that sound impressive on big speakers, or headphones, also work on average equipment. A classic mistake is to use a deep bass synthesized sound for a dramatic heartbeat. While sounding effective on big speakers with extended bass response, the same sound may sound like low level clicks on a transistor radio.

There is a definite difference between 'Hi-Fi' loudspeakers and 'monitoring' speakers (the speakers that come with PCs are neither). Hi-Fi speakers are designed to make everything sound good and well balanced. Monitoring speakers are designed to be analytical. They are your equivalent of the doctor's stethoscope. You want to be able to hear things, and correct them, before your listeners hear them. This is why sound balancers monitor at higher levels than ordinary mortals. They wish to hear inside the balance. They will also 'dim' the speakers (reduce the level by 12–15 dB) from time to time to check realistic levels on both their monitoring speakers and the 'grot' speakers.

11

Mastering

11.1 Line-up and transmission formats

The mastering process is concerned with producing the final product that is actually going to be used. This could be for use by yourself; say, a CD of audio illustrations for a talk, or it could be a multitrack tape for a *Son et Lumière*. It could even be a disc of MP3s for your car with extra processing to overcome the noise of the car's engine on the motorway.

It could be a programme for broadcast. Here the final recording has to be prepared in the correct way to fit in with the requirements of the broadcaster as to levels and line-up information.

The requirement will differ between organizations and reflect their internal practice. For example, if your customer requires programmes provided on CD to be recorded with a peak level of −4 dB, then all you need to match this requirement is to normalize your programme to −4 dB and burn it to CD.

Increasingly broadcasters are accepting material on CD-ROM as a .WAV file. They usually require that the professional sampling rate of 48 kHz is used for these. Wave files can also contain formatted text information. Adobe Audition 2.0 provides dialogs for entering this (Figure 11.1).

There is also a European Broadcasting Union standard for including text in the wave file to provide production information (Figure 11.2).

Line-up

Quarter inch tape and digital audio tape (DAT) usually need to have a line-up tone, or tones, at the front of the programme. Here BBC practice is for 20–30 seconds of, nominally, 1 kHz tone on both tracks recorded at 11 dB below peak programme level. Analog tapes are also required to have a section of 10 kHz tone, so that head alignment errors can quickly be detected. Programmes on CD often require line-up tone as this removes the ambiguity between the various metering systems available. A BBC practice is to put the programme as track 1, make track 2 two minutes of silence with the tone on track 3. This copes with the bad habit of some CD players of dropping out and reverting to track 1 if they are left set up for a long period.

The 'Generate' menu title covers useful tools here.

PC Audio Editing with Adobe Audition 2.0

Figure 11.1 Wave text information

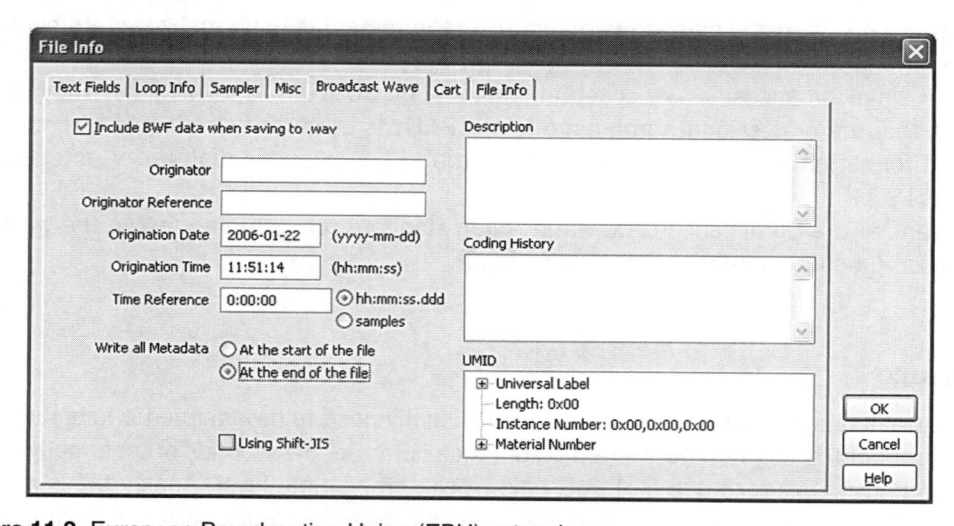

Figure 11.2 European Broadcasting Union (EBU) extensions

11.2 Generate silence

This option does just that, it generates a set number of seconds of silence. It is different from the silence in the Effects section which mutes the highlighted section of the recording. Here, the silence is generated as a new recording – any audio recording after the cursor is pushed onwards to make way for the silence, that is, rather than being replaced by the silence. This means that you can insert a silence between a line-up section and the programme material.

11.3 Generate noise

This is a specialized option. Generate noise means just that. You have a choice of three types of hiss: white, pink and brown.

How can noise have a color? Well, the analogy is with the light spectrum. The three noises are hisses containing all the frequencies of the sound spectrum, just as white light contains all the colors of the rainbow. As we all know, there are different types of white: warm white, cold white, etc., so there are different colors of hiss.

White

White noise has equal proportions of all frequencies present; each discrete frequency has the same energy present as any other frequency. Because the human ear is more susceptible to high-frequencies, white noise sounds very 'hissy'.

Pink

Pink noise has a spectral frequency similar to that found in nature. Each octave has the same energy as any other octave. As you go up the octave range, each octave has 'more frequencies' in it; so for every octave increase the energy for each discrete frequency is halved. If you look at the spectrum analysis of white noise it appears to have a flat frequency response but pink noise has one that falls 3 dB for every octave.

It is the most natural sounding of the noises. By equalizing the sounds you can generate rainfall, waterfalls, wind, rushing river and other natural sounds.

Brown

Brown noise has much more low end, and there are many more low frequency components to the noise. This results in thunder- and waterfall-like sounds. Brown noise is so called because, mathematically, it behaves just like Brownian motion. This is how molecules move in a nice hot cup of tea!

As has been suggested, these three types of hiss can be dragooned into helping a sound effects mix. With a little top cut, brown noise can take on the role of 'city skyline'. This was the background rumble that, for decades, was the sound of silence on television. When they had mute film, a touch of skyline gave a sense of something going on. Legend had it that it was the sound of Victoria Falls played at half speed.

This sort of rumble added to a sound effects mix can give it a depth it might not otherwise have. Like a spice in cooking, it improves without being distinguishable.

Pink and white noise are also very useful test signals to check out audio systems, including loudspeakers, resonances and frequency response dips show up quite clearly. The high-frequency losses of recording systems such as analog tape recorders are also easily checked. However, be warned: errors of a fraction of a decibel can be heard and this may well be within the maintenance tolerance of the machine.

11.4 Generate tones (Figure 11.3)

This can generate quite complex tonal mixes but, for our purposes, a simple pure sine wave tone at a specified level is all that is required. For line-up, the phase difference should be zero as tone may be used as a system phase check.

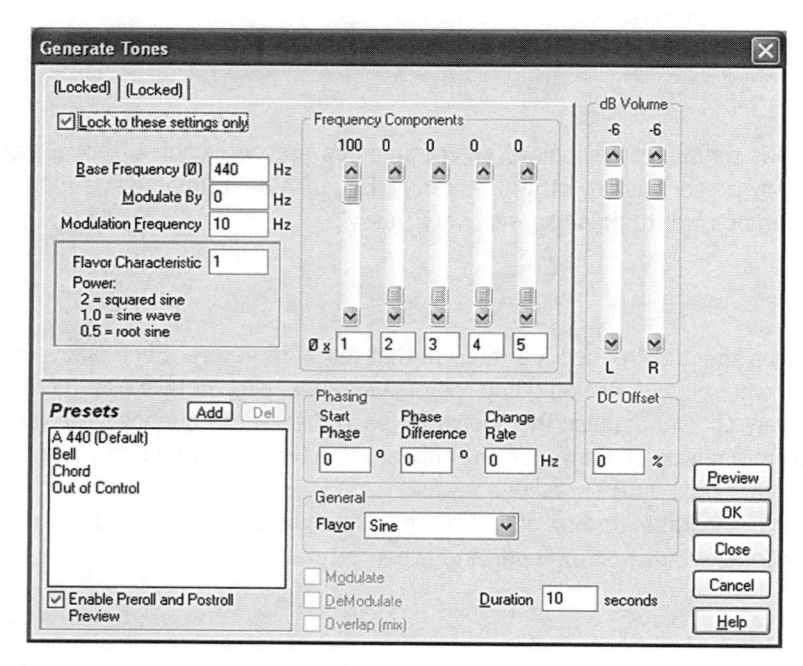

Figure 11.3 Generate tones

The Generate tones dialog has the option of using beginning and end settings by unticking the 'Lock to these settings only' box when the tabs showing 'Locked' become labelled Initial Settings and Final Settings. This could be used to generate a continuous frequency response test run for 'squeaking' the programme chain.

11.5 Changeovers

If ¼" tape is being used, then the maximum time available on a 10½" NAB reel is 30 minutes. Programmes longer than this have to be on more than one reel, and each reel has to be cut for a changeover in such a way that it can be performed perfectly without rehearsal. While this is a rapidly disappearing requirement it is still worth considering how best to cut a changeover.

Ideally this should take place at a change of item, preferably where there is a slight pause. (If the changeover is botched then the pause just gets slightly longer – with presented programmes it is regarded as good practice to end the first reel with the presenter's cue and start the second reel with the insert, or the music. The logic is that, if the changeover is missed, it does not sound as though the presenter has gone to sleep but is obvious that it was the technician!) The pause should always end the first reel NOT start the second.

11.6 Mastering process

The mastering stage is the last chance you have to check that the levels are consistent and that the overall recording is as loud as it can be. Picking off isolated peaks and reducing their level and then renormalizing can give a substantial extra level with no audible consequences. Commercial music CDs are often put through a multiband compressor which divides the audio into separate frequency bands and separately compresses them. Adobe Audition 2.0 has introduced a multiband compressor which can be applied to individual .WAV files or in real time within a multitrack mix.

DAT and analog masters are created by the simple process of dubbing from the computer to the external recorder. Mastering CDs can be as simple, if an external CD recorder is used. However, these often can only use the more expensive consumer blanks, and share, with the other media, the need to copy in real time. CD writers for computers are relatively cheap and can burn CDs at many more times than real time speed. Windows XP contains basic software to do this. Version 2 of Adobe Audition 2.0 includes relatively sophisticated software to burn audio CDs (see Chapter 12 'CD burning').

11.7 CD labelling

Surprisingly, the most fragile surface of a CD is the label side, not the playing side. Scratches on either side can cause errors. Many CDRs come printed with labels for handwriting their contents. This is perfectly possible, but care must be taken to use a pen certified as suitable. A ball-point pen will just damage the CD and some ink solvents will dissolve the protective layer. Some CD blanks have a thin layer of ceramic on their label side that, not only protects the disc, but makes them suitable to be directly printed using a suitable ink jet printer.

Using ordinary stick-on labels may, or may not, work, depending on the glue used on their backing. However, they can end up damaging your drive, rather than the CD, as it may be being spun at 40 or 50 times the standard speed and the off-centre weight can disrupt reading the disc and strain the bearings. Labels can also dry out in the heat of the drive especially within laptops causing the disc to dish.

Purpose-made labels for CDs can be purchased. Different brands are usually associated with different proprietary applicators that will centre them properly. While these seem expensive, the applicators also come with software for designing and printing onto the labels.

You can also buy transparent labels to use with unprinted disc blanks to make the CDR label look more like a conventional CD. However, getting these onto your CD without air bubbles is more of an art than a science. Some people, who wish doubly to protect their CDs, put the transparent labels on top of the paper printed ones and also make them impervious to liquid spills which make the ink jet printing run. Ensure that the labels you use are suitable for your printer as ink jet and laser printers have different requirements. Some ink jet printers can print directly on to CD surface although special blanks are needed. Even with these it can take some time for the ink to dry.

11.8 Email and the World Wide Web

The requirements for the Internet are constantly changing but the constant for all but those with expensive lease lines is that there is a limit to the amount of data that can be sent every second. What is now often called POTS (Plain Old Telephone System) can at the very best with a V90 modem get 6–8 Kbytes/second of data, and even that is in one direction only – from the Internet Service Provider to the user. Upload speeds in the opposite direction are usually half that.

Integrated Services Digital Network (ISDN) can offer those sorts of speeds reliably in both directions or twice that if both channels of an ISDN circuit are combined. (Usually at the penalty of being charged as two calls. While weary broadcasters are apt to translate ISDN as the 'It Sometimes Doesn't Network', it actually stands for 'Integrated Services Digital Network'. It is capable of handling any data, be it telephone, fax, graphics or even stereo music. The main complications are due to a plethora of standards, many of which are not, quite, compatible with each other.) Lease lines offer virtually any speed that you want but at a price.

Asymmetric Digital Subscriber Line (ADSL) is available on both standard telephone lines and also on cable. This allows you to be on-line all the time and have access to much faster data communications. Different cable and telephone companies differ in the 'contention ratios' they offer. The bandwidth you offered is shared and the amount you actually get from moment to moment is dependent on how many people are using the network.

However well you are equipped, you want audio that you put onto the Internet to be accessible to users who are not so well equipped. This means that there is a need to make files as small as possible with a minimum loss of quality. A CD quality recording takes up 10 Mbytes for every minute. This makes it impossible to play in real time from the Internet except on the very fastest connections, as well as occupying a great deal of space on the hard drive of the server where it will be stored.

There are simple ways of reducing this but with quality penalties. Mono, rather than stereo, halves the size instantly, as does halving the sampling rate. Reducing the bit rate from 16 to 8 halves the size yet again. By the time you have done all this the quality is getting decidedly iffy.

What is needed is some way of compressing the audio data so that it occupies less space. Immediately ZIP files come to mind. They are almost universally used as a way of transferring other files. It makes the files smaller and they take less time to be sent. A ZIP file uncompresses its contents to exact copies of the originals. This works very well for text, but very badly for .WAV files. It can even happen that the zipped version of a .WAV file is actually bigger than the original.

What is needed is a different way of recording the data so that it sounds as good as possible, while throwing away as much data as possible. These are known as 'lossy' forms of compression, unlike the lossless ZIP file. Minidisc uses such a system, so that a 140 Mbyte Minidisc holds the same amount of audio as a CD containing 720 Mbytes of audio. This uses psycho-acoustic tricks so that it only records what we hear, not the background material that is hidden by the foreground.

The Microsoft ADPCM (Adaptive Differential Pulse Code Modulation) form of the wave file (it too has the extension .WAV) uses about a quarter of the data by sending information about the differences between samples rather than absolute values. A 1 minute stereo file that starts out at 10 336 Kbytes reduces to 2600 Kbytes. The quality stands up very well – it is mainly the high-frequency transients that get subtly lost. Converting to mono and halving the sample rate would reduce that to a quarter, one-sixteenth of the original size.

Real Audio, and Real Media, files are designed to be sent to a modem. They can be 'streamed'. Their associated playback software, which can be integrated with Internet browsers, can play the files as they come in rather than having to wait for the entire file to be downloaded (there is a pause while an audio buffer is filled). They can be optimized for the intended delivery modem. Our 10 Mbyte speech file can be delivered as a 118 Kbyte mono Real Audio file. The penalty is a phasiness that sounds as though the speech was recorded in an enclosed telephone box. A larger Real Media file of 367 Kbytes sounds better.

Cool Edit Pro version 1 was able to save in these formats, but version 2 and Audition do not. Neither version has a direct facility for loading them, as they are intended for delivery, not for editing. If you do need to transfer these files into the editor – maybe you are doing an item about the Internet – you can record the output of Real Audio/Media by setting Adobe Audition 2.0 to record the output of the sound card handling the Windows output, or, often more conveniently using a specialized utility like 'TotalRecorder' http://www.highcriteria.com/.

The best-known format is MPEG (Motion Picture Experts Group) which comes in various flavours with extensions of MP1, MP2, MP3 and MP4. MP3 has become increasingly used as it is very efficient in providing low data, good quality sound in stereo. It can handle different sample rates and can be set to use different bits/second. The 10 Mbyte wave file reduces to 939 Kbytes retaining 44 kHz sampling and using 120 Kbits/second encoding. This gives good quality stereo for an end user. Reducing the bit rate to 48 Kbits and halving the sampling rate reduces the audio file to 353 Kbytes. The quality is still comparable to the original provided you are listening on PC style speakers or in a car with speakers mounted in the doors. Adobe Audition 2.0's MP3 coder can record down to 20 Kbits/second.

Audition includes the MP3Pro variant which is backwards compatible and can be played by ordinary MP3 decoders. However, an MP3Pro decoder can take advantage of extra low bit rate information that allows it to add back high-frequency information. This extra information is a bit of a 'guess' but can substantially improve the perceived quality.

Also provided by Adobe Audition 2.0 is Microsoft's WMA (Windows Media Audio) format which can be used for streaming. It can provide files of similar length to MP3 but its advocates say that it can produce better quality at similar bit rates to MP3 or Real Audio and Real Media. Microsoft claim that WMA can achieve similar quality at up half the bit rate of an equivalent MP3 file.

Adobe Audition 2.0 is able to save many of the Microsoft audio formats using the encoders that your windows system has acquired either from the Windows installation disc or when downloading other programmes like the Windows Media Player. These are using the ACM waveform option in Adobe Audition 2.0's Save dialog. The options button then gives you access to whatever MS formats you have on your machine (Figure 11.4).

All these compression formats make assumptions about your ears and the programme material. In practice, some music and some voices compress better than others. For example, a light male voice often sounds less 'phasey' than a deeper male voice at high compression ratios.

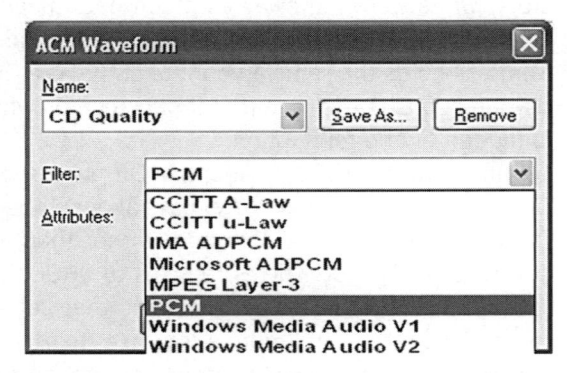

Figure 11.4 Using the ACM waveform option

Whatever your final format, it is especially vital that the sound file is properly prepared. Do everything at full 32- or 16-bit resolution and your normal sampling rate. Only when everything is ready do you reduce the sound to the lossy compression format.

You may wish to try adding a little more mid-to-high frequency presence to the sound to help it punch through. Use the hard limiter to lose isolated peaks. Maybe a little audio compression will help, try single and multiband. Save the audio to your chosen format and listen to it critically. Repeat until you are satisfied.

BEWARE when you SAVE AS to, say, an MP3 format, Adobe Audition 2.0's filename extension will change at the top of the screen, but its buffer will still contain full resolution, uncompressed audio. You have to close the file and reload to hear the result.

Adobe Audition 2.0 can load and save a large number of different formats. Some are obsolete but still found. Others are used by other types of computer (such as AIFF for Apple Macintosh computers). The exact formats that you can load will combine those supported directly by Adobe Audition 2.0, plus those supported by Windows and any added by your sound card software.

11.9 Analytical functions

Digital audio files consist of a large quantity of numbers. As such they can be analysed statistically and produce, at worst, pretty pictures and, at best, useful diagnostic information about an audio problem. This may be particularly valuable for older users who are losing their high-frequency hearing and allow them to *see* problems that others can *hear*.

Spectral view

Normally when using Adobe Audition 2.0 the waveform view is used. This shows how the amplitude of the signal varies with time. But there is an alternative view, the spectral view. This analyses the spectral content of the signal. Much of the time this is good for making a pretty picture on your screen and little else. However, it can be helpful in tracking down audio problems. Spectral view displays a waveform by its frequency components, where the height represents the frequency and, as usual, time is represented horizontally. This will show you which frequencies are most prevalent in your waveform. The greater a signal's amplitude component within a specific frequency range, the brighter the displayed color will be. Colors range from dark blue (little audio in this range) to bright yellow (frequencies in this range are high in level). You can change the colors in the Option/Settings dialogue/Spectral 'Colors' tab.

As a diagnostic tool it will show you frequency response problems such as a suck-out at particular frequencies (perhaps resulting from a head alignment problem in an analog original). It will also show, as in Figure 11.5, the presence of low-level tones and help you identify their frequency. Know the frequency, and you are half way to finding the source of the problem. It can also be very useful in showing up clicks in audio from transferred gramophone records.

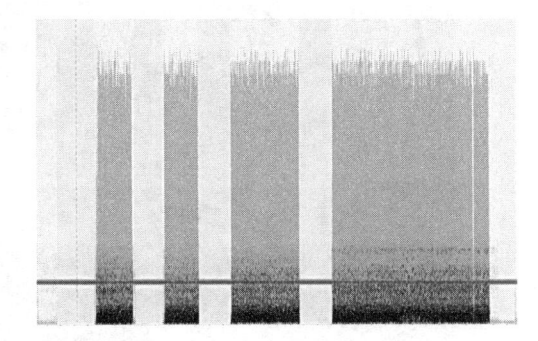

Figure 11.5 Spectral display showing low-level 3 kHz

11.10 Spectral editing

Audition 2.0 has an extraordinarily powerful spectrum edit capability which virtually allows you to 'retouch' audio files to remove spurious noises and modify syllables. There are extra tools for this which can be selected either by icons at the top of the screen (Figure 11.6) or by the menu option 'Edit/tools' (Figure 11.7). There is a different selection of tools in the multitrack view. The normal mode for the mouse is as a 'Time Selection Tool' (keyboard shortcut 'S'). The second option is 'Marquee Selection Tool' (keyboard shortcut 'M'). The mouse pointer changes from the standard 'ibeam' (often used for text selection elsewhere) to a cross. This allows you to 'rubber band' rectangular marquees within the spectral display.

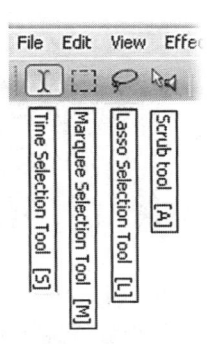

Figure 11.6 Selection of tools in the multitrack view

Figure 11.8 shows an audio file with blips of tone within it indicated by the horizontal white lines. Figure 11.9 shows four of the blips surrounded by the selection marquee. Shift

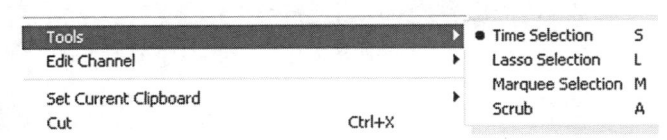

Figure 11.7 Edit/tools menu option

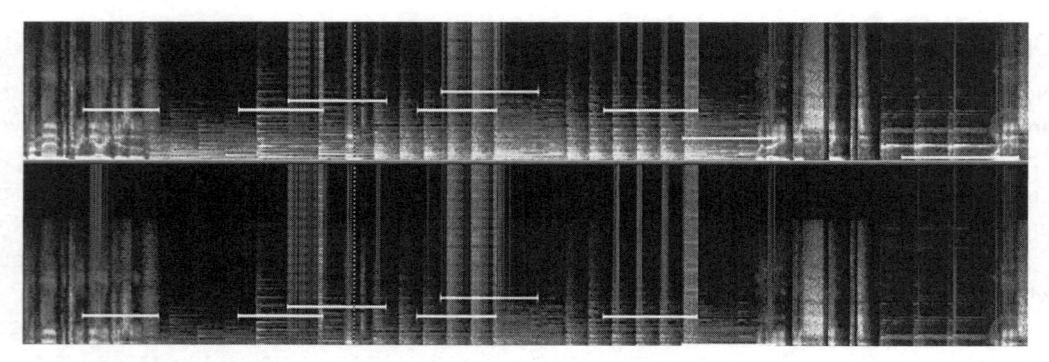

Figure 11.8 Audio file with blips of tone

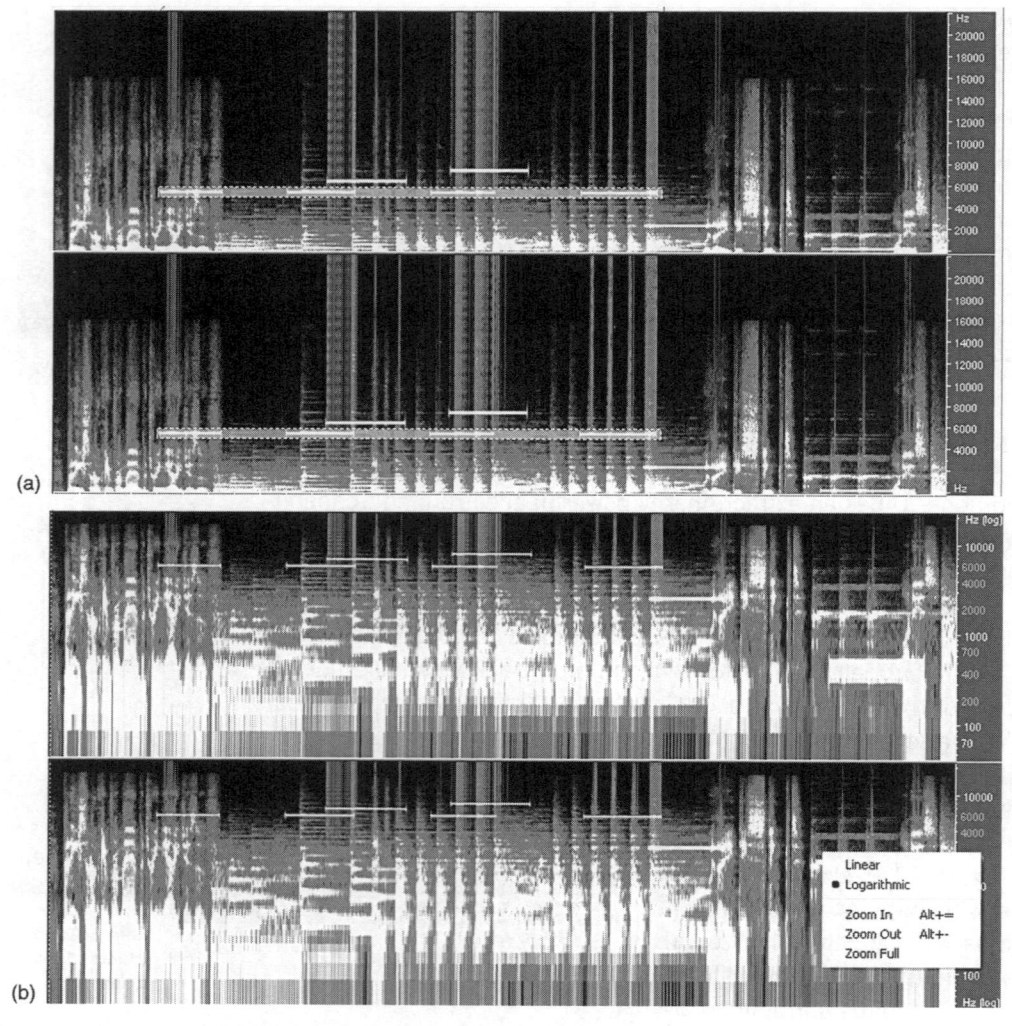

Figure 11.9 Blips surrounded by the selection marquee (a) linear view, (b) log. view

clicks of the mouse will snap the nearest part of the rectangle to the mouse position. This can be easier if you are using a laptop touch pad rather than a traditional mouse. Normally both tracks of a stereo file are treated the same but if you have disabled one side by clicking the time selection mouse at the top or bottom of the waveform display then only the enabled track is selected. So now what do you do? The answer is almost anything. The full gamut of effects processes can be applied, from noise reduction to muting. You could, for example, beef up a cymbal's 'tiss' or a reader's sibilants by adding white noise. People will be finding increasingly imaginative possibilities. Yes, it will be time consuming but this is the audio equivalent of retouching an image. For the purpose of this example I have merely used the Control/X cut option to mute the marqueed area.

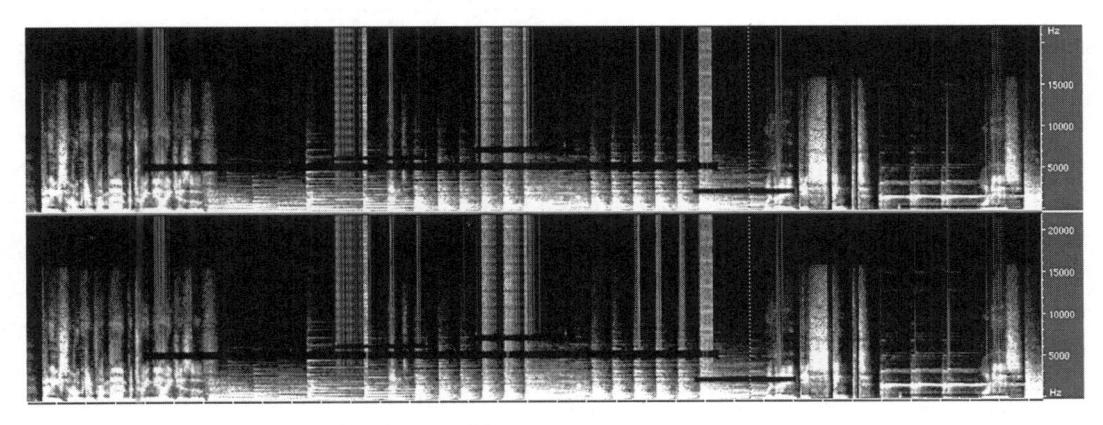

Figure 11.10 Results applied to the tone blips

Figure 11.10 shows the result applied to all the tone blips. You can expand the display vertically by using the vertical zoom and then dragging the frequency labels to where you need to be. This allows your marquee to be drawn more tightly and to retouch lower frequency sounds (Figure 11.11). Alternatively, right clicking the frequency labels gives you the option of changing the spectral display to a logarithmic one which expands the lower octaves (Figure 11.12).

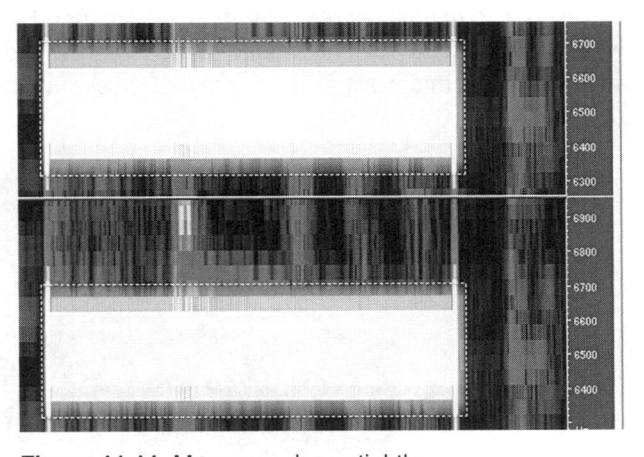

Figure 11.11 Marquees drawn tightly

The Lasso Selection Tool (keyboard shortcut 'L') works in a similar way except that you can make a freeform shape. *This tool also allows you to make multiple selections at the same time using the 'shift' key,*

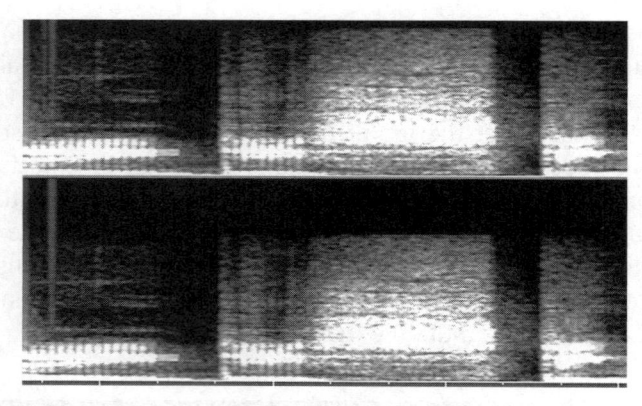

Figure 11.12 Changing the spectral display to a logarithmic one

enabling the same process to be applied to all of them at once. The mouse pointer becomes a graphic of a lasso (Figures 11.13–11.16).

There is a fourth tool option which is 'scrub'. The mouse pointer becomes a combined delta loudspeaker graphic. In play, pressing the left mouse button down allows you to play the audio back and forth in an attempt to emulate the old scrubbing of tape spools to make an edit. How successful this is will be is down to your graphics card, your processor speed and available memory. I confess that it does nothing useful on my 2.8 GHz laptop. It is like moving the audio via an elastic band. Alternatively, holding down the 'Ctrl' key while scrubbing loops the audio directly beneath the tool.

There are two more spectral views. These are not editable but are aids to analysis. View/spectral pan display (Shift/A) represents where the stereo images are panned to. Some experience is needed to read the display. A panned mono source will produce a thin line. A wide stereo occupying the whole sound stage will produce a wide display. Figure 11.17 shows the normal waveform display of a sequence where mono speech starts centre,

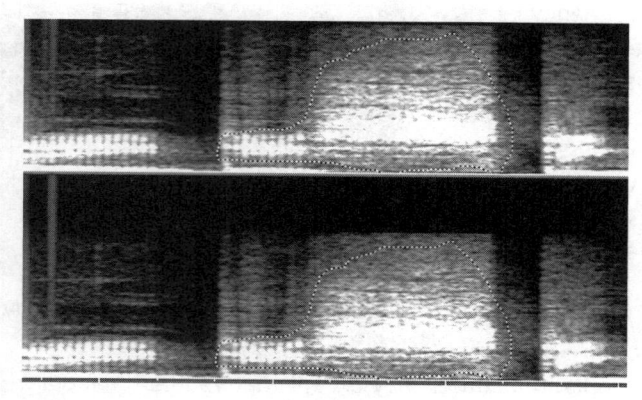

Figure 11.13 Lasso applied

Mastering

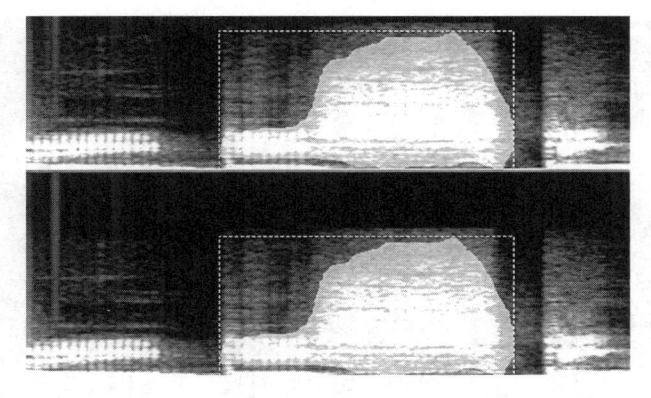

Figure 11.14 Rectangle appears around 'illuminated' lassoed area

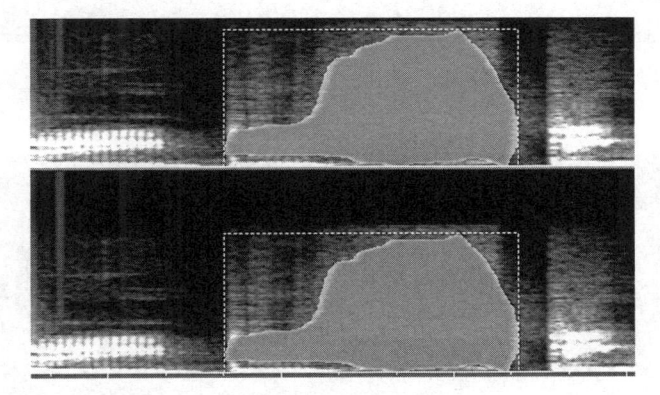

Figure 11.15 Delete pressed, lasso 'dark illuminated'

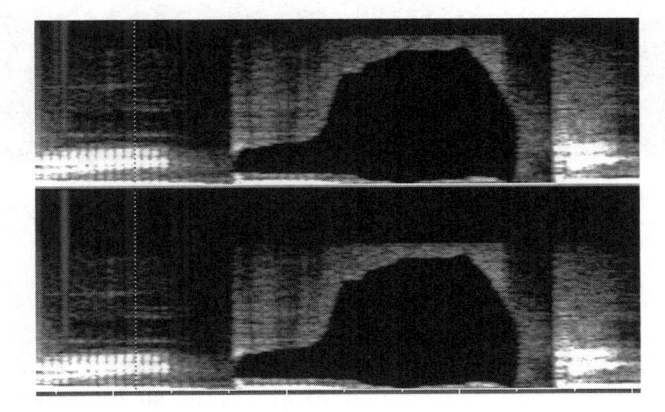

Figure 11.16 Cursor selected away, audio deleted

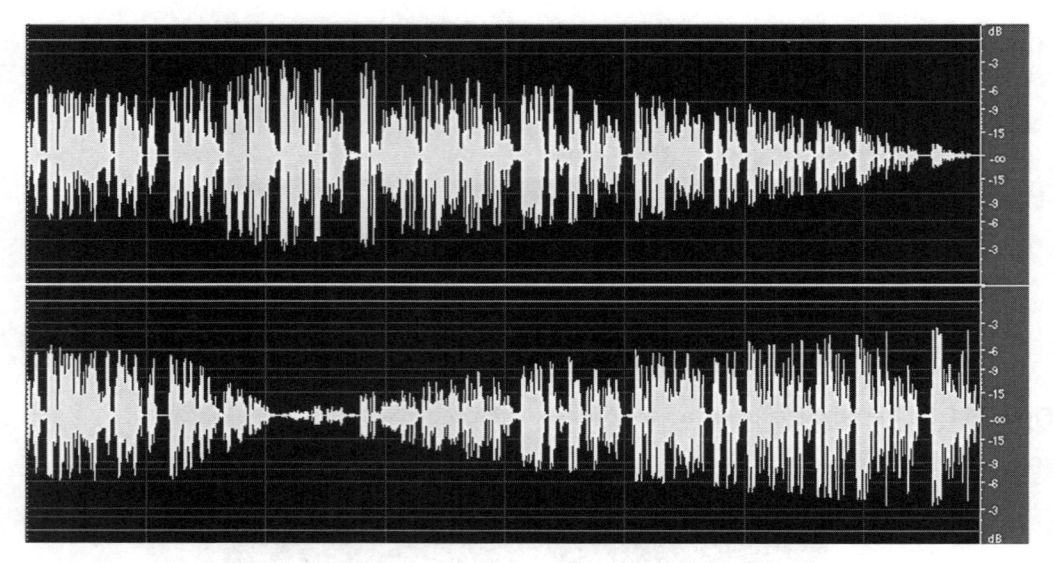

Figure 11.17 Normal waveform display

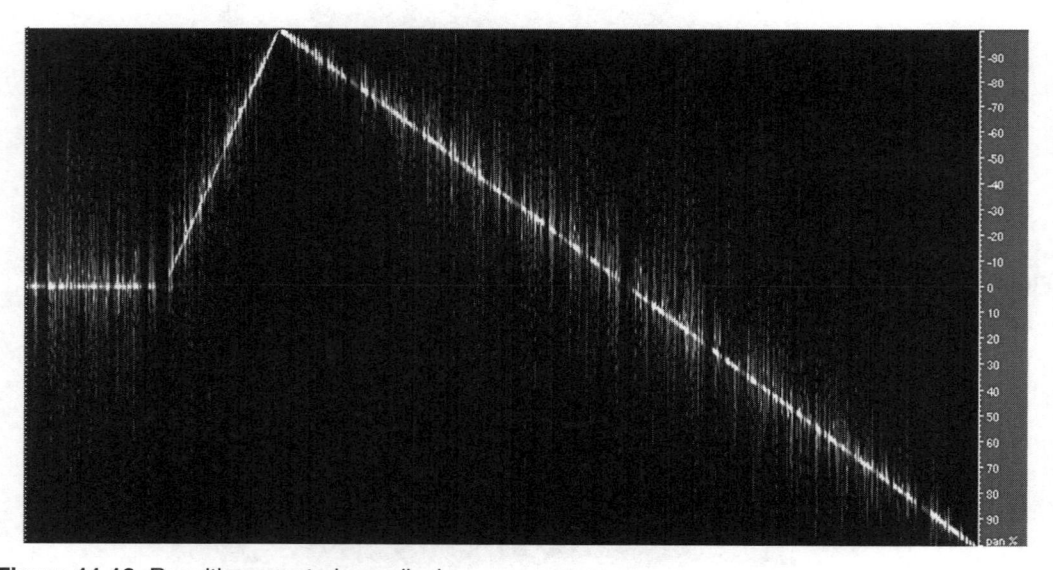

Figure 11.18 Resulting spectral pan display

then moves to the extreme left and then pans to the extreme right. Figure 11.18 shows the resulting spectral pan display.

The Spectral phase display shows the 'stereo-ness' of the file; a narrow display indicates mono, a wider display indicates wider stereo. Figure 11.19 shows a stereo recording gradually being narrowed until, at the end, it is mono.

Mastering

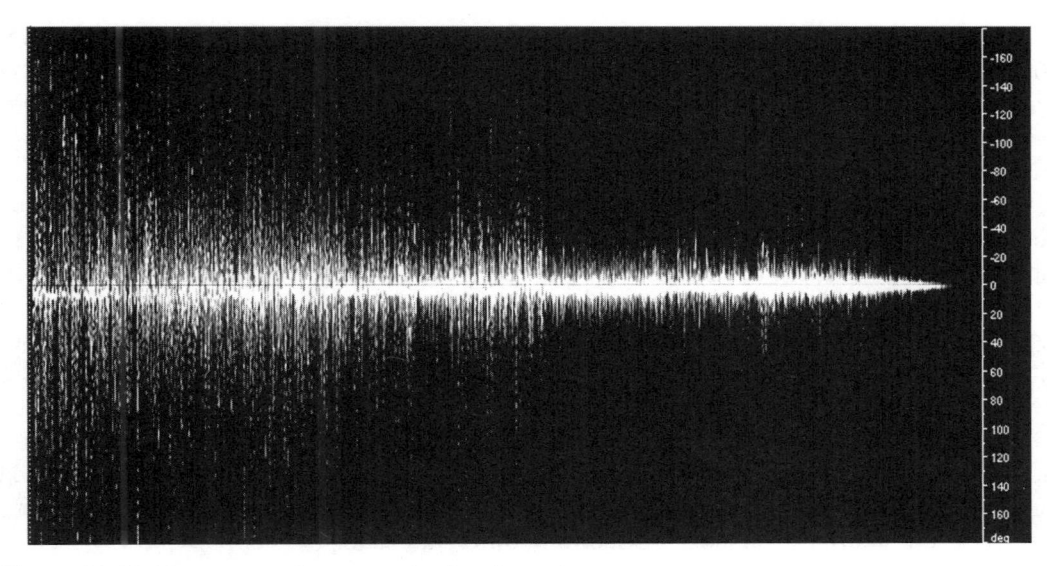

Figure 11.19 Stereo recording narrowing to a 'mono'

Frequency analysis (Figure 11.20)

The Frequency analysis window contains a graph of the frequencies at the cursor or at the *centre* of a selection. This window 'floats', meaning that you can click in the waveform on the main Adobe Audition 2.0 window to update the analysis while the Frequency analysis window is on top.

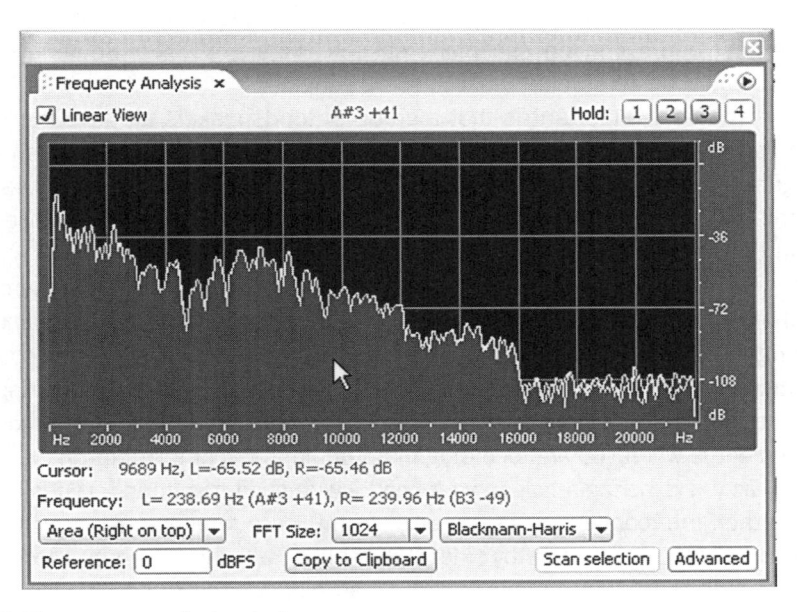

Figure 11.20 Frequency analysis window

The information in this dialog is like one 'slice' or line in the Spectral view of the waveform. The most prominent frequency is interpolated and displayed in a window below. You can move the mouse over the graph area to display the frequency and amplitude components of that frequency. Where there is a single instrument playing, you can often see its harmonics quite clearly.

You can control how finely the frequencies are analysed. Provided this is not too rigorous for your computer, there will be enough processing power for the graph continuously to update itself while the file is being played. This dynamic display can often reveal low level, or high level, tones, etc., that cannot be spotted otherwise. For example, high-frequency noise may be present where there is no audio activity. This should be filtered for the benefit of people who can hear it. Similarly, a recording made from FM radio may show signs of breakthrough from the 19 kHz tone that synchronizes the stereo system.

You can also generate a step-by-step animation by clicking on the main waveform window and then holding down on the right arrow key. As the cursor scrolls across the display, Adobe Audition 2.0 displays the spectral information in the Analysis window. When you view stereo data, the left and right channels are shown in different colors.

Phase analysis

What is phase?

At its simplest this is the direction your loudspeakers move when a positive audio cycles is applied to them. Do they move forward or do they move away from the listener? The absolute phase – does this correspond with the direction the original microphone's diaphragm moved – is not thought to be important, although some people dispute this. It is often surprising just how asymmetrical audio can be especially speech (Figure 11.21). So it is possible that absolute phase could be audible, at least to some people.

What is undisputedly important is that a group of loudspeakers all have the *same* phase. In other words, in a stereo set up, the two loudspeakers move together when fed with an identical signal (mono panned centre) (Figures 11.22 and 11.23). If they move in opposite directions this will blur the stereo image and also reduce the bass response as the two speakers will tend to cancel out low frequencies.

You can tell if your speakers are in-phase by feeding a centre mono signal to them. Do not use tone – it is hard to locate – but speech or music with lots of transients – a piano is ideal. If the phase is right then the sound will come from the centre half way between the speakers. If the phase is reversed then it will be hard to place the image. On a very good monitoring set-up, the sound will sound as if it is inside your head. This is a good way of assessing how good the monitoring environment is when you are recording in an unfamiliar room. Out-of-phase mono sounds quite painful – as if it is boring a hole in your head – with good monitoring. If it is difficult to tell the difference then the room acoustics are poor. In this case temporarily move the speakers so that they are close together and side by side. Out-of-phase will show itself by a loss of bass.

Professional monitoring equipment often has an easy way of switching the phase of one leg of a stereo signal but much semipro equipment does not and most computer sound cards

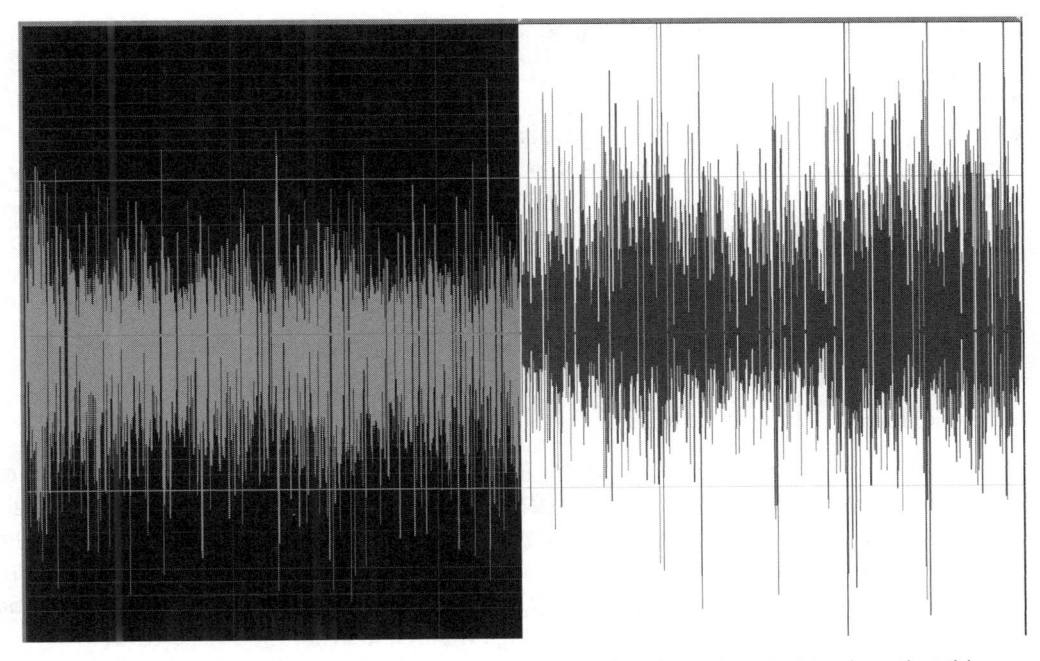

Figure 11.21 Asymmetrical speech. The highlighted area has been inverted to show that this particular speech recording (single male voice reading a story) is biased in one direction even though there is no DC offset

do not. It is well worthwhile to make a test file that can be played through the system. While you are at it, a channel identification can be added.

Making a phase check test file

- Record in mono onto a stereo wave file so that left and right channels are identical.
- Record words similar to:

    Phase and channel identification ...
    Identifying left channel only ...
    Identifying right channel only
    Identifying centre mono both channels in-phase ...
    Identifying centre mono right channel out-of-phase ...

- In Adobe Audition 2.0, select the right-hand channel of the phrase identifying the left. Press Delete. You now have that phrase on the left-hand channel only.
- Select the left-hand channel of the phrase identifying the right-hand channel. Press Delete. You now have that phrase on the right-hand channel only.
- Leave the phrase identifying mono centre alone.

- Select right-hand channel of the phrase identifying out-of-phase. Click on the menu item Effects/invert.
- Save the file.
- You can now play this file and check your loudspeakers. If you burn it to CD you can use this to check playback equipment and Hi-Fi equipment.
- If you find that the phase is wrong – you get a better image with the phase reversed sound – then reverse the connect to *one* loudspeaker only. It does not matter which you change.

Adobe Audition 2.0 Phase Analysis

As well as providing a spectral view of phase, Adobe Audition 2.0 implements a simulation of a long standing hardware way of analysing phase. This was to feed the left and right channels of a stereo signal to the X and Y inputs of an oscilloscope. The X input moves the spot on the screen left and right. The Y input moves it up and down. This means that if there is an X input only then a horizontal line is produced. Y only produces a vertical line. Mono produces a line at 45°. A stereo signal produces a complex image that spreads around the screen.

Adobe Audition 2.0 also has an MS option. What this does is feed the sum of the left and right signals to the Y input and difference between the channels to the X input. This has the effect of rotating the display through 45°. A mono signal gives a vertical line. A left-hand-only signal gives a 45° line going up to the left. A right-hand-only signal gives a 45° line going up to the right. This means that the display aligns visually with what you hear. This is the option that I prefer.

Interpreting the display

If you are copying analog tapes or cassette, azimuth error is a common problem. Normally the record and replay heads are set to be exactly at 90° to the travel of the tape. If one or the other are out of alignment then high-frequency losses will occur. This can be corrected by realigning the playback head to the setting used by the recording head. This is why it is always advised that cassette recordings are transferred using the same machine that recorded the take as most use the same head for record and playback.

The reason that high frequencies are lost is that the playback head's 'gap' needs to be of the order of half the wavelength of the highest frequency to be played back. If it is slanting with respect to the recording then the top of the gap is scanning a different part of the recorded waveform to the bottom. This makes the gap appear as if it were wider than it really is.

A quick fix for a mono recording is to play it back with a stereo head and only use one track. This halves the effective gap width error (see Figure 11.24). Adjusting the head for best high-frequency output by ear is harder than you might think. However, the Adobe Audition 2.0 Phase analyser can give you an objective view of this. Play the mono recording through a stereo machine as usual and look at the phase analysis. Azimuth error will not show as a

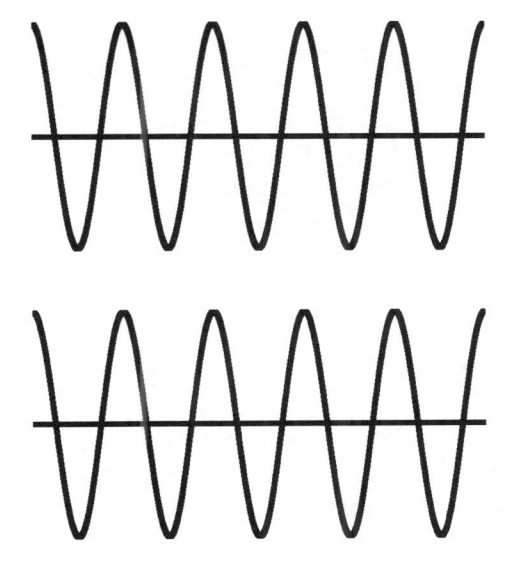

Figure 11.22 The left hand pair of waveforms are in-phase

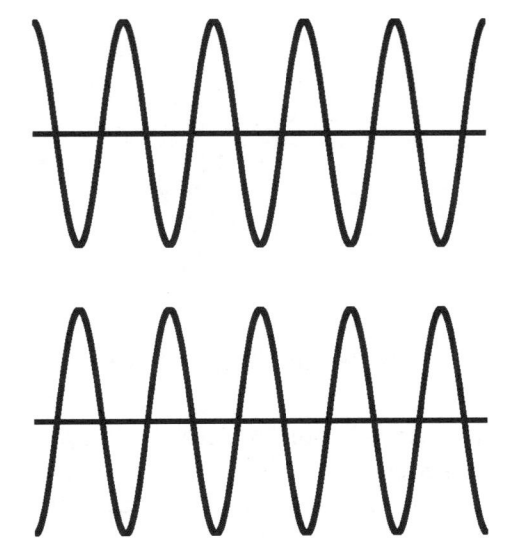

Figure 11.23 The right-hand pair is out-of-phase as the bottom waveform has been inverted

straight line, but will have width; pure tone will show as an ellipse. Figure 11.26 shows a dramatic error of 15° at the frequency used. This technique can be used on a stereo recording provided that you can find some centre mono such as narration or presentation.

Why does it matter?

When you are dealing with stereo or multi-channel audio the relative phasing between feeds is important for exactly the references we used to identify the error. The sound stage image becomes blurred and bass sounds lose their impact. While the untrained listener will not be able to say why, they will often find 'something wrong' with the playback and leave the room or switch off the recording, usually attributing their dislike to the performance in an undefined way. If your recording is likely to be broadcast then mono listeners will hear the sum signal (the vertical part of an MS display). An out-of-phase centre signal will be inaudible. Effectively they hear the *Karaoke* version!

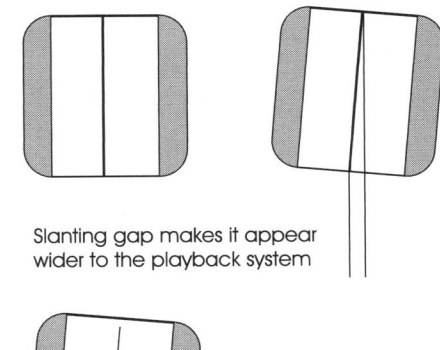

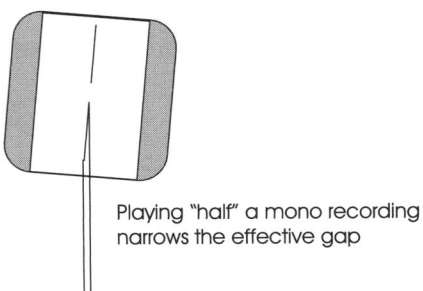

Slanting gap makes it appear wider to the playback system

Playing "half" a mono recording narrows the effective gap

AZIMUTH ERROR

Figure 11.24 Azimuth error

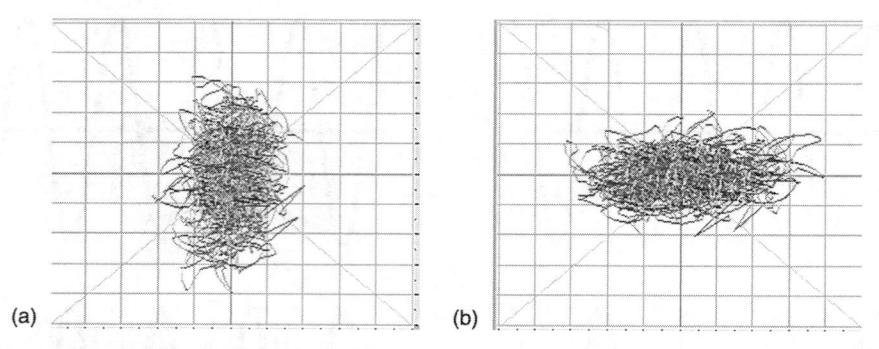

Figure 11.25 MS display: (a) in-phase and (b) out-of-phase stereo audio

You can very quickly develop 'an eye' for using the display to check your recordings for phase errors as, on average, the sum signal will be slightly larger than the difference. Figures 11.25 and 11.26 show an identical stereo signal in-phase and out-of-phase. The normalize button amplifies the display so that it is visible even on low level displays. The samples drop down controls the number of samples used. The more samples used, the denser the display.

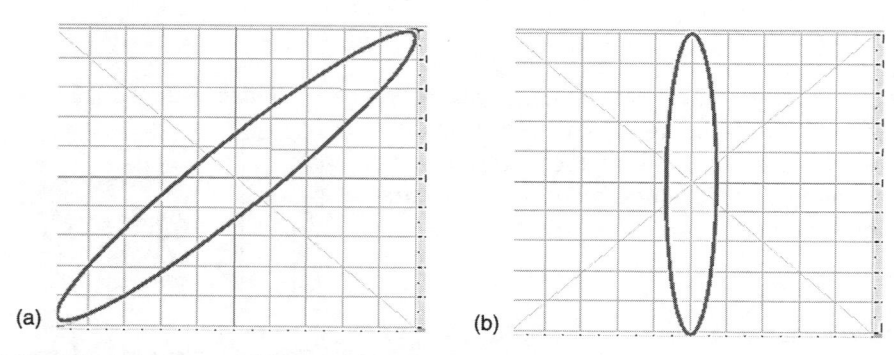

Figure 11.26 15° phase error: (a) LR and (b) MS

Amplitude statistics (Figure 11.27)

WIndow/Amplitude statistics produces information about the whole waveform or that section that is selected.

Use the Statistics dialog to get the following details about the current waveform.

Minimum/maximum sample value

Minimum, in this context, does not represent the 'quietest sample'. It couldn't, as every cycle of audio goes through zero. What it does represent is the peak negative-going excursion of the waveform. In both cases, clicking the arrow will take you to where this maximum or minimum is.

Mastering

Peak amplitude

This combines negative and positive, and expresses the peak level in decibels. A value of 0 dB indicates a fully modulated wave file. Negative values indicate how much the level would be increased if normalized. Clicking the arrow will take you to the point the peak amplitude occurs.

Possibly clipped samples

A sample which has the value of the highest possible, or lowest possible, value can indicate full modulation.

But, because no higher number can be represented, it may indicate that clipping – digital overload – has taken place. Clicking the arrow allows you to examine the waveform to see if there really is a problem by taking you to the point.

DC offset

This reports the DC offset to within 0.001 per cent and represents a test of how well the manufacturer has set up your sound card. Each sound card input may have a different value. Ideally they should all be zero.

Minimum/maximum/average/total RMS power

This is roughly equivalent to the area of the waveform display. A very peaky waveform may consistently touch 0 dB but it may have less power than a continuous sound of lower level. The arrows will take you to minimum and maximum points. These may well be in different places from the minimum and maximum amplitudes.

Histogram tab (Figure 11.28)

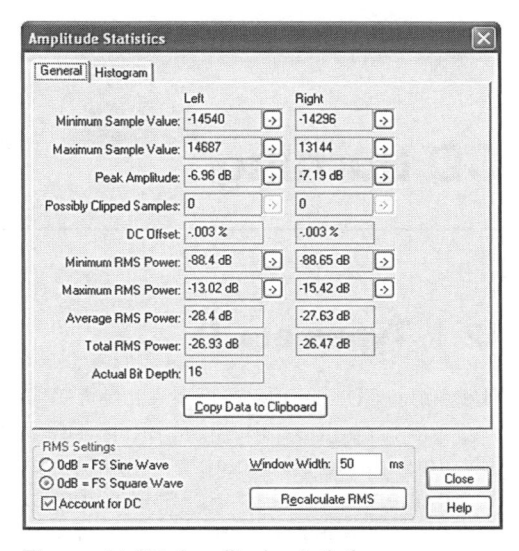

Figure 11.27 Amplitude statistics

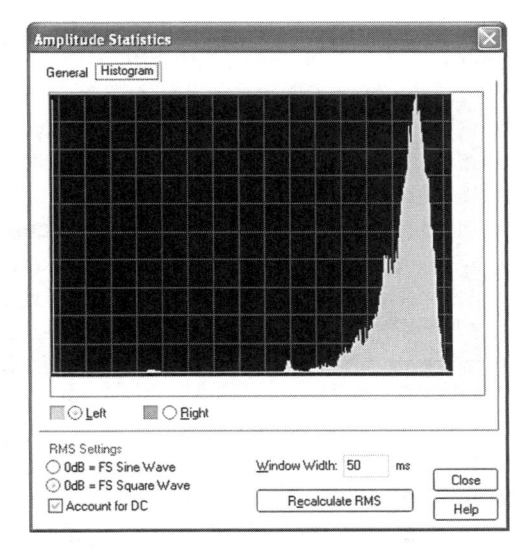

Figure 11.28 Histogram tab: amplitude statistics

This is a way of seeing the distribution of amplitudes in your recording. The graph shows which amplitudes are most common. The horizontal ruler measures the amplitude in decibels, while the vertical ruler measures percentage. The more audio there is at a particular amplitude, the more the graph will show for that amplitude.

12

CD burning

12.1 Types of CD

The audio CD, designed in the late 1970s, is now quite an old format. Today, its specification quirks can be quite frustrating. The audio CD was never designed as a filing system and fetching data to bit accuracy is not straightforward. CD-ROM adds the necessary file system information but at the expense of reducing data capacity. Yet, it remains a robust format and has spawned a large number of other formats, including DVD, using the same sized optical disk.

The audio CD is what we would now call a streaming format. It was designed to provide a continuous bit stream to a CD player. It was also designed to allow errors. It is not a filing system, that role is taken by the CD-ROM. This can be confusing if you are new to making your own CDs. You can make either from the same blanks. They look the same but one will be played by a CD player and the other spat out, played silently or, at worst, played with tweeter searing loud noise that in no way resembles your audio. CDs play from inside to out and start with 'Table of Contents' to tell the player where the tracks are.

It all comes down to error correction. A 74-minute CD blank can record 720 Mbytes of audio but just 650 Mbytes of data. The audio tracks are allowed to have errors which the CD player will correct sometimes but more often as not it makes a guess at the missing value (interpolation). Otherwise, it just mutes what might well be a loud click. The CD-ROM uses the 'missing data' to apply more robust error correction techniques. For more information on error correction see the section 'Errors' in Chapter 2 'Some technical bits', Page 10.

The CD-ROM can contain a very large number of files. These can be any type of data including audio of any format. For example, using very low bit rates, you can get 500 or so MP3 music files onto a single disc. CD audio discs are much more restricted. They have an absolute maximum of 99 tracks. Each track must be at least 4 seconds long. The first track should start 2 seconds into the disc. The tracks must be stereo and use 44 100 kHz sampling rate with 16-bit resolution.

Each track can have up to 99 index marks, the first one being the beginning of the track. The original idea was that a track would contain a single classical work, with each movement marked by index marks. Although this is part of the original CD specification, many CD players do not make use of this facility. The corollary of this is that not many commercial CDs have them either. Even so, they are used by many sound effects collections which cram many more than 99 sounds onto each CD. So, for example, a track of 20 door slams would have each slam marked with an index mark.

12.2 Audio CDs

Because of the streaming nature of CD audio, the tracks do not have file names and were not designed so that points within the track could be found with bit accuracy.

The first has been addressed partially by an addition to the CD specification which allows text to be associated with each disc and each track so that information can be added rather like on a Minidisc. Again many CD players ignore this information.

The second has made CD copying and recording problematical. The 'easy' way of copying a CD is merely to play it and use the CD-ROM drive's audio output to record. This can only be done in real time and includes the limitation of the analog section of your sound card. Reading the CD digitally has to be done in blocks. But when the CD-ROM drives go back for the next block it cannot find where it finished to sample accuracy. This can lead to what is often called 'jitter' as the resulting files can have bits of audio missing along with sections where there is an overlap. The normal way of dealing with this is for the software deliberately to fetch overlapping blocks and then line them up to find identical sections and electronically splicing the blocks together on the fly. This software can be in the computer or contained within the CD-ROM drive's firmware.

Not being able to find exact points within a CD track also causes the traditional problem with recording CDs. CD burning is not something that can usually be interrupted and this means that many computers cannot be used for anything else while the CD is being made. However, the good news is that the CD can be burnt in less than real time. Computer CD recorder drives that can handle $\times 24$, and faster are common. In itself, a fast drive is not enough; the computer system itself must be fast enough to be able to provide the data without a break.

Everything is fine provided a continuous bit stream can be sent to the recorder. However, computers can be doing other things and there may be a delay before the next block of data is sent. To counter this CD recorders have memory buffers of several megabytes so that they can continue to work on the buffered data during any gaps from the computer. As you can record CDs at much faster than real time even quite large buffers of several megabytes are often not large enough. While a buffer may be able to contain the equivalent of 40 seconds of audio, if you are burning the CD at 16 times real time then it is only proof against a break in data of 2½ seconds. Running out of data in the buffer is known as a 'buffer under run'.

With older CD burners, the only option is to ensure that the computer is doing as little as possible in the background with software like virus checkers switched off. Ideally the audio should be on a hard disk that has been freshly defragmented to speed up the supply of data. Over time, data files are fragmented around the surface of the hard disk. This means that the head has to dodge about collecting the data. Unfragmented files allow the head to run continuously from beginning to end of the file. The Windows software utility to defragment files is usually found from the Start menu under:

Programs/Accessories/System Tools/Disk Defragmenter.

Routine defragmentation of files is good practice anyway and helps prevent your computer slowing down over a period of weeks. If the buffer empties – even only momentarily – then the CD recording process fails and you are left with a useless disc often dismissively referred to as a 'coaster' or 'table mat' as it is good only for protecting your desk from your cup of coffee. However, recorder speeds of up to ×4 should be well within the capabilities of all but the most ancient of machines.

Fortunately technology has come to the rescue and CD recording drives are coming with improved techniques of dealing with this, such as Burnproof® technology. With this, the CD burner monitors the input buffer. If it falls below a certain capacity, say 10 per cent, it stops the recording process in an orderly way. Once the buffer has filled again, it can match the data in the buffer against the last block it recorded, playback what it has already recorded then drop in to record and continue without a gap. Actually there will be a tiny gap, but it is within the tolerance specified from the very beginning of CD manufacture so all CD players should have no problem. This technology also works for data being recorded as a CD-ROM.

12.3 Multi-session CDs

Most CD blanks, CDRs, can only be written to once. There is no way of erasing them and starting again. There is a half-way house where you can make 'Multisession' CDs. This allows you to add tracks in separate sessions. In order to achieve the table of contents for the CD is written in a different place. The effect of this is that most audio CD players will not recognize the disc while most computer CD-ROM drives will. The capacity of the disc is reduced as there is recording space lost each time for internal housekeeping. When you have finished adding tracks you can 'fix' or 'close' the disc. What this does is to copy the latest Table of Contents to the place on the disc where ordinary CD players expect to see it. At this point no more tracks can be added but ordinary CD players can play the tracks. This made a lot of sense when the cost of a CD blank was measured in tens of pounds or dollars but nowadays with blanks being so cheap creating multi-session CDs is losing its usefulness. The exception to this is if you want to create a mixed-mode CD which combines data with a CD-audio disc that will play on a CD player and show data on a computer.

12.4 Rewritable CDs

Rewritable and erasable CD blanks, CD-RWs, are easily available at quite low prices. At first glance, these seem to be a panacea. However, they are still more expensive than CDR and have several disadvantages. The major one is that they cannot be played by most ordinary CD players. They have a lower reflectivity compared with standard pressed CDs and need a stronger laser to read them. Many DVD players can read them as a by-product of being able to read DVDs.

CD-RWs have to be erased before reuse although this usually only means wiping the table of contents and is quite quick. You cannot erase a single track as such. Using CD-RWs as 'large floppies' is a very attractive idea and there is software that allows you to do this. How successfully it does this seems to be very machine dependent. Some people have used CD-RWs in this way with no problems at all while others have had nothing but 'blue screens of death' as their computers crash. The capacity of the disc is much reduced by the need to format it. This takes time and a large amount of data is taken by the formatting information and cannot be used to record. Quite often other machines using nominally the same software cannot read the discs so the portability of CD is lost. Similar software is available for DVD-RW which of course has much higher capacity.

As with multi-session CDs, many people take the view that CDRs are so cheap that the extra hassle of CD-RW is not worth it. Others compromise and only use them written as if they were CDRs only making use of their erasability.

12.5 CD recording software

Normally basic CD software requires you to specify if you are burning an audio CD, a CD-ROM or a mixed format. Tracks are dragged into a window and the CD burnt. If you are planning to produce actual audio CDs then more sophisticated software is needed. Simple software works on the basis of one wave file per track.

This need not necessarily be so; Audition can also have tracks 'marked up' in a single file and use individual files for all of the tracks of a CD. For CDs of events, some burners can vary the length of each inter-track gap or have none at all. You can even continue the audio through the inter-track gaps. For a CD of a concert you can arrange for each song to begin each track but for speech links and so on to be in the gaps between the tracks (the CD player counts down the seconds in the gap). The CD will play through continuously, but a track skip will take you directly to the next song. If you do a lot of this sort of thing, then you will need something Sony's CD-Architect which can create complex structures very easily (Figure 12.1).

If you have access to professional or semi-professional CD players which will make use of them, you can also add index marks. These can be useful to you for archival purposes, allowing you to mark individual items without starting a new track. You can mark individual contributions within a feature or program. Unlike tracks, index marks do not appear in the table of contents and the CD player literally has to scan through the track to find them, so index searches are slower than track searches.

Traditionally CD burning software offers various options as to how the disk is recorded. You decide whether you want to record the whole disc at once or go for the multi-session option. If you do, you often have to remember to 'Import' the last session so that the new recording is added to the previous Table of Contents.

Adobe Audition 2.0 CD software

Adobe Audition 2.0 provides audio-only CD burning. Audition's CD burner does not burn indexes, at present, but Audition's index markers within files are recognized by other software such as Nero so that everything up to the actual burning process can be prepared on Audition. A single wave file can be split into several CD tracks, not necessarily in their order in the wave file. Remember MP3 files count as data and cannot be placed on an audio CD. The booby trap is that it might look if you can, but Audition will convert the MP3s to WAV files before burning thus substantially reducing the number of tracks that can be put on one disc.

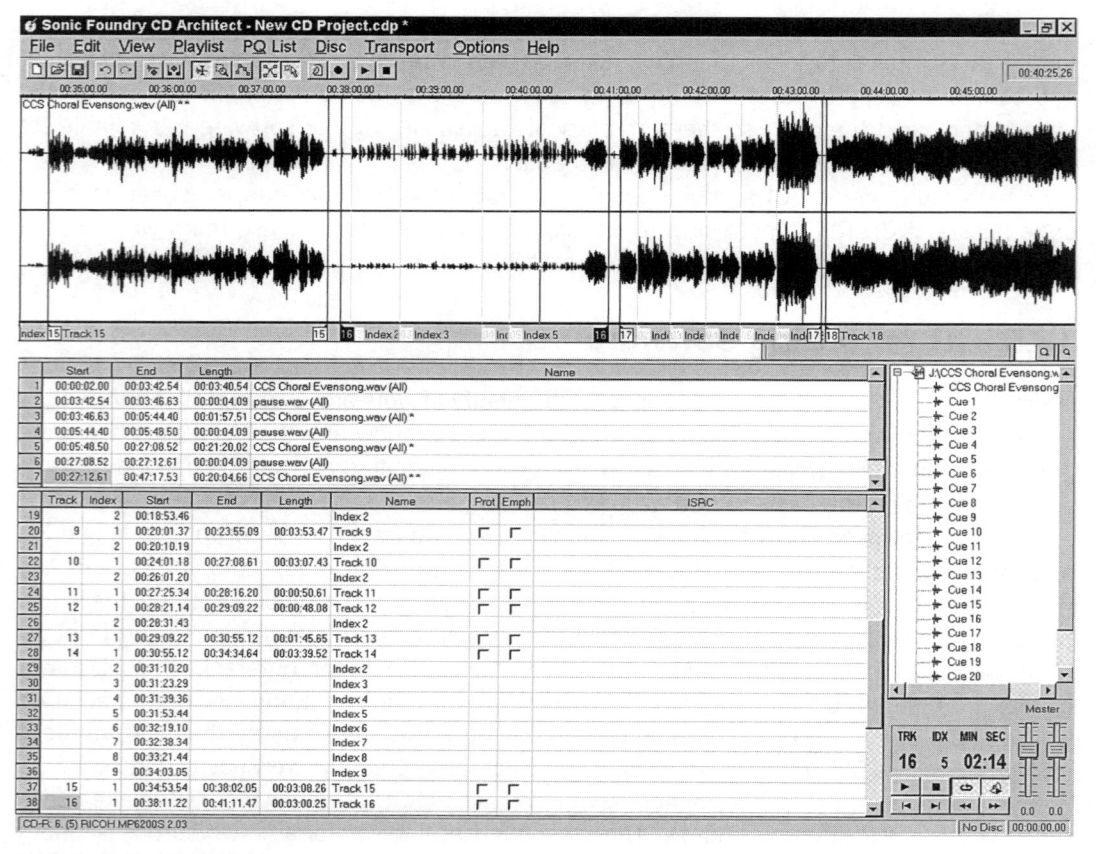

Figure 12.1 CD Architect

The CD burner is integrated into Audition as the third option to Edit and Multitrack views (Figure 12.2).

CD burning dialog in use

On the left of the screen is a list of files or a single file. These can be loaded into the editor or imported into the CD burner (File/Import). If the file has internal markers for tracks then a plus symbol will appear on the left. If this is clicked then a list of the marked tracks will be shown. These can be dragged onto the main panel individually. Alternatively the file itself can be dragged. This will provoke a dialog asking how you want to handle any markers within it (Figure 12.3). Some CD burner programs can read the available duration information of the disc. Adobe Audition 2.0 only has selectable assumptions of 74 or 80 minutes (CD View menu). The normal Free Space indicators for time and megabytes at the bottom of the screen use this.

CD burning

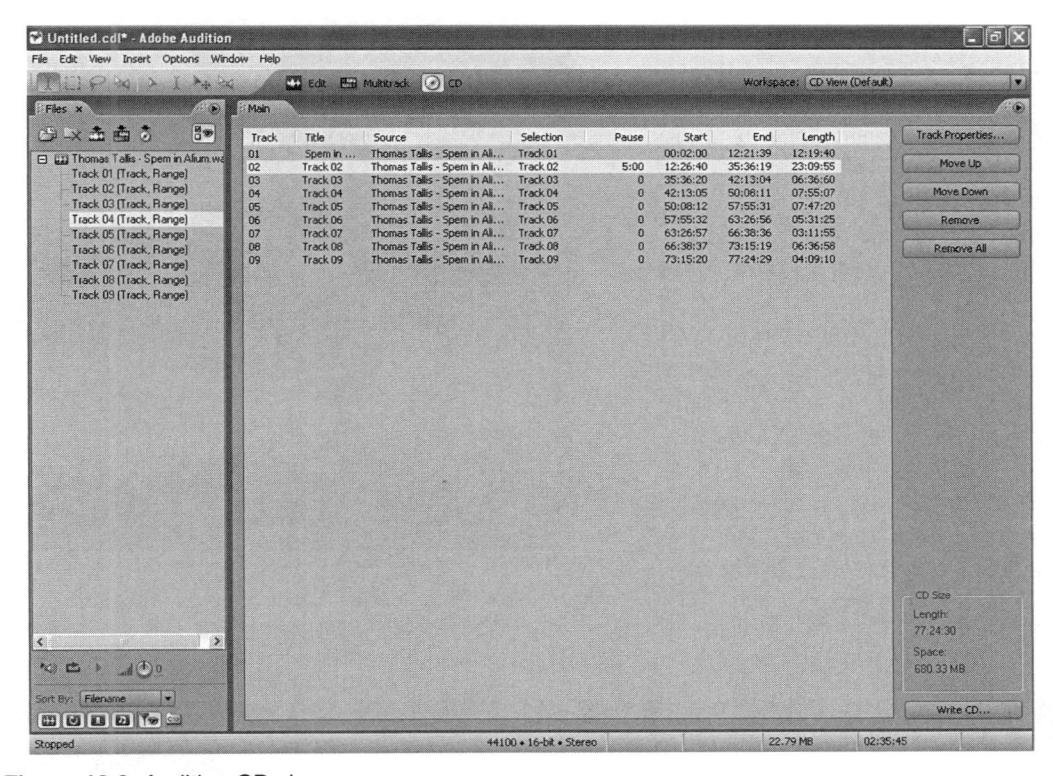

Figure 12.2 Audition CD view

Files shorter than 4 seconds are ignored as tracks must be 4 seconds or longer to conform to the CD specification. The file uses a different marker from ordinary 'Cue' markers. The track marker is created using Ctrl/Shift/F8. This will create a track that corresponds to the current selection or a track marker at the current cursor if there is no selection. (This is different from previous versions of the program which used Shift/F8.) This now selects the workspace session window. Ctrl/F8 still selects an index. You may wish to consider changing the shortcuts through 'Edit/keyboard shortcuts and Musical Instrument Digital Interface (MIDI) triggers' (Alt/K).

Both point and region markers have their uses. If you have a continuous recording, say of a live performance then all you may

Figure 12.3 How to handle markers

want to do is to mark particular event in the recording so that listeners can find their way around easily. This is easily done by marking a track at the current cursor position (Figure 12.4). However, it may be that you want to extract clips from a recording. Mark them as regions by track marking a selection then these clips will become available as separate tracks. You can mix regions and cue points in the same wave file. The point markers can be converted to regions in the marker list window. (Alt/8 toggles) using the Merge selected icon at the bottom of the panel (second from right) (Figure 12.5).

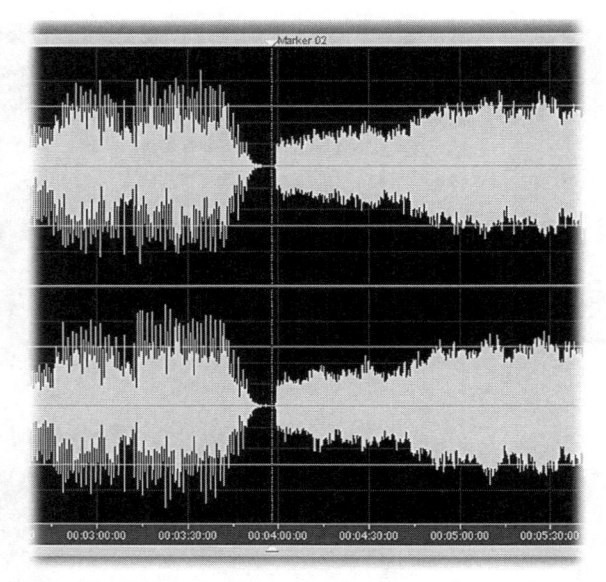

Figure 12.4 Point marker at current cursor position

Figure 12.5 Markers panel: Bottom icons; Edit Marker Info (toggle), Auto Play (toggle), Add Marker, Delete Selected, Merge Selected and Batch Process Marker regions

You can see the point markers as they have zero length. Select the ones you want to merge into regions and click 'merge selected'. The conversion does not assume that you want to start at the beginning of the file so if you are making CD-tracks then make sure you put a point marker at the beginning and end of the file (Figure 12.6).

You will end up with one less track than you had point markers which are analogous to 'posts' supporting the panels of a fence (Figure 12.7). The start and finish values can be tweaked either by editing the times using the marker list editor, or more easily by sliding the markers

CD burning

Figure 12.6 Point markers

Figure 12.7 Point markers posts supporting the panels of a fence

within the file. Figure 12.8 shows a close-up of the markers. The start marker is headed by a triangle with a right facing hypotenuse whereas the end marker is the other way round. In the illustration we see the junction between two tracks so the end marker for the previous track is shown before start marker for the next track. The playback cursor is topped by an equilateral triangle. If the transport 'go to next marker' button has been pressed, this often

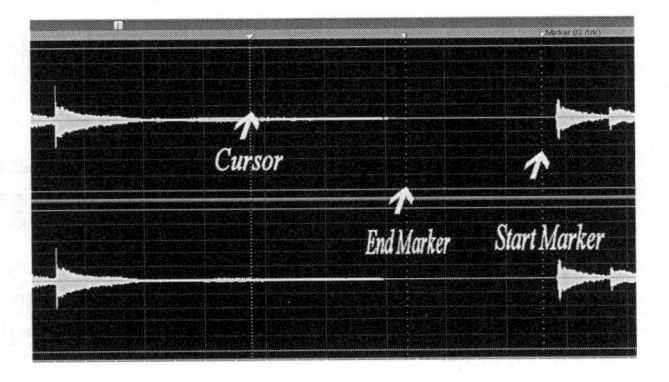

Figure 12.8 Marker close-up

end up on top of and hiding a marker. Normally all three are set to different colors.

Selecting a track and clicking the Track Properties button (or Track Properties on the right click menu), allows you to vary how the tracks are placed (Figure 12.9). You can also set the Title and Artist used by CD-Text, if you are using it. Normally a track will use the default

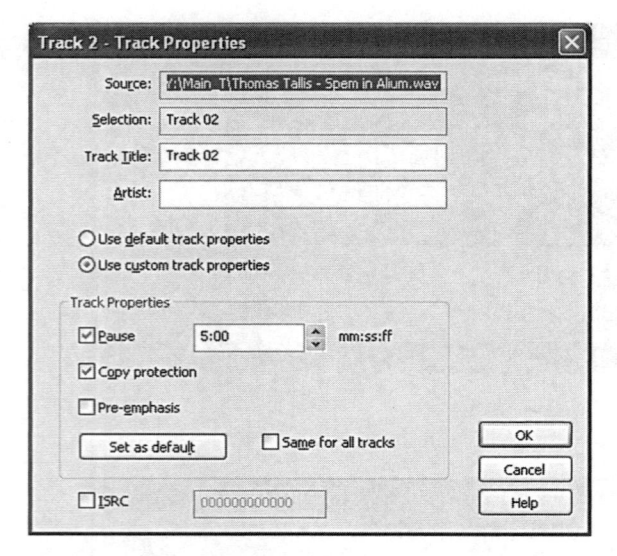

Figure 12.9 Track Properties

properties but you have the option of an individual setting with the Use Custom Properties option.

You can select whether there will be a pause at the beginning of the track and how long it will be. The copy protect and pre-emphasis flags can set although neither has much value these days. Most CD copying software will ignore the copy protect flag so merely takes on the role of asserting copyright for you. CD has an option of using a top boost or pre-emphasis. The idea is that on playback a top cut is applied which further improves the signal to noise ratio. However, with modern audio tastes for a lot of high frequencies in the mix this is often counterproductive. Further, the CD burner will not apply the pre-emphasis for you; you have to apply it yourself. Ticking the 'Same for all tracks' box will set the other tracks to the same setting without changing the default. You can make your new setting the default by clicking the 'Set as default' button. For professional use, you also have the option of giving individual tracks an ISRC number. Theoretically this should conform to the International Standard Recording Code format but there is nothing to prevent you having your own coding system for personal use.

Clicking the 'Write CD' gives you more options including selected which CD burner to use if you have more than one (Figure 12.10).

'Write mode' gives you the option of 'Write', 'Test', 'Test & Write' (Figure 12.11). When you first use the system rehearsing the burn using the test mode can save blanks. In practice, once the system has proved itself, you will leave the 'Write' option set.

You can set to make more than one copy and to eject the disc once it is done. Depending if your drive can handle CD-Text you can disable or enable it; adding in a title and an Artist description for the overall CD.

'CD Device Properties' (Figure 12.12) allows you change the buffer size from a drop down list. Depending on the CD burner, you may be able to specify the write speed although the single option 'Optimal' may be all that is available. A tick box selects 'Buffer Underrun prevention' on or off.

You can also add a Universal Product Code/ European Article Number (UPC/EAN) number to give the CD a catalog number. Theoretically this should conform to the UPC/EAN standards, sometimes known as bar codes. However, you can use your own numbering system for personal use if you wish.

If the disc is for someone else, you may want to check it before sending it to them. Do not use the CD burner you used to record it! If you have a separate CD-ROM then use that or,

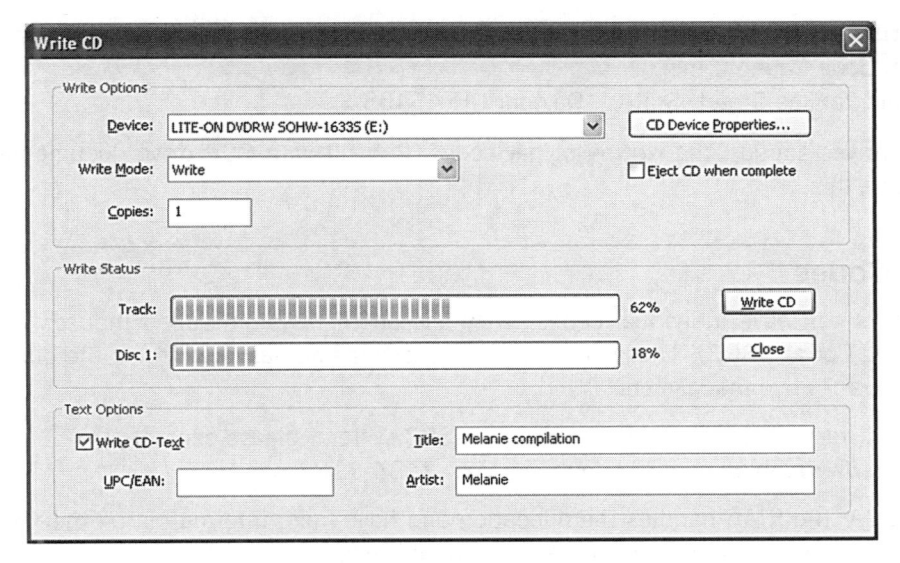

Figure 12.10 Options in clicking Write CD

Figure 12.11 Options in Write mode

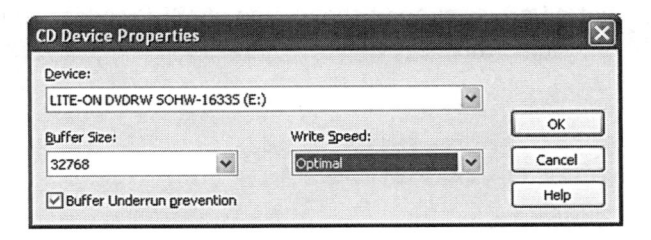

Figure 12.12 CD Device Properties

better, use a domestic CD player. If you are paranoid about the quality you can rip the newly created CD and compare the resulting files with the originals.

12.6 Recording codes

UPC/EAN Code

EAN or UPC codes identify products and manufacturers. Sometimes known as bar codes, they are usually only required for professional CD mastering. Codes are assigned on a country-by-country basis.

In the UK:
Association for Standards and Practices in Electronic Trade
EAN UK Ltd.
10 Maltravers Street, London WC2R 3BX

In the USA:
Uniform Code Council, Inc.
8163 Old Yankee Street, Suite J, Dayton, OH 45458

Not all drives support the writing of bar codes. Consult your CDR drive documentation to determine this.

ISRC Codes

The tracks can have individual codes. They use another system called Industry Standard Recording Codes (ISRC). These are designed to provide an easy way to log the play of specific CD tracks by broadcasters.

In the UK, for further information about the ISRC system, please contact:
PPL; http://www.ppluk.com or telephone: 020-7534-1122

In the USA, the RIAA handles the allocation and have more information on their web site, including a description of how the code is formatted:
http://www.riaa.org/Audio-Standards-3.cfm or telephone 1-202 775-0101

Not all drives support the writing of ISRC codes. Consult your CDR drive documentation to determine this.

13

Making programmes: acquiring material

13.1 Responsibilities

All the equipment, bells, whistles and toys are of no use whatsoever unless there is something worthwhile to record, something your listeners will want to hear. Acquiring that material is the first task.

At the more complex level, you may be editing something that has been put together in a studio, or maybe a recording of a live event. In the context of this book, this means that the recording has already been done; you are presented with a stereo mix on analog tape, a DAT or even a multitrack mix on an eight-track video cassette-based digital multitrack; it may even be on a removable hard disk or a flash card.

If you have a suitable sound card then you may be making the original recording using Adobe Audition. In practice most people, most of the time are using something external to originate material. It may be off-air from radio, from a CD or any number of delivery formats.

Your first responsibility is to ensure that you can handle the delivery format. An eight-track ADAT tape is fine only if you have an ADAT machine and your computer is set up to record its tracks, either by analog or using the optical ADAT eight-track digital interface.

Even with DAT, you should have specified the sampling rate you require. This should normally be the rate you intend to use for the final edited version: 44.1 kHz for compact disc related work or, maybe, 48 kHz for material destined to be broadcast. Do not use lower sampling rates, or bit rates lower than 16, even if material is destined as a reduced data Internet sound file. You never know when there will be another use for the audio and the better the input to a poor reproduction system the better it will sound.

However, much of the material put together with a PC audio editor is likely to be less ambitious, and self-acquired; probably using a portable recorder, preferably digital with an ability to transfer digitally to your computer.

There are now a bewildering array of devices that will record audio. Some are purpose built, others have a primary rôle where audio is incidental; digital cameras, mobile phones, MP3 players, etc. For the professional DAT is still used but is not an option for new gear. Minidisc has had a revival in the form of Hi-MD. The new machines can reformat old-style minidiscs and double their capacity; 1 Gbyte Hi-MD discs are available that allow up to 94 minutes of linear 16 bit audio. As well as Sony ATRAC data compressed system they will also record MP3 audio and the discs are also being promoted for pictures I have always liked Minidisc physically. It fits in the pocket easily and is big enough to write a sensible label. At the moment, machines suffer from having to use Sony proprietary software to copy the files.

My hunch is that the medium of the future is the flash card as most devices use it as a solid-state hard drive and thus they can be connected to a computer (usually via USB) and files copied using the operating system desktop. They are small enough to be posted in an envelope but do suffer from the 'It fell down the back of the chair' syndrome; they are easily lost and have little room for writing labels. These are exciting times for recording technology and any further speculation here is likely to date rapidly.

When choosing a recorder beware of rights management software which is often configured to take your rights to your own recording away. So obsessed have manufacturers become about copyright theft that they sometimes give little consideration to the makers of original recordings. I know of cases where original recordings have become unusable because of rights management software. When they were copied the first time the software has logged the unique ID of the hard drive. Later the hard drive has failed or the owner has bought a new computer. The software refuses to make a 'second' copy and the original recordings have become completely unavailable.

A common way of obtaining your material is the interview; so here are a few guidelines to get the best out of them. These guidelines are as applicable to the promotional sales interview with the chairman of the company, as for a news interview for local or national radio.

13.2 Interviewing people

Preparation

Editing starts here, so consider the following; so as not to waste your own or other people's time and resources:

What is the purpose of the interview?

Are you interviewing the right person? (Do they have the information you want? Do they have a reasonable speaking voice?) Remember that the boss's deputy or assistant often has a better finger on the pulse of day-to-day problems so it is often useful to interview both, if you can.

Prepare your subject area for questions, rather than a long list of specific questions.

Choose a suitable, and convenient, location.

A senior manager's office is usually suitable, if it has a carpet and soft furnishings; but it is extremely important to get the person out from behind the desk. While you will tell them that acoustics is the reason, equally vital is that people speak differently across desks. Sitting beside them on a sofa will usually get a much more human response.

Type of interviews

- *Hard*: To expose reasoning and to let the listener make up his/her own mind. Interviewee comes in cold, with no knowledge of questions. Commonly used with politicians or those in the public eye.
- *Informational*: Aim is to get as much information as possible, so this is likely to be a more friendly, conversational style of interview. You may prepare your interviewee with a warm-up by outlining the areas of questioning.

- *Personality*: Here you are interested in revealing the personality of the interviewee.
- *Emotional*: Probably the most difficult type of interview requiring the greatest tact and diplomacy from the interviewer. Such interviews are of the 'How do you feel?' variety used when interviewees may be under great stress following a tragedy (although try to avoid the actual question 'How do you feel …?').

Technique

Make sure that you actually know how to use the recorder! Try it out *before* the day of the interview. Test the recorder before you leave base. Test it when you arrive and are waiting for your interviewee. Test it when you take level. Always keep the test recordings.

Always record useful information such as the date, who you are going to interview. This will help speed the identification of a recording when the label has fallen off and when it has been transferred to computer (some DAT and Minidisc machines will record a time of day and date code on the recording. This can be useful, provided you remembered to set the clock when you took the machine out of its box).

NEVER say anything rude about the person or company. 'I'm off to interview that prat Smith of that useless Bloggo company' might get you a laugh in the office, but loses its edge when you find yourself accidentally playing it back to Mr Smith after he has given you level. Machines that only offer headphone monitoring are a help here but spill from headphones can be very audible.

If using a handheld microphone, then sit, or stand, close to interviewee. Again; get them out from behind their desk! Side by side on a sofa usually works very well. The microphone should be 9–12 inches from interviewee and a similar distance from you. If you have to compromise, favour the interviewee.

Make a short test recording 'for level'. The nominal task is to set the recording level so that it does not distort or is so low in level as to cause noise problems when amplified for use. Do not ask a question for level that is going to be asked during the interview.

Traditionally, interviewers were supposed to ask what their interviewee had for breakfast. These days it is likely to produce either a monosyllabic 'nothing' or a long involved account of the special Muesli they eat. No, this is the opportunity to record the interviewee's name and title. Get them to say it themselves and use this as a way of checking the level on the recorder. As well as giving you a factual check – are they Assistant Manager or Deputy Manager – it gives you a definitive pronunciation of their name; is Mr Smyth pronounced Sm-ith or Sm-eye-th? In some styles of documentary this can be used for them to introduce themselves rather than this being done by the presenter.

It is good practice to leave the level set from your previous test recording so that the level is roughly right if you run into a major news story on your way to or from the interview and need to 'crash start' a recording.

Another function of the level test is check for unnoticed problems. An air conditioning noise that you do not notice 'live' may be very obtrusive on playback. Listen for, and anticipate, external noises, for example, children playing, dog barking, phone ringing, interruptions.

There can also be unexpected electrical interference. Faulty fluorescent tube can cause interference as can radio and computer equipment. Beware especially of mobile phones. As well as the obvious annoyance of them ringing during an interview, audio equipment is very prone to interference from the phone's acoustically silent 'handshake' signals with the network. Switch off your own mobile and ask your interviewee to do the same.

Relax the interviewee, if necessary, by discussing areas of questioning.

Keep eye contact and make appropriate silent responses, such as nodding or smiling.

Ask short, clear questions, one at a time. Remember we want to hear the interviewee not the interviewer.

Listen carefully and keep your questions relevant to what is being said. Ask open questions – who, what, where, why, how? BBC Radio Four's first *The World at One* presenter, William Hardcastle maintained that he added to this list the non-alliterative 'Is it on the increase?'

Pre-recorded interviews

Record background atmosphere for 15–30 seconds before and after interview to use when editing. In my experience, leaving a 2-second gap between 'Right we're recording' and the first question will give you the cleanest pause.

If you are using two microphones with a stereo recorder, then deliberately record the interview 'two-track' so that the interviewee's voice is on one track and the interviewer's on the other. Keep to a convention so that questions are always on the same track. 'The presenter is always right' was the tongue-in-cheek convention used in BBC Radio. In the days of quarter inch tape, this also had the advantage that the answers could be heard on an old half-track mono machine which only read the top (left-hand channel) of the tape.

After the interview play back the last few seconds to ensure that the piece has recorded OK. Ask the interviewee if there is anything that they feel was left out.

Vox pop interviews

In the trade, interviews done with people randomly selected in the street are known as vox pops (from the Latin *vox populi*; voice of the people). They present their own unique problems.

Always test the equipment before leaving base. Follow a closed question (e.g. 'Do you think …?') with an open question (e.g. 'Why do you think that?) to get interviewees to justify their opinions.

Prepare several forms of the same question to put to interviewees.

Choose a suitable location away from traffic, road works, etc.; unless, perhaps, the item is about the traffic or chaos caused by the road works. Approach interviewees with a brief explanation of why you're conducting a survey of public opinion. Keep the recording level the same and adjust the microphone position for quieter or louder interviewees. Record background actuality at beginning and end of tape.

(A comprehensive guide to interviewing is contained in *Radio Production*, fifth edition, by Robert McLeish, Focal Press, Oxford, 2005.)

13.3 Documentary

Documentary material will also use 'actuality'. This is material recorded, on location, of activity which can be used both for illustration and for bridging passages of time.

Record lots of it! It is too easy to be so closely focused on the interview material that the non-vocal material is forgotten. Do not be in the situation of the producer who recorded a feature about a mountain climbing expedition and forgot to record any sounds of people climbing! (In this particular case the situation was saved by a single 2-second sequence being looped, repeated and used in a stylized way.)

Actuality is different from sound effects. Actuality is the real thing, recorded at the actual location. Sound effects come from a library and are used to fake actuality. This is fine for drama but has only limited use in factual items. Using an FX disc of a Rolls Royce driving off would be fine as an illustration of a car driven by rich people, but not ethical if cued as 'Joe Smith driving off in his Rolls Royce after my interview'.

Music can enhance an item, but beware of copyright and performance rights. While the sound of a busker playing in the background is unlikely to be a problem, using a complete performance of a piece of music usually is.

13.4 Oral history interviews

Here you are interviewing someone, usually elderly, about their memories of decades before. This is possibly going to be the first and only time they have been interviewed. You want them to be relaxed so that they will speak freely. You need to be relaxed as these interviews are likely to be long. Sat side by side on a sofa can work, but some people can be unhappy about sitting so close to a stranger. Here, I have used a tip from Charles Parker, who with the Radio Ballads, pioneered British documentary radio. He suggested actually sitting or kneeling at the interviewee's feet. This has the effect of lowering your perceived status and emphasizes that their memories and views are important. With one knee on the floor, the other becomes a convenient place to rest the elbow of the arm holding the microphone. Acoustically it is good as the microphone naturally falls about 18 inches from you and the interviewee with both bodies providing sound absorption.

13.5 Interviews on the move

Perhaps not so much an interview as two or more people reacting to an environment. This could be observing badgers at night, a historical house or a carnival. The object is to capture as much of the atmosphere as possible while not making it totally impossible to edit.

In some ways, an ideal medium would be a four-track recording. Two tracks would be fed from a stereo microphone to record atmosphere; the other two would be fed by two mono microphones, one for the presenter – possibly a personal 'tie-clip' mic – and the other, hand-held, for whoever else is talking from moment to moment. An alternative might be two small portables – Minidisc or DAT – one recording atmosphere and the other recording dialog.

The major disadvantage of such an arrangement is that the technology is in danger of taking over. The person holding the mics and carrying the recorders may be being used for their knowledge of the subject rather than their technical skill. Here, a separate recorder operated by the producer can help out with collecting actuality.

However, it is very possible to get a dramatic, and memorable, recording using a single hand-held stereo mic but it is vital that the mic is not much moved about with respect to the speakers as, otherwise, their voices will go careering around the sound stage. In noisy environments, the mic should be as close as feasible to the voices in order to get as much separation as possible, so that what is being said can be understood. Plenty of actuality should be obtained for dubbing over the edited speech. It is easy to add sound, difficult or impossible to take away. A sensible discipline is also required so that the unrepeatable sounds of an event are either not spoken over, or a commentary is given that is articulate and will not need editing.

When editing a feature about, say, an outdoor market, the buzz of its activity is vital. While some material may be recorded in shops away from the noise of the market, probably the best material, editorially, will be acquired in the crowd. A beat pause before each question, or response, from the presenter is all that is needed to form the basis of a cross fade on atmosphere to a different section. Where the edit causes an abrupt change of character of the background then a short section of separately recorded actuality can be added to cover the change (see Section 7.7 'Chequer boarding' on Page 70).

13.6 Recorders

The original portable acquisition format was direct cut Edison cylinders. Later, 78 rpm lacquer discs were used by reporters during WWII. They were heavy, more transportable than portable, and could not be used 'over the shoulder'. With the perfection of tape recording by the Germans during that war, the 1940s and 1950s saw the rapid appearance of more and more useful portable tape recorders.

Tape (like disc) had the advantage that the recording medium was the same as the editing medium, which was the same as the final user or transmission medium. Removable digital media have the potential to be the same. However, at the time of writing, recorders using this sort of technology tend to be bulky and heavy and there is no agreed standard. Recorders are becoming available that use small media with sufficient capacity not to need to compress the audio. They are either removable or can download their recordings via a USB or Firewire port. Download times of 40 times real time are achievable. The only disadvantage is that, either the recorder has to be brought back to the computer at base or it has to have removable media, such as a hard disk or a flash card that plugs into a suitable port on the base computer.

Small recorders using Minidisc are very attractive. Ordinary domestic recorders can produce superb quality at prices a fraction of professional gear. Also becoming readily available are recorders using memory cards. A problem with these as well as minidisc is that they often use 'lossy' data compression techniques. This can cause problems if the edited recording are further going to be data compressed, especially using different systems. This is a hazard that especially faces broadcasters. If this is important to you, then shop around. Hi-MD has non-lossy modes as do many memory card recorders. Although cheap, the trade-off is that domestic recorders use fragile domestic plugs and sockets. This is not a problem if your environment allows you to take care of the machine, or is sufficiently well financed as to regard them as disposable.

Alternatively, an option is to purchase a case designed to protect the recorder. These make the whole package larger but they have professional XLR sockets and also a large capacity battery for longer recording times. There is also the incidental, but not inconsequential, effect of making the reporter look more 'professional'. This can make the difference between being granted an interview or rejection. There are professional versions of both minidisc and memory card recorders. These have XLR sockets and are generally more robust than their domestic brethren.

Whatever, and wherever, you are recording, you need to get the best out of your microphone and a basic review of acoustics and microphones may be helpful.

13.7 Acoustics and microphones

Acoustics

When making a recording, realize that the microphone will pick up sounds other than the ones you really want. So what does a microphone hear?

1 The voice or instrument in front of it.
2 Reflections from the wall.
3 Noise from outside the room.
4 Other sounds in the room including voices and instruments.
5 General noises, if outdoors.

Separation

Much of the skill of getting a decent recording is to arrange good separation between the wanted sound and the unwanted sound. The classic ways of improving this are:

Multimiking

There are many ways of reducing the spill of unwanted sound while retaining the wanted. It can seem that the answer to most problems is to have as many microphones as possible. In fact, the opposite is true. If you have six microphones faded up, each microphone will 'hear' one direct sound from the sound source it is set to pick up. However, it, AND THE OTHER FIVE MICS, will also get the indirect sound from its own sound source AND every other sound source.

In my example, you will be getting six lots of indirect sound for every single direct sound. As a result, multi-mic balances are very susceptible to picking up a great deal of studio acoustic.

With discussions, a lot of mics very close together on a single table can provide some extra control and give some stereo positioning. Avoid the temptation to put contributors to such discussions in a circle with comfy chairs, coffee tables and individual stand mics. The resulting acoustic quality will be excellent; unfortunately the discussion is likely to be boring and stilted. This is because once you go beyond a critical distance of about 4 feet, people stop having conversations and start to make statements.

A studio designed for multi-mic music recording will have a relatively 'dead' acoustic with reflections much reduced. The disadvantage of this is that the musicians and singers cannot hear themselves properly. In a very dead studio, a singer can go hoarse in minutes! You have to substitute for the lack of acoustic by feeding an artificial one to headphones worn by the artist. Needless to say, such a studio would be a disaster for a discussion.

Booths/separate studios

Gives good separation but with the penalty of isolating the performers from each other.

Multitrack recording

Laying down one track at a time while listening to the rest on headphones gives good separation but it can be difficult to realize a sense of the excitement of performance.

Working with the microphone closer

Because the direct sound becomes louder, you can turn down the fader and therefore, apparently, reduce the amount of indirect sound from that microphone. This has the disadvantage that the microphone is much more sensitive to movement by the artist and other unwanted noises can become a problem: guitar finger noise; breathing becomes exaggerated; spittle and denture noises can become offensive.

It is a matter of opinion just how much breathing and action noises are desirable. Eliminate them totally and it no longer sounds like human beings playing the music. Too much and it becomes irritating.

Using directional microphones

Microphones are not particularly directional and are likely to have angles of pickup of 30° to 45° away from the front.

High directivity mics will give poor results with an undisciplined performer who moves about a lot. The main value of directional microphones is not the angles at which they are sensitive but the angles where they are dead. It is more critical to position a microphone so that its dead angle rejects the sound from another source. Here a few degrees of adjustment can make a lot of difference to the rejection and no audible difference to the sound you actually want.

Orchestral music

Classical and orchestral music tends to be performed in a room or hall with a decent acoustic with a relatively small number of microphones; typically a main stereo mic plus some supplementary mics to fill in the sound of soloists, etc. Not only will the musicians be happier – many non-electric instruments are hard to play if wearing headphones – but also the larger number of musicians involved often makes it impractical to find that number of *working* headphones, let alone connect them all up and ensure that they are all producing the right sound, at safe levels.

Microphones

There is no such thing as the perfect microphone: a single mic that will be best for all purposes. While expense can be a guide, there are circumstances where a cheaper microphone might be better.

There is often a conflict between robustness and quality. As a rule, electrostatic microphones provide better quality. They have much lighter diaphragms that can respond quickly to the attack transients at the start of sounds. They have better frequency responses and lower noise figures. However, they are not particularly robust. They can be very prone to blasting and popping, especially on speech, where a moving coil microphone can be a better choice. Electrostatic microphones are reliant on some form of power supply which will either come from a battery within the mic, or from a central power supply providing 'phantom volts' down the mic lead. These fail at inconvenient times.

Quality

Yes, of course, we want quality but what do we mean by this? We want a mic that can collect the sound without distortion, popping, hissing or spluttering. With speech, there is a premium on intelligibility that is only loosely connected with how 'Hi-Fi' the mic is. An expensive mic that is superb on orchestral strings may be a popping and blasting disaster as a speech mic.

Directivity

Directional microphones are described by their directivity pattern. This can be thought of as tracing the path of a sound source round the microphone; the source being moved nearer or further away so that the mic always hears the same level – at the dead angle the source has to be very close to the mic. At the front, it can be further away.

Beware that sound picked up around the dead angles of a microphone can sound drainpipe-like and can have a different characteristic from the front.

As a rule, the more directional a microphone is, the more sensitive it is to wind noise and 'popping' on speech. When used out of doors, heavy wind shielding is essential. Film and television crews usually use a very directional 'gun' mic with a pickup angle of about 20° either side of centre. They are invariably used with a windshield; the large furry sausage often seen pointing at public figures in new conferences.

The most common directivity mic is the *cardioid* (Figure 13.1), so called because of its heart-shaped directivity pattern. Cardioid microphones have a

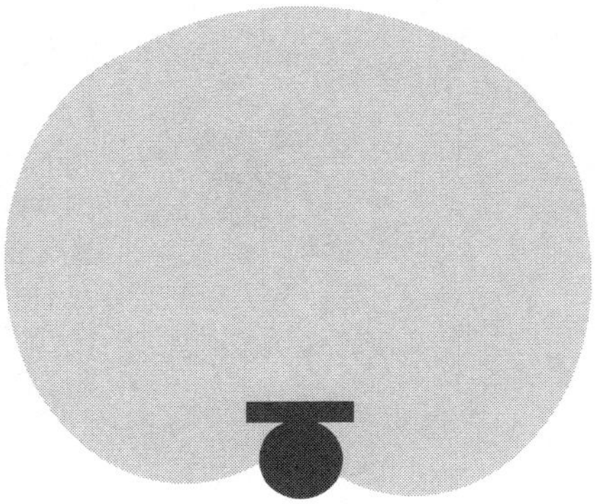

Figure 13.1 Cardioid microphone

useful dead angle at the back. Placing this to reduce spill is more important than ensuring that it is pointing directly at the sound it is picking up.

A so-called *hypercardioid* is slightly more directional (not more heart-shaped!) (Figure 13.2). It has the disadvantage that it has a small lobe of sensitivity at the back and is dead at about 10° either side of the back of the microphone. However when used for interviews using table stands the dead angle is usually about right for rejecting other people around the table.

Another common type of microphone is called *omnidirectional* as it is sensitive equally in all directions (Figure 13.3). Omnidirectional microphones are particularly suitable for outdoor use as they are much less sensitive to wind noise and blasting. They can be better for very close working, where directivity is not important, as they pop or blast less easily; an important consideration with speech and singing.

Figure 13.2 Hypercardioid microphone

Barrier mics (PZMs) can produce good results with little visibility as a result of being attached to an existing object (Figure 13.4). This can be anything from the stage at an opera, to a goal post on a football field. They work best when attached to a large surface, such as a wall, floor, table or baffle. They can be slung over stages using transparent sheet plastic as the baffle material to reduce visibility. The resulting pick-up pattern is hemispherical; an omnidirectional pattern 'cut in half' (Figure 13.5).

A more specialist directivity pattern is the *figure-of-eight* (Figure 13.6). This microphone is dead top and bottom and at the sides, but 'live' front and back. This directivity pattern is most often found in variable-pattern microphones. Figure-of-eight mics can give the best

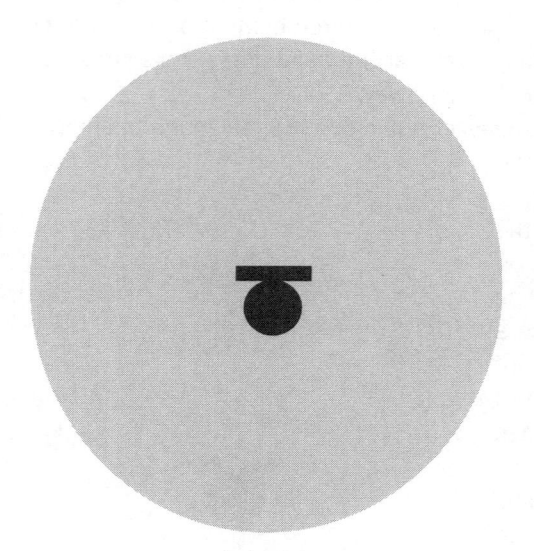

Figure 13.3 Omnidirectional microphone

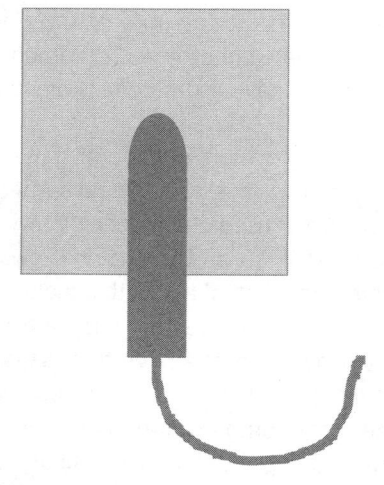

Figure 13.4 Barrier microphone

Making programmes: acquiring material

Figure 13.5 Hemispherical pick-up pattern microphone

Two cardioids added give Omni response

Two cardioids subtracted give figure-of-eight response

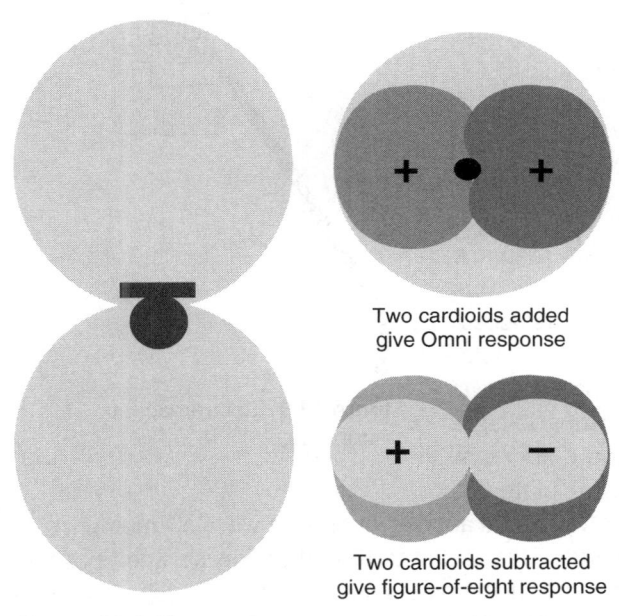

Figure 13.6 Figure-of-eight microphone

Figure 13.7 Variable pattern microphone

directional separation provided that they can be placed so that the back lobe does not receive any spill, or can be used to pick up sound as well. Ribbon mics have the very best of dead angles that can be very useful when miking an audience which is also being fed sound from loudspeakers.

Variable pattern mics

These, usually very expensive, microphones offer nine or so patterns. They work by actually containing two cardioids back to back (Figure 13.7).

If only one cardioid is switched on then, as you would expect, the mic behaves as a cardioid.

If both cardioids are switched on and their outputs added together, then an omnidirectional response is obtained.

If they are both switched on and their outputs subtracted then a figure-of-eight pattern results.

The other intermediate patterns are obtained by varying the relative sensitivities of the two cardioids.

Beware, these derived patterns are never as good as the same pattern produced by a fixed pattern mic. The cardioids retain their physical characteristics so an omni mic pattern derived from the two cardioids does not have the resistance to wind noise and blasting of a dedicated omni mic. The dead angle rejection or a derived figure-of-eight is not as good as a ribbon. Most microphones are 'end fire': you point them at the sound you want. These mics are 'side-fire' with their sensitive angles coming out of the side of the mic.

Transducers

Microphones are a type of 'transducer'; a device that converts one form of energy into another. In this case, acoustic energy is converted into electrical. Most microphones fall into two categories: electromagnetic and electrostatic.

Electromagnetic

Most electromagnetic microphones are moving coil. They are often called dynamic mics. They are like miniature loudspeakers in reverse; the acoustic energy moves a diaphragm attached to which is a coil of wire surrounded by a magnet (Figure 13.8). This causes a small audio voltage to be generated.

A specialized form of electromagnetic mic is the 'ribbon' mic that once was very widely used by the BBC. These have an inherent figure-of-eight directivity. They consist of a heavy magnet and a thin corrugated 'ribbon' of aluminium which moves in the magnetic field to generate the audio voltage. (These are 'side fire' mics.)

Electrostatic

Electrostatic mics, also known as 'capacitor', or 'condenser' microphones, fall into two types. Many require a 'polarizing' voltage of about 50 V for them to work. This is not a problem in a studio that can provide the necessary power as 'phantom' volts down the mic cable. Some mics contain 12 V batteries which operate a voltage boost circuit to provide their own polarizing voltage.

Electrostatic mics work by making the diaphragm part of a 'capacitor'. This consists of two plates of metal placed very close together. When a voltage is applied, a current will flow momentarily and the plate acts as a store for this charge. This is due to the attraction between positive and negative electricity across the narrow gap. If the plates are moved closer together, then this attraction increases and the plates can store more charge. If they move apart then the attraction reduces and the amount of charge that can be stored is reduced. The consequence of the plate distance changing is a current through the circuit.

If one of the 'plates' is the capsule of a microphone and the other is the diaphragm, made of a very thin film of conductive plastic, then we have the basis of a high-quality microphone. The very lightness of the diaphragm gives it the ability to follow high frequencies much better than the relatively heavy assembly of a moving coil microphone. As the diaphragm moves out, so the storage of the capacitor is reduced and charge is forced outwards. As the diaphragm moves inwards, the storage increases and the current flows in the opposite direction as the capacitor is 'filled up' by the polarizing voltage.

However, particularly for battery operated mics, an 'electret' diaphragm can be used. This does not need the polarizing voltage, as a static charge is 'locked' into the diaphragm at manufacture. It is the electrostatic equivalent of a permanent magnet. While the technology is much improved, these tend to deteriorate with age.

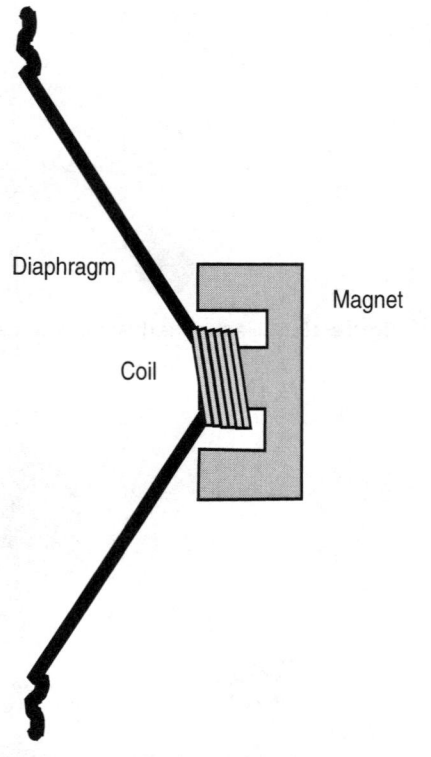

Figure 13.8 Electromagnetic microphone

There is a variant on the electrostatic mic where the capacitor, that is the diaphragm, is part of a radio frequency oscillator circuit. The changing capacitance changes the frequency of the oscillation, producing a frequency modulated signal that is converted into audio within the microphone. RF capacitor mics have a good reputation for standing up to hostile environments and are frequently seen as 'gun' mics used by film and TV sound recordists.

Robustness

Is the mic likely to be dropped, blown on, driven over? Does it blast on speech or singing?

Environment

Are you indoors, or outdoors? Is the mic going to get wet or blown upon? Non-directional (omni-directional) mics are least prone to wind noise. This can be further improved by a decent wind-shield. In practice, using your own body to block the wind can be the most effective.

Moving coil (dynamic) microphones are the most resistant to problems with moisture or rain. Umbrellas are of little help in rain because of the noise of the rain falling on them! The windshield is usually enough to prevent rain from getting into the mic. If you ever need to record under water, or in a shower, then a practical protection is a condom.

Size and weight

Presenters and singers do not like carrying heavy mics. Mics slung above people must be well secured by two fixings, each of which is capable of preventing the mic falling onto the audience or artists.

Visibility

This is affected by size and weight, but is also affected by the color of the microphone. Television producers of classical music programmes want minimum visibility. This may be helped by mics being painted 'television gray' and by small physical size.

Operas on stage are often miked by six or so PZM/barrier mics laid on the stage isolated from stage thumps by a layer of foam plastic (stage mice).

Power requirement

Total reliance on phantom power fed down the mic cable can mean losing everything if the power supply goes down. Batteries fail unpredictably. (Even if you thought you just put a new one in; in the stress of a session, it is easy to put the old one back in. Get into the habit of scratching a mark on the old battery, before taking the new one out of the wrapper.) Unpowered electromagnetic mics have definite advantages here.

Reliability, consistency and 'pedigree'

Microphones made by reputable manufacturers, bought through recognized dealers, are likely to be more reliable and long lasting than an unknown make bought through a dodgy outlet. They are also likely to be consistent in quality so that one XYZ model mic sounds just like another model XYZ mic.

13.8 Multitrack

While most computers are equipped with single stereo sound cards, it is equally possible to fit cards that can handle more than two channels at a time. The main use for these is in music recording but they can also be useful for speech, drama and documentary. However, multitrack is very intensive in its use of data. To state the obvious, a CDR with 80 minutes' capacity of stereo will only have room for 20 minutes of eight-track audio.

Many multi-channel systems have quality benefits even if you rarely need multitrack recording. They use 'break-out' boxes so that the critical audio components are outside the hostile environment of the computer. These boxes can be placed conveniently on the desk while computer can be floor standing.

If a long recording run is not required then Minidisc, hard disk, flash card or even tape cassette-based multitrack recorders can be an option. There are also video cassette-based eight-track systems which can provide about 40 minutes' continuous running. These are all galloping towards obsolescence and being replaced by hard disk based machines.

While there may be a temptation to record a discussion multitrack with a view to pulling out extracts for a programme, in practice it is quicker (and cheaper) to have the discussion mixed straight down to stereo by an experienced sound balancer. You also save by not having to allocate time for mixing down the material from multitrack to mono or stereo.

14

Making programmes: production

14.1 Introduction

This is where the disparate strands of material you have been gathering get pulled together and made into your final programme, or programme item. The script is written or a running order prepared. Each item is put into the correct order, checked with the existing material and recorded. Traditionally this was done as a studio session. While there are still plenty of occasions when this is still the case, the digital audio editor can make this less necessary.

With many speech-based programmes, the main purpose of the studio is to provide a quiet, and relatively dead, environment for your presenter to record their voice links between items (aka inserts) and voice-overs on top of actuality. Given that your presenter is in this sterile environment, it makes sense for them to be able to hear the insert material in context so that they can react to it.

However, small, cheap and light, portable recorders make the studio increasingly unnecessary. Its remaining function of providing a sound console seamlessly to mix the material together is subverted by the digital audio editor.

14.2 Types of programme

There are many types of recorded programme, from a simple talk to a full-scale drama or music recording. A modern PC has within it the resources to cope with all of these. The only programmes not within its scope are, arguably, those that need two-way communications. The interview on the telephone or circuit is still best done in a studio with its specialist communications and talkbacks. While a PC can handle a modern Integrated Services Digital Network (ISDN) line, it is the sophisticated communication between the producer and the interviewer, as well as the two-way discussion on the circuit, that is important. While it is not impossible for a computer to do this, the requirement is sufficiently specialized as not to be something that you can buy easily or cheaply.

Incidentally, when time is available, there is a technique for recording telephones in good quality even when international or mobile phone calls are involved. Put simply, each end of the conversation is recorded locally, in good quality, on a portable recorder or, of course, in a studio. The two tapes, or discs, are then combined in the digital editor after they have been posted, or freighted, to bring the two halves together. With digital recorders, little slippage of synch will take place but, even when it does, a little sliding of tracks can easily be done.

A track can always be split while it is silent because 'its' person is listening to the other end. The best results are achieved if the room atmospheres at each end are kept running continuously. The main difficulty with this technique is that people tend to speak differently on the phone. On a poor line they shout. This sounds very odd when both voices sound as if they are in the same room. The key to this is to arrange for the person's own voice to be fed to their headphones at a level that leaves them comfortable without shouting.

14.3 Talks

The most straightforward, if not easiest, production is the straight talk. Once common on radio, this now struggles to keep in view. However, it also manifests itself as commercial promotions – a sales message on cassette or, even, the chairman's morale boosting address to the company's employees.

This merges into the illustrated talk which can soon become a magazine programme.

From the production point of view talks tend to be serial in nature. They start at the beginning and finish at the end. Talks will be all the more effective if they are not overburdened with detail. The BBC Radio Talks Department, in the 1950s, thought that the ideal talk would handle one concept in 15 minutes, provided that the structure was of the form:

In this talk I shall tell you X.
I am telling you X.
I have just told you X.

Or as the late Alistair Cooke described it, 'first you must say what you are going to talk about, secondly you must talk about it and then you must say what you talked about'.

This may be effective in communicating X; it may work for training, but it will be difficult for it to be entertaining. Cooke pioneered the reaction against this with his discursive series 'Letter From America'. Somewhere between will be what you want. Do you want the audience to retain the facts, or to get a good impression of your chairman?

There are two likely locations for this talk; a special recording or one made at a public meeting. In the first case, you are in control. In the second, you are not. Do not be tempted to use your best music mics, as they are almost certain to be a popping disaster. A good moving coil mic that is known to be insensitive to making popping and blasting noises is best. It is also unlikely to have an extended bass response that will pick up every low frequency distraction, from distant slamming doors to air conditioning noises.

Will the speaker stand or sit? When recording a public meeting, ignore what the speaker's office tells you: use two microphones. One should be set up for the speaker sitting down. The other should be rigged for the speaker standing up. If at all possible, use floor stands for the mics so as not to pick up table thumps. Place the mics about 2 feet (60 cm) from the speaker's head. Any closer and the balance will become very sensitive to head and body movements. The further away they are, the less likely the speaker is to move them. (There is a perception problem here; if a mic is very close and it is pointing at their lips, they have an urge to move it so that it is 'pointing at them'. This leaves the mic pointing at their eyes.

In fact, this is not a problem, only their desire to move the mic is. 'Miking' the eyes can, therefore, make sense).

If a mixer is not available, then splitters can be purchased which will allow the two mono mics to be fed separately, one to the left-hand channel and the other to the right-hand channel of a portable recorder.

If you want audience reactions – applause, etc. – then you need an additional pair of mics left and right of the audience. These should be fed to the mixer, if you are using one, or to a second recorder (two Minidisc recorders are more generally useful and cheaper than a decent mixer). If both of the recorders are digital then there will be no problem with synchronizing the recordings once transferred to the computer.

Place the microphone for secondary speakers slightly to one side but pointing at them. They will tend to talk towards 'the boss' no matter how many times they have been asked to talk to the audience! You will obviously need a mixer for this sort of event.

For the specially recorded talk, you are more in control. Someone sitting at their desk, with a table stand holding the microphone, will produce adequate results from an audio point of view but, very likely, produce a very formal delivery. This may be what is required.

Another option is to record the talk as if you were recording an interview, but without any questions being asked. The speaker has eye contact and is more likely to talk to you, representing the idealized single listener. The vital thing to remember with all recorded (and broadcast material) is that you are communicating with individuals. There may be thousands of them but they are not a mass audience; they are a large collection of individuals listening by themselves.

14.4 Illustrated talks

Best results are obtained if the illustrations can be played in as the talk is recorded. This allows the speaker to react to the material more naturally than if they record 'cold'. Whether or not the inserts, as played in, are recorded at the time is down to production convenience. A minimalist set-up might be two Digital Audio Tape (DAT) or Minidisc recorders; one being the play-in machine, perhaps feeding a couple of small powered speakers as used for PC sound.

14.5 Magazine programmes

Structurally, the magazine programme is a series of premixed items linked by, usually, one or two presenters. In many ways it is rather like an illustrated talk but on a larger scale. The individual items are often 'presented' by reporters and contributors other than the main programme presenters.

14.6 Magazine item

Magazine items often take the form of illustrated talks with a presenter/insert/presenter/insert format but they can also be simple features.

14.7 Simple feature

A simple feature differs from an illustrated talk by being continuous with an attempt to make the item seamless. The presenter will talk over actuality, and interviews will be chopped up and presented as extracts.

14.8 Multi-layer feature

The multi-layer feature is the most complex form. Various threads are interwoven to make a seamless whole. Before computer audio editing these were time consuming, labour intensive and complex to make. The multitrack non-linear editor doesn't remove the complexity but makes it much more easily handled. Expect at times to be simultaneously mixing a dozen tracks.

14.9 Drama

Drama brings with it the widest range of techniques. It can be formal and theatrical when reproducing a stage play. It can also be shot like a film using portable equipment on location.

Drama is often studio bound for the simple reason that, with no cameras, the actors are traditionally not expected to learn their lines but to read them off scripts. There is often a technician in the studio to do 'spot' effects such as pouring tea and knocking on doors. This is the radio equivalent of the Foley artist used in cinema films.

Each speech in a drama script should have a left-hand margin with the character name at the start of each speech which should also be numbered, starting from 1 on each page. This allows speedy and simple communication by referring to 'Cue *N* on Page *Y*'.

Drama often, but not inevitably, leads to a lot of post-production. Background effects are best done at the time. It is especially valuable to the actors if they can hear what they are talking against. This is the advantage of location drama, as the sounds are real. If the actors are on a street, they raise their voices quite naturally as a bus passes. This is also a problem with location drama, as you will probably have no control over that bus passing. It can also result in tiring the listener, unless there are contrasting scenes away from the hurly-burly.

Modern technology allows drama, set in modern times, to be made quite effectively with portable equipment and a PC for post-production. The biggest problem can be making it clear to passers-by what is happening. People are now quite *blasé* about camera crews but can be quite confused when there is no camera. I have even heard the suggestion that a balsa wood mock-up camera would be an asset! The idea is to prevent your best take being ruined by the microphone picking up passers-by speculating, amongst themselves, as to what all these strange people are doing!

14.10 Music

This encompasses the whole range from the solo singer, to the multitracked band, to the symphony orchestra. There is also the range from electronically synthesized and sampled music, usually using Musical Instrument Digital Interface (MIDI), to acoustically produced sound from traditional instruments. Music technology is a whole industry and is beyond the scope of this book except to say that, surprisingly, little has to be added to a PC already equipped for what we are describing, to take on the full gamut of music.

Audition has the capability of importing MIDI files as tracks to the multitrack mixer. Some basic simple editing is also possible. This means that your music lovingly prepared on a sequencing package like *Cubase* can be run into your synthesizer as part of the Audition session. An alternative solution is to synchronize Audition to your sequencer using MIDI Time Code.

15

Archiving and backup

15.1 Selection for archiving

Archiving often involves difficult decisions. If you keep everything, then you soon will be buried in old recordings. To be any use, they need to be catalogued; this is a lot of work.

Yet, so often, the interview that is thrown away is the one that turns out to have commercial value because of later fame or notoriety acquired by the person interviewed.

15.2 Compact disc

The ease with which computers can burn CDs has simplified the matter. A CD blank can contain about 80 minutes of audio in standard CD audio format. The discs themselves take up relatively little space, especially if kept in folders rather than conventional 'jewel cases'. Plastic sleeves punched for insertion into standard lever arch folders allow the discs to be kept with any paper documentation.

Keeping your material as CD audio means that it is easily accessed using ordinary CD players. The audio can be reloaded into the computer faster than real time ($\times 12$ or more). However, editing information is lost and there can be a small loss of quality for each generation of copying due to the compromise nature of CD audio recording, which was never designed as a data storage format.

The Track At Once option of much CD burning software has the advantage that it allows you to burn multisession CDs. This could be useful for archiving, as you can add a track and remove the CD. On another day you replace the CD and add another track. While you are doing this the CD will be readable on your computer but NOT on an ordinary CD player. When you have finished adding tracks the CD is 'closed', or 'fixed'; no more tracks can be added but the disc now becomes readable on an ordinary CD player. However, adding a track to a multisession CD can fail and you are likely then to lose the entire contents of the CD.

A better strategy for archiving is to allocate hard disk space for your archive material and then burn the CD in one go. Make two copies for real security. Ideally, archived material should be kept as computer files in CD-ROM form as this is error free.

15.3 CD-ROM

CD-ROM is designed as a data storage format at the expense of a little storage capacity; a little over 700 Mbytes of data for an 80-minute CDR blank. For most medium-length items, this is enough to contain all the material used in the editing session, including the edit decision lists (session files in Audition speak). This means that you can access the complete editor information and can 'unpick' edits or access the original uncut material. The data CD can also contain word processor and text files and even pictures associated with the item. This may include transcripts if they were prepared for editorial, copyright or World Wide Web use.

On the other hand, the convenience of having a CD audio version is very great, so a good compromise is to burn one of each. The audio CD becomes your every day 'library' copy. You can take it home, or to someone's office, and listen to it on any CD player. The CD-ROM is kept on the shelf and used only to transfer material to be reused. If the CD audio version becomes damaged then a new one can be made from the data files on the CD-ROM version. *In extremis*, a new data CD-ROM of the audio can also be made from the audio version but this will, of course, have none of the extra non-audio information that was on the original CD-ROM.

However, there is a spectre that has already loomed over the horizon; the life of CDRs. I have already had experience of well-stored CDRs burned 6 years ago sprouting data errors. This gives the CD audio disc an additional advantage; they are still playable with errors. CD-ROMs give file errors and refuse to give up their data.

15.4 Digital Audio Tape

Digital Audio Tape (DAT) cassettes can be used for archiving. Audio cassettes have 2 hours of capacity at 44.1 or 48 kHz. They have twice that at their half speed setting. This records at 32 kHz sampling rate. Using DATs manufactured for computer backup, you can get 4 hours or more onto a single cassette at full quality (8 hours or more at long-play, or LP, speed). However, the tape is thinner than audio DAT machines like and actually using them for computer backup is a much better option.

DAT can be useful as a way of archiving 48 kHz masters in a way that can be played as audio on conventional machines. DAT has a number of disadvantages:

- The tape is very fragile and is prone to humidity problems. When a DAT gets mangled in a machine you have lost everything. While rescuing it is technically possible, this is a very expensive and time-consuming activity with no guarantee of success.
- While it may sound trivial, a substantial disadvantage is the small size of the cassette. There is little room to write an informative label on them and they have also been known to disappear down the backs of chairs! The CD with jewel case or folder has a definite advantage here.

15.5 Analog

It is possible to archive material onto analog media such as compact cassette or reel-to-reel. This has a number of disadvantages and loses all the advantages of digital. It may give a short-term compatibility with an existing system. Sooner or later, the existing system will have to be transferred to digital; it makes sense to do it sooner rather than later. There is a problem. While this is good for 5 years, there is no digital medium that is known (rather than claimed) to have a life measured in decades. Analog is a proven medium. At present, the pressed CD looks as though it has a good life but this far too expensive for routine archiving.

15.6 Computer backup

Just as all your word processor files and web transcripts can be transferred to the same CD-ROM as your audio, so this can be done using whatever computer backup system you have – you do have a computer backup system don't you?

All computers crash. All computer systems lose data. No one believes in backups until the day they lose a day's, a week's, a month's, a year's work. Companies have gone bankrupt because of poor, or no, backup strategy.

A comprehensive backup strategy will also take into account the possibility of theft and fire. Programs are (usually) replaceable. Data files represent investment of time and money. They cannot be replaced unless you have a backup.

Even if you do have a backup, do you know how to restore your files? A cynic would say that the world is full of good backup systems which are rubbish at restoring. Beware, while it is easy to protect against the theft of information by password protection and encryption, this protection is lost if the password is written on a Post-it® note stuck to the side of the monitor. Equally, if the computer is lost in a fire then so will the Post-it® note! You must have a safe place to keep the password.

Choosing a password

Passwords should be easy for you to remember but difficult for others to work out. If you must write a password down then leave it in a secure place. Avoid the obvious; 'secret' is one of the first things that a malicious hacker will try. They will also try the names of your spouse and any children.

For the best protection, you should use at least six characters and avoid dictionary words. For extra safety include non-alphanumeric characters such as '&$*@%'.

With some programs the password may be case sensitive so that 'rad$38' is treated as a different password to 'Rad$38' (a capital 'R' is used in the second but not the first). Here, the Caps Lock key being on can cause initial panic when the password you are so confident about is not accepted!

If you are an employee, then your company should have an admin. system where your password(s) can be kept securely against the day when you meet your demise in a car accident (or, with long-term data, they may need to recover the data even if you left the company 2 years prior to the disaster).

Storage systems

There are a plethora of hardware data storage systems, some based on tape cartridges and some based on disc cartridges. Because of the possibility of theft of the computer or fire, the backup must be on some form of removable medium. Tape cartridge systems have the disadvantage of a relatively short cartridge life. The tape wears. Some recommend cartridges being replaced after anything from 20 to 100 uses (this includes DAT).

With the dramatic fall in price of hard disks it is worth considering using standard high capacity hard disk drives mounted in removable caddies for your backup. An alternative, for home use, can be relatively cheap devices that allow you to connect Universal Serial Bus (USB) disc drives to your Wi-Fi network. These can then be in a different room and out-of-sight and accessible to your laptop and computers on the network.

Your backup regime should rotate round at least three removable media – the so-called grandfather, father, son strategy. Again, because of fire, you should store them away from your computer, in another room or even another building, especially with mature commercially valuable data. One way of achieving this is to backup, via a network connection, to a remote server. This can use a local area network or, if you have fast enough access, the Internet. For the amount of data involved in audio files you will need, at least, Integrated Services Digital Network (ISDN) or Asymmetric Digital Subscriber Line (ADSL) speeds and, even so, this will involve overnight backup runs. There are firms that provide off-site storage on their own computers. Talk to your system administrator about preferred ways of off-site backup.

Despite all this, for many purposes, taking a backup home with you overnight can be an effective way of meeting reasonable backup safety criteria. However, beware falling foul of the Data Protection Act.

A related problem is how to move material from one place to another. Sending a CD or a flash card in the post may be adequate. An audio file can be 'emailed' from computer-to-computer. Some radio stations use their administrative computer network to send audio data, often at less than real time, as and when there is spare capacity.

16

Tweaks

Computer programs always have options that allow you to tweak how they work so that you can make them easier to use. Adobe Audition 2.0 is no exception.

16.1 Mouse options

Your choice of mouse will make a difference to Adobe Audition 2.0 as it exploits a wheel mouse so that the wheel can be used to zoom in and out in the single wave view and to move up and down the tracks of the multitrack view. Within the Effects transform windows you can use the mouse wheel to alter settings. This is done simply by holding the mouse cursor over a slider and rotating the wheel.

The way the right mouse button works in Edit View can be customized – Edit/Preferences/General tab has the options for 'Edit View Right clicks' of Popup Menu or 'Extend selection' (Hold Control for popup menu).

The default action is that when you right click in the wave editor you will get the popup shown in Figure 16.1 giving useful options. The downside of this is that the massively useful right click to extend a selection is accessed by having to press control as well. The options in the popup are available directly from the top menu so many people prefer to use the original *Cool Edit* option of reversing the functions so that a right click extends the selection and you press control to get the popup.

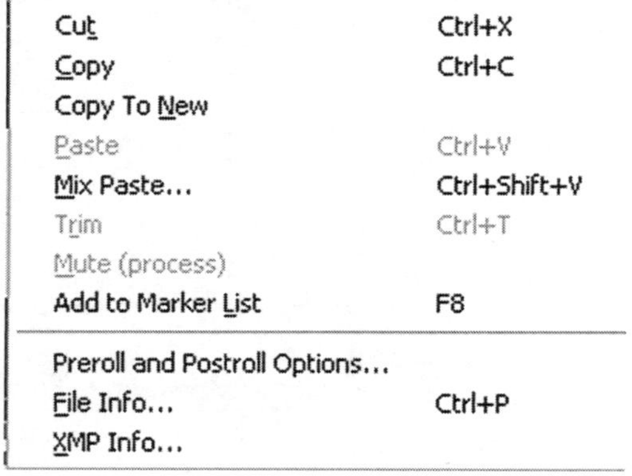

16.2 Hardware controller

The disadvantage of all computer-based editors is that you only have 'one finger'; the mouse pointer. With the availability of Universal Serial Bus (USB) and Firewire ports it

Figure 16.1 Right Click Popup

is becoming increasingly practical to have an external control surface to provide you with extra controls that are dedicated to your program. Windows 98 onwards provides 'Human Interface' drivers that are allowing a degree of standardization.

Controllers can actually be full-scale mixers with faders and equalization (EQ) controls but these are understandable expensive although they are beginning to come down. Some use Musical Instrument Digital Interface (MIDI) rather than USB.

Syntrillium, who produced *Cool Edit Pro*, Adobe Audition's predecessor, designed an optional extra of a dedicated USB controller called 'Red Rover' (Figure 16.2). This is still available and works with Adobe Audition. This is largely aimed at users that want to record themselves. For example, a guitarist would have problems with the electric pick up also picking up the electrical interference from the computer monitor, or even the computer itself. Moving a few metres away can eliminate this. Red Rover provides transport controls and level indication. These can also be useful when editing. Additionally cue markers can be added on the fly and the metronome switched on or off.

In the Multitrack View, it provides an easy direct control of the master volume. Another knob is used to select any single track. That track's volume can also be changed. Its Mute, Solo and Record Enable buttons are available along with indication of their current settings.

The 'External Controllers' tab in the 'Edit Preferences' (F4) allows you change the settings of hardware controllers. At the time of writing, Red Rover has just the one option that allows you to change the step value of the gain control knobs between 0.1 and 0.5 dB in 0.1 dB steps. However, the 'Add Cue' and 'Metronome' buttons are reassignable.

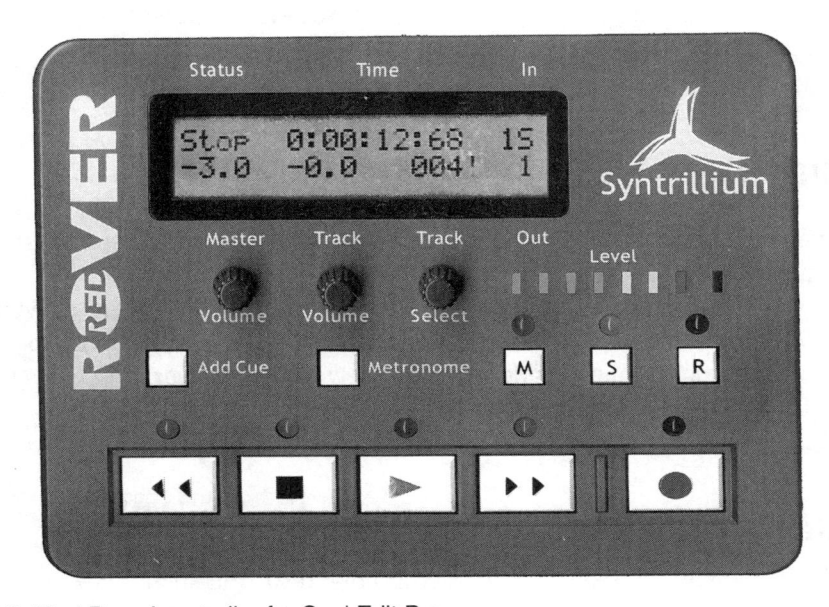

Figure 16.2 'Red Rover' controller for Cool Edit Pro

16.3 Keyboard shortcuts

Adobe Audition 2.0 comes with some keyboard shortcuts already defined. You can change these or add more with the 'Edit/Keyboard Shortcuts and Midi Triggers' dialog. Set up the actions you use most as simple key press combinations.

If you do a lot of editing it would be useful to give yourself some spool controls on your keyboard. The settings I use are illustrated in Figure 16.3. This allows me to play a file and then go into continuous fast forward with a single press of the '3' key on the numeric keypad. Hitting '1'

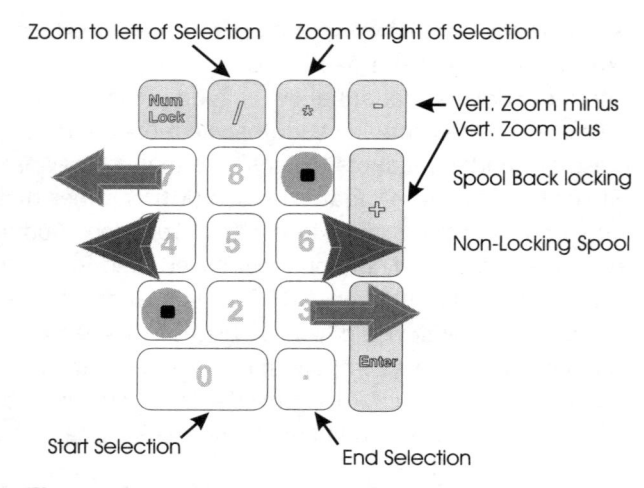

Figure 16.3 In a multitrack session, you can override the defaults for a particular track. (See assigning inputs and outputs to audio tracks. Figure 7.22)

reverts me to play. '7' and '9' do the same for reverse wind. '4' and '6' become momentary reverse and forward spool buttons.

I also allocate the '+' and '−' keys on the numeric keypad to operate the vertical zoom controls.

You may also wish to consider reallocating the 'Selection anchor left when playing' and 'Selection anchor right when playing' controls from '[' and ']' to '0' and '.' on the numeric keypad ('.' is shown as 'decimal' on the shortcut list). These keys are easier to find than '[ ]' which sit on little used keys.

16.4 MIDI triggering

Shortcuts can also be set to MIDI commands. These could be from a specialist MIDI controller box, a sequencer like Cubase. or even a MIDI keyboard.

In effect, you can turn Adobe Audition 2.0 into a simple sampler (trigger playback from a MIDI keyboard), or even save your files with a foot pedal if you really wanted to.

To assign a MIDI trigger, click inside the Press New MIDI Note box, and press the desired key on a MIDI keyboard, or adjust a MIDI controller (such as a foot pedal).

To enable MIDI triggering

From the Options menu, choose 'MIDI Trigger Enable' (F6). You should now be able to trigger Adobe Audition 2.0 from your MIDI device.

16.5 SMPTE synchronization

Through Society of Motion Picture and Television Engineers (SMPTE) time code, you can effectively control Adobe Audition 2.0's transport from a device such as a MIDI sequencer, or with the appropriate hardware such as a VCR, or tape deck. When slaved to SMPTE, Adobe Audition 2.0 will synchronize with frame accuracy to the master device which is generating the time code. It needs about 5 seconds pre-roll to lock.

Adobe Audition 2.0 can also generate time code to synchronize external device (see also Appendix 3 'Time Code').

16.6 Favorites

The Favorites menu can be built to make a collection of your most commonly used actions. Use Edit Favorites to create, delete, edit and organize items appearing in the Favorites menu.

Edit Favorites can instantly call up any customized Adobe Audition 2.0 Effect Transform or Generate effect, Script or even many third-party tools. The menu can also contain sub-menus for easy organization.

Figure 16.4 shows an example how Favorites can be laid out using submenus. Figure 16.5 shows the edit dialog that created that menu. The '\' character creates a submenu. Further entries in that submenu are created by repeating the text up to the '\' and then adding a new name. A menu division is created by using multiple dashes as a name. The figure at the end is to make the dashes different texts so as not to confuse the engine that makes the menu.

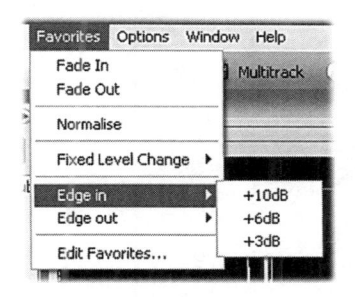

Figure 16.4 Showing resulting menu

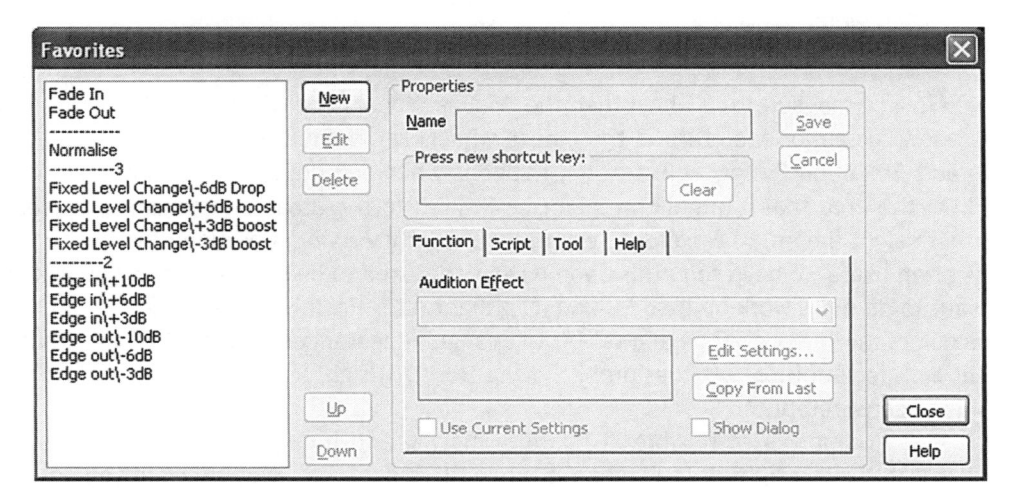

Figure 16.5 Favorites dialog

16.7 Preferences (F4) (Figure 16.6)

This tabbed dialog is the key to customization of Adobe Audition 2.0. This varies from what colors are used by the display to the arcane settings of MIDI and SMPTE parameters. The General tab shown contains most of the settings that can annoy. It is here that the action of the right click in the waveform editor can be changed to the way it worked originally. Using the right mouse button to extend the selection is such a frequent event that having to press control is a major slow down.

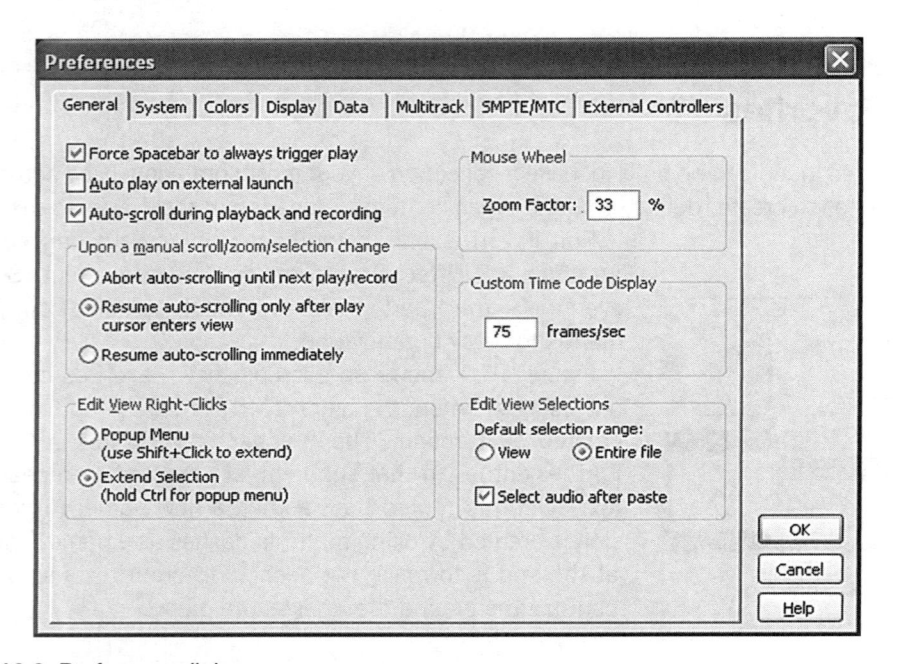

Figure 16.6 Preference dialog

If you have a mouse with a wheel then the 'Mouse Wheel Zoom Factor' controls how much the window zooms for each turn of the mouse wheel. The default is '33 per cent'.

'Default Selection Range' selects what happens when you double-click on a waveform to select it. The area that is highlighted can be limited to the area you can currently see on screen or select the entire waveform, even if you're only viewing a portion of it.

It is often useful to have something you have just pasted to be highlighted as you will usually want to do more work on it so ticking 'Highlight after Paste' makes sense.

The colors tab (Figure 16.7) allows you to change appearances. Personally I find that the default look for Audition is rather drab. Moving the 'UI Brightness' slider fully right brightens things considerably.

Now that Audition is so flexible in floating, moving and hiding panels, the 'Workspace' drop down in the top right of the screen is a boon (Figure 16.8). Not only can you get back to the original settings but also you can create and save your own.

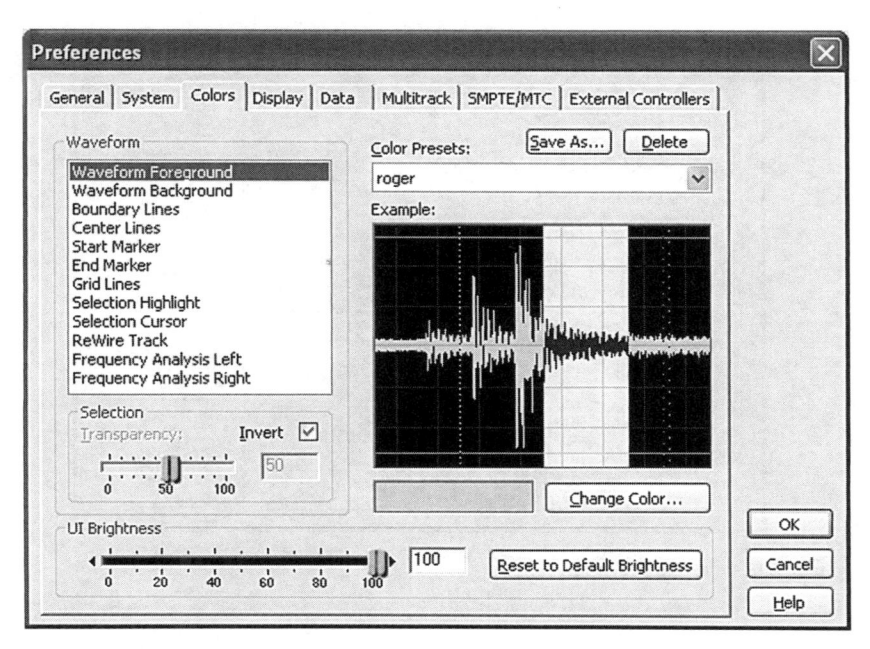

Figure 16.7 Color tab

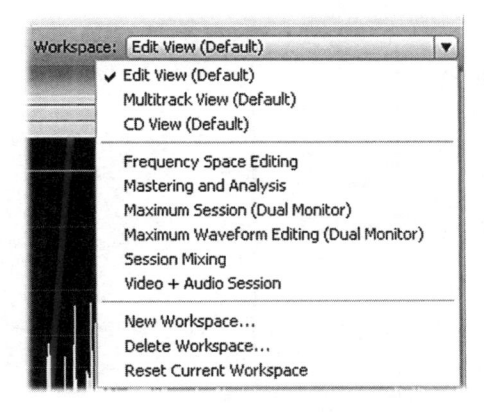

Figure 16.8 Edit/Audio Hardware Set-up

When you set inputs and outputs for recording and playback, Adobe Audition can use two kinds of sound card drivers: DirectSound and Audio Stream In/Out (ASIO). Some cards support both types of drivers.

ASIO drivers are preferable because they provide better performance and lower latency. You can also monitor audio as you record it and instantly hear volume, pan and effects changes during playback. The main advantage of DirectSound is that you can access one card from multiple applications simultaneously.

Click the Edit View, Multitrack View or Surround Encoder tab (Figure 16.9).

For Audio Driver, choose a driver for the sound card you want to use. (Choose an ASIO driver if one is available; otherwise, choose the DirectSound driver, Audition Windows Sound.)

Click Control Panel, set driver properties and then click OK (optional).

Do any of the following (see Figures 16.10–16.12):

- Under Edit View, choose stereo ports from the Default Input and Default Output menus.
- Under Multitrack View, choose stereo or mono ports from the Default Input and Default Output menus.

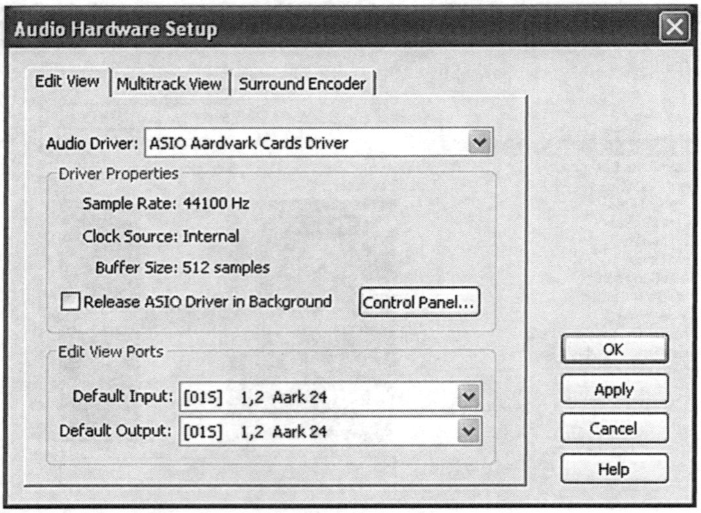

Figure 16.9 Edit View Hardware Set-up

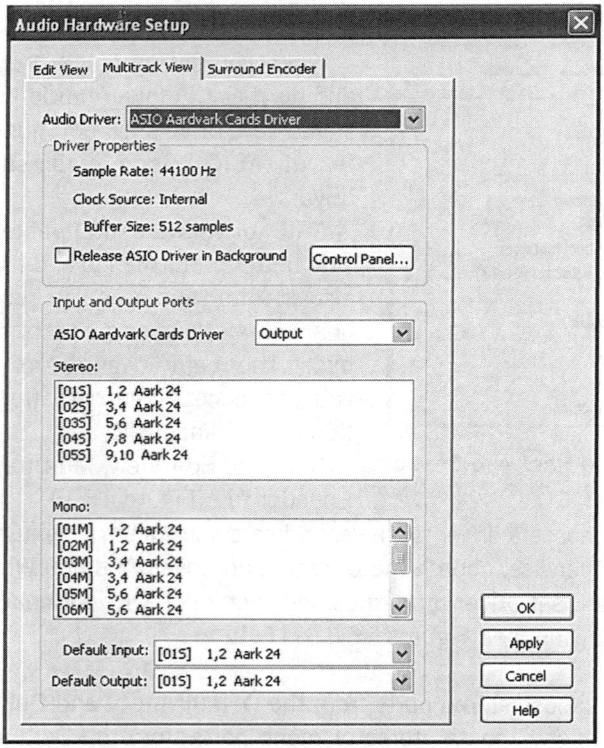

Figure 16.10 Multitrack card shown in Multitrack View

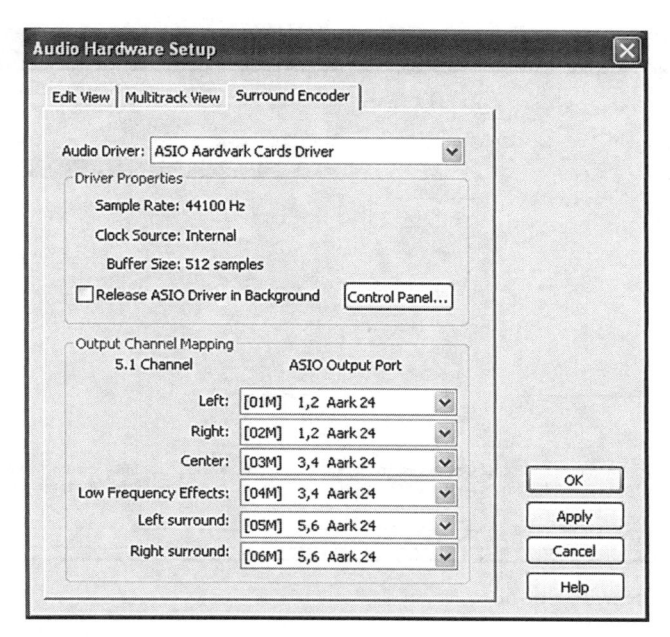

Figure 16.11 Multitrack card in Surround View

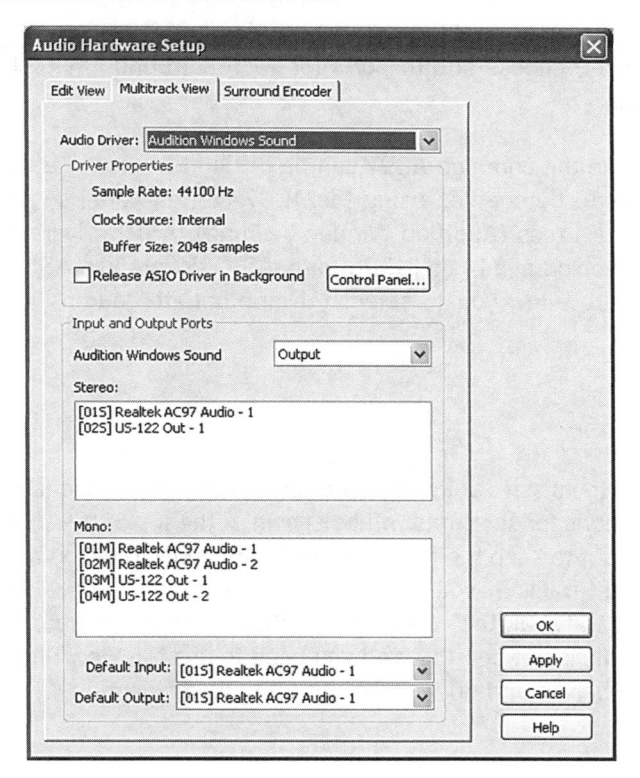

Figure 16.12 On Board AC97 with USB card

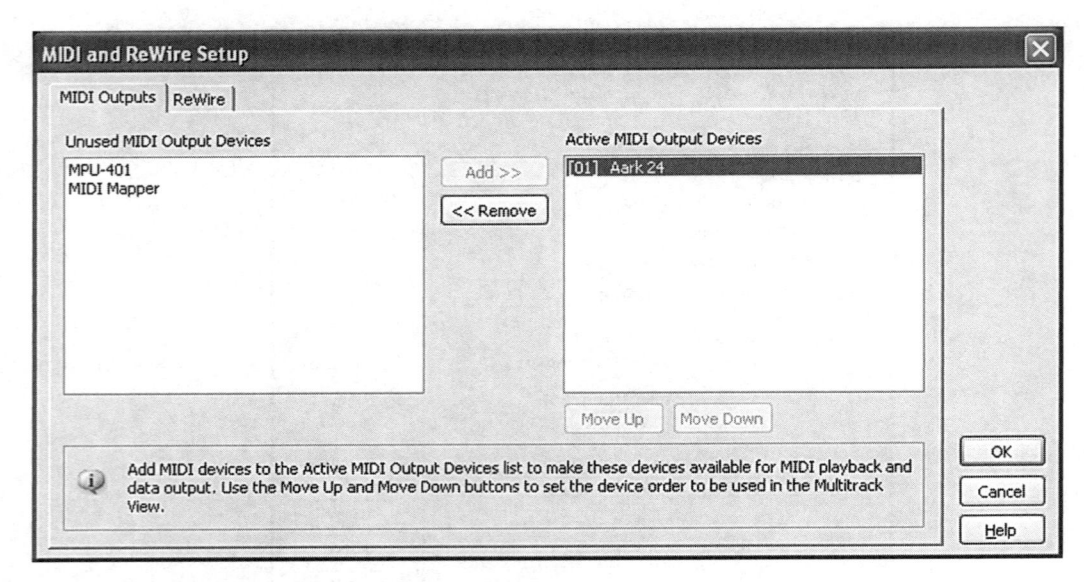

Figure 16.13 MIDI and Rewire Set-up

In a multitrack session, you can override the defaults for a particular track. (See assigning inputs and outputs to audio tracks. Figure 7.22)
- Under Surround Encoder, choose output ports for each surround channel in the Output Channel Mapping area.

On board sound cards, like the common AC97 can be difficult to combine with other cards. The Tascam US 122 has to be enabled using the AC97 control panel so that they appear together as DirectSound sources (Audition Windows Sound). ASIO drivers seem not to be supplied with the AC97 so you need to try third-party ASIO drivers like ASIO4ALL.

MIDI and Rewire Set-up allow you to select the appropriate sources and destinations (Figure 16.13).

16.8 Icons

Some people love icons, others stare at them uncomprehending. If you like icons then you can choose which groups of icons for shortcuts will be shown at the top of the screen. At the very least, you will probably want to keep the file group as this includes the icon to toggle between single wave view and multitrack. If you hate icons then you can remove them altogether. Normally the easiest way to tweak the icons is right clicking on the toolbar but, if you have no icons selected, then this is not possible and you need to use the View/Shortcut Bar menu item. You can toggle the bar on and off without changing your selection.

17
Using the CD-ROM

17.1 Adobe Audition

Most of the CD-ROM is taken up with the install files for Adobe Audition 2.0. This is a 30-day time-limited version which can be upgraded to the full-working version and is identical to the 'tryout' version found on the Adobe web site www.adobe.com.

Installation notes

1 Double-click the Adobe_Audition_2.0_Tryout.exe file, and click Unzip when prompted.
2 Navigate to the folder where you unzipped the files and double-click the file called Audition 2.0 Setup.exe.
3 The set-up will guide you through the rest of the installation process.

17.2 Bonus material

Additionally, exclusive to this book there is additional material for you to work with.

Low Bit-rate demo

This contains three folders: Compressed Music, Compressed Story and Story. Within each of these folders are two folders called Dither and No-Dither. These contain files of the form 10000000.wav, 11000000.wav, 11100000.wav, 11110000.wav, 11111000.wav, 11111100.wav, 11111110.wav, 11111111.wav (the story folder does not have 10000000.wav files). These correspond to simulations of 1–8-bit recordings with and without dither applied (random noise). Further explanation can be found in the 'Notes' text file in the 'Low Bit-rate demo' folder.

Music mix exercise

The 'Stems' folder contains 15 tracks which are the constituent parts of a song especially recorded for this book in a professional studio. You are left to produce your own mix. If you wish, you can re-record tracks or add others. At very least you will probably want to produce a stereo mix (the demo is mono).

RNID: Tinnitus demo

This is the demonstration of what Tinnitus sounds like on the Royal National Institute for the Deaf (RNID) web site mentioned in Chapter 2. There is a brief account of how the material was obtained at the start of the file. My thanks go to the institute for their permission to reproduce the file.

Contact information: RNID Information line, 18–23 Featherstone Street, London, EC1Y 8SL, UK.

Telephone: 0808-808-0125, Textphone: 0808-808-9000, Fax: 020-7296-8199.

Email: informationline@rnid.org.uk, Web site: www.rnid.org.uk

Story Edit

This a short ghost story recorded especially for this book. Alec Reid, the author and reader of the story deliberately did not prepare before recording the story so that it is particularly rich in fluffs and 'overlaps'. It is included to give you practice in this basic form of speech editing. Additionally, the script is included for you to print out for reference if you wish. I have also included JPG files of the script with the edits marked up. The convention used is one of many that are used and is chosen only because it is the one I personally use.

Vox pop

This is some raw 'Vox pop' interviews recorded by Torben Horler for this book, to provide you with some speech material practice editing.

18

Hardware and software requirements

18.1 PC

Audio editing on a PC became a practical proposition for professional use once the equivalent, or better than, a Pentium II running at more than 200 MHz was reached. These days, this is regarded as being very slow but even so will give more than adequate speed of processing and is able to perform multitrack mixing with simple software. With the increase of processor speed software writers have taken advantage of this to 'bloat' their programs so that they are much cleverer but commensurately slower. Audition 2.0 is slower than Audition 1.5 but machines are that much faster and Audition 2.0 is that much cleverer; 2–3 GHz should be your target for modern Windows XP software.

Consider getting a 'full tower' case as this will then conveniently sit on the floor releasing space on your desk. Floor standing towers can also more easily be strapped down if theft is likely to be an issue. Tower case will have plenty of room for adding extra drives. These can mount up very quickly. A reasonably full quota might be: CD-ROM drive, CD-RW drive, DVD drive, removable hard disk system. Two (Integrated Drive Electronics) IDE drives to separate programs and system from data and, maybe, two more drives dedicated to audio use.

18.2 Sound card

This should be capable of at least 16-bit audio at sampling rates of 44.1 kHz (CD standard) and 48 kHz (DAT standard). Lower rates will be useful if you need to audition files intended for the Internet as well as for reproducing noises made by other programs and by Windows operations. Higher sampling rates (88.2 and 96 kHz) have become fashionable. They double the length of audio files. People still argue a great deal as to whether this actually provides a better sound.

Fitting two cards can be convenient. A basic games type card will handle the Windows and games sounds and a second high quality one for your audio editing, maybe with several inputs to allow multitrack inputs.

If you have access to high quality analog sources such as 15/30 ips Dolby Spectral recordings (SR), or high quality microphones in quiet studios, then a higher bit rate is highly desirable; even if your finished recording is going to be 16 bit. This is because headroom will have to be allowed at the time of recording to prevent digital overload. Quite likely a few levels will have been increased during post-production. This means that you can end up delivering the equivalent of a 12-bit recording! (see Section 3.4 on 'Analog transfer' on Page 22).

Some sound cards have microphone inputs. These are usually of relatively poor quality. They are best faded out on the Window mixer to prevent them adding noise. The useful inputs are the 'line' level ones. These will match the line outputs of audio gear. Multitrack cards often have the option of handling professional levels which are 14 dB higher than domestic levels. They may also be able to handle 'balanced' feeds as well as the domestic-style unbalanced type (see Page 22).

As well as conventional analog inputs and outputs, it can be useful if your sound card will handle digital inputs and outputs so that your original digital recordings can be 'cloned' on to your computer and your final mix returned without quality loss.

In practice, many of the domestic machines – often used by many professionals – have only optical digital outputs rather than the electrical outputs required by many sound cards. However, you can buy converter boxes which will do the job very adequately but at the expense of yet another box cluttering up your desk.

Optical connectors use modulated red light rather than electricity to transfer the data. This light is visible to the naked eye when a connector is carrying an output.

Be aware that there are two physically different optical standards. One is called TOSLINK and the other is the same shape and size as minijack audio connectors. The socket is usually dual function and can be used for electrical analog audio or for optical digital (see Page 17).

Your card should be switchable between the domestic and professional data formats. While they are nominally compatible there can be circumstances where a domestic format input can interpret and professional style output as being copy protected.

Sound cards are traditionally slotted into the back of the computer, after removing the cover. This is probably as good a place as any; the card is kept safe and not likely to be dropped. However, various boxes that plug into the Universal Serial Bus (USB) or Firewire input of the computer can be used instead. These have the advantage that they can be plugged and unplugged without switching off the computer. While some versions of Windows 95 claim to be able to use USB, you really need Windows 98SE or later for success (see Page 19).

Most editing is done away from the recording site and the audio brought to the computer on removable media. However you may need one or more microphones. For simple one-microphone use for voice-overs or story readings then there are plenty of devices that provide a clean amplifier for a microphone including phantom power if required. They often include equalization and compression as well. Some sound cards with breakout boxes have high quality microphone inputs – usually about 4 will be provided with a software mixer provided if you don't want to use Adobe Audition 2.0 directly. Once you want more than four microphones then you are really talking about a studio set up and a proper mixer will not only be more convenient, but also provide useful things like talkback and auxiliary feed to the artist's headphones.

18.3 Loudspeakers/headphones

Ordinary PC 'games' loudspeakers are not adequate to assess sound quality. You should pay the extra for music quality active speakers ('active' means that the speakers have their amplifiers built in for driving directly from the sound card output). If you have the space this can be a proper Hi-Fi system with the computer's sound card feeding a line level 'aux.' input.

If you are editing audio at work then much of the time, you may be operating in a shared office using headphones. These should be of the best quality with a decent bass response. Try always to check mixes on loudspeakers, if at all possible, as headphones give a very different impression to loudspeakers. As a generalization, a mix that sounds good on loudspeakers will sound good on headphones. The reverse is not always true. It is all too easy to have too little separation between a voice and music or effects behind it.

Headphones are much more critical of poor editing than loudspeakers. The standard advice to broadcasters used to be, not to worry if an edit was audible on headphones but not on loudspeakers. However, with the popularity of MP3 players with radios, it is no longer possible to assume that all your audience is listening on loudspeakers.

Sometimes the audience is known to be listening on headphones – walk-round cassette guides to exhibitions, for example. Here headphones will give you accurate results. Use your best headphones for production, but do check what the mix sounds like on the, probably cheaper, headphones used at the show.

18.4 Hard disks

Ideally you should have a second separate hard drive for your audio data. Even better, for the fastest operation, yet another separate disk for handling the audio editor's temporary files. This reduces head clacking where the hard disk head is flipping between the source file and the destination file during copying. This is eliminated if the source file is on a different drive from the destination. How fast this can happen is controlled by the spin speed of the hard drive. The 7200 rpm is standard with 5400 rpm being regarded as redundant like the 3600 rpm, 4500 rpm of the past. Presently, top of the line SCSI drives can be found running at 15 000 rpm with a corresponding high price.

Traditionally the advice was that these drives should be of audio-visual (AV) quality especially if you intended to 'burn' CDs. Early, non-AV quality, disks recalibrated themselves at inconvenient times, to compensate for temperature changes, interrupting a continuous flow of audio (or video). This is likely only to be a problem if you are intending to use very old equipment; modern drives should have no problem in delivering the data when required, but, as always, you should confirm with your supplier that the drives supplied are suitable for the task.

Modern IDE/EIDE (ATA (Advanced Technology Attachment)) drives are entirely adequate for audio purposes although many people still think that the SCSI system is well worth the extra cost for AV work. SCSI also has the advantage of being able to handle up 15 devices, which can include scanners as well as hard drives, rather than the four IDE drives that most computers support.

Nominally called the IDE bus; it's also known as the ATA specification (ATA Bus). The IDE bus is a Parallel bus. A new form is being introduced called Serial ATA (SATA), as a result the parallel form is now commonly called parallel ATA (PATA).

SATA provides greater scalability, simpler installation, thinner cabling and faster performance (up to 3 Gbytes/second). SATA maintains backward compatibility with PATA drivers. Users of the SATA interface benefit from greater speed, simpler upgradeable storage devices and easier configuration.

If you are likely to be working on several projects at once then some form of removable storage can be helpful. USB key fob 'dongles' are becoming cost-effective.

While less convenient, another alternative is to use standard hard drives connected by USB and 300 Gbytes is quite common these days.

18.5 USB

The USB has become very common and there is a great deal of equipment that can use it. From the user's point of view it has the major advantage that you do not have to open up your computer to connect devices. Even better, they can be connected and disconnected without switching off or rebooting your computer. The disadvantage can be that you end up with a clutter of devices on flying leads scattered around your computer.

As well as things like CD burners, sound cards, digital cameras, scanners and external drives, basic things like a mouse, a keyboard and a printer can be added. Theoretically, over a hundred separate devices could be hung on to a single USB port daisy-chained one after the other. In practice USB manufacturer don't implement this facility. The amount of data that can be sent at any time is limited. Although this is still vastly more than the ordinary serial port, it will limit data hungry devices.

USB2 connections are faster but are still limited. New computers usually have three or more USB sockets but if you are going to use USB devices then you may consider purchasing a USB hub so that you can add more. Hubs can come as separate boxes or, these days, many monitors have hubs built in. USB keyboards usually have chaining sockets to take a USB mouse plus, say, a Zip drive.

Many modern computer motherboards have USB ports built-in but for those that don't. PCI cards containing two or more ports can be added to an existing computer.

18.6 Firewire

Firewire is an alternative to USB. It is much faster but, at the time of writing, there are fewer products on the market for it. Again PCI cards can be bought to add ports to existing computers. There is no reason why both USB and Firewire cannot be used on the same computer. Daisy chaining is quite often implemented. Firewire sockets (also known as IEEE 1394) come in two versions. The 6-pin connectors are similar in size to USB and have power to drive downstream devices. There is a smaller 4-pin version designed to connect cameras to computers. These do not have power as the cameras will have their own batteries. The 4–6-pin converters are available. My laptop has three USB sockets and one 4-pin Firewire and I usually connect my scanner via the Firewire to release a USB socket.

18.7 Audio editor

The audio editor is the key to the whole operation. There are any number of different ones made by a host of manufacturers. These range from full-scale Digital Audio Workstations (DAWs), often using proprietary hardware, to very simple 'freebies' with the minimum of facilities.

Most of the examples in this book are taken from a good example of presently available software. *Adobe Audition 2.0* is a combined linear/non-linear editor potentially able to handle any number of audio tracks, limited only by the hardware capabilities. It implements most of the facilities you are likely to meet. Not only is it a good editor in its own right but is excellent for training in the different aspects of audio editing.

18.8 Linear editors

Editors fall into two methods of working: linear and non-linear. Linear editors can handle one stereo, or mono, audio track at a time. They operate like word processors; any editing is destructive, a deleted section is actually deleted from the audio file (although, as with word processors, there is likely to be an 'undo' feature available). When saved, any cut material is lost, unless a backup copy of the original has been kept.

As because of the large amounts of data involved with audio, cut and paste are instantaneous only for the smallest of sections. Copying a 40 minutes stereo file can take several minutes. While the process may be much quicker than with pre-digital technology, minutes spent staring at a progress bar crossing your screen are frustrating and unproductive (see 'Blue bar blues' on Page 42).

18.9 Non-linear editors

Non-linear editors do not change the audio files being edited in any way. Instead they create 'Edit Decision Lists' (EDLs). They don't play an audio file linearly from beginning to end. Instead the files are played out of sequence – non-linearly – with edits performed by skipping instantly to the next section. It is a sophisticated version of programming the playback of a CD. You set the CD only to play certain tracks but the omitted tracks are still on the disk. EDLs allow the playback to skip from instant to instant within and between tracks.

The most obvious advantage is that material is never lost. But this can be a disadvantage as well, because your hard disk can soon become cluttered with unwanted material. In practice, some culling with a linear editor is useful before using a non-linear one.

You can have any number of different EDLs for the same audio, thus you can have any number of different versions. You can decide to keep a version that you are reasonably happy with, but continue to see if you can further improve the item. This is ideal if the same item has to be repackaged for different programme outlets. A programme trail can be one EDL and the broadcast programme another. Even in the domestic environment different versions can sometimes be useful, even if it is just a shorter version for your mother to listen to!

While this is possible with linear editors, it is at the expense of much duplication of audio data, which can soon fill a hard disk. Only the non-linear editor can give you the opportunity to 'unpick' edits for your new version.

Non-linear editors can also handle more than one audio file at a time and play from different sections of a single audio file simultaneously. In this way cross fades and mixes can be performed, without permanently changing the original audio material.

Non-linear editing operations, in general, can be faster as the computer is not having to move multi-megabyte chunks of digital audio data around the hard disk. Instead, the changes in the EDL are measured in just tens of bytes.

18.10 Multitrack

You don't have to handle complex EDLs yourself as the non-linear editor presents itself, on screen, as if it were a multitrack tape recorder. The lists are managed by the editing software not by you. However, it is far more flexible than a physical multitrack tape recorder.

The non-linear editor shares, with multitrack tape, the ability separately to record individual tracks in synchronization with existing ones. You can also 'drop-in' to replace a section. But, unlike hardware multitrack with its physical restraints, you can also 'slide' tracks back and forth with respect to each other. Track 'bouncing' (copying from one track to another) is virtually instantaneous. You can add any level changes and cross fades in a way that would need full-scale automation in a studio using physical multitrack. When you have finished, you can usually get the editor to save space by performing the virtual edits on the audio files so that they become physical edits.

Just as over the years, automobiles have become more and more like each other, so have multitrack audio editors. As an operator, moving from one to another is rather like driving different makes of cars. You may occasionally wash your windows at someone rather than flashing your headlights and you may need to read the manual to be able to tune the radio! But basic driving – the main function is relatively easily achieved.

18.11 Audio processing

The standard tricks of an audio editor are cut and paste, as in a word processor, plus the ability to cross fade between two pieces of audio.

In addition, audio editors for professional and semi-professional use come with a barrage of special effects. Some you will have daily use for, others may 'get you out of a jam' some time. Many editor programs have the facility for their user to buy 'plug-ins' that will add specialist, or enhanced quality effects. Some of the most sophisticated of these can cost more than the editor they plug into.

The most important processing options are:

- *Normalization*. This standardizes the sound level of each item, although not necessarily the loudness.

- *Reverberation.* Sometimes known as 'artificial echo', this will add a room, or hall, acoustic to your recording. This is most often used with music recording. With speech programmes you are less likely to need it, except to get you out of a jam by recreating an acoustic. There is also the annual ritual on radio of adding reverberation to witches' voices for Halloween.
- *Compression.* This reduces the range of volume between the quietest and loudest sound. Used with care, it can enhance your recording. Used carelessly it can make your item offensively difficult to hear. The overall effect is to make your material louder while remaining within the constraints of the maximum digital level available. It has the disadvantage of bringing up background noise and making voices breathy.
- *Noise reduction.* This should not be confused with the noise reduction used on analog tape systems like dbx, Dolby A, B, C, S or SR. These are 'companding' systems where programme material is compressed for recording but expanded back to its original dynamic range on playback. (Dolby A and SR are used in professional studios; Dolby B, C and S are domestic systems.) A digital system does not need such techniques. However, you may have a digital recording copied from a poor original, either from the archive, or one afflicted with technical problems at the time of the original recording. While intended to remove such recording problems as tape hiss, your editor's noise reduction system may well turn out to be useful in removing acoustic noise such as traffic rumble or air conditioning whine. I have even managed to remove aircraft noises and telephones ringing. Audition even has a spectral editing facility where you can 'lasso' a sound within the spectrum and process or remove it.
- *Declicking.* Old 78 rpm recordings and vinyl Long Play (LPs) are plagued by clicks as well as noise and will benefit from electronic declicking. However, don't be too enthusiastic about cleaning up old recordings as they often need to sound old in the context of a programme item.
- *Filters.* These are useful for filtering out constant sounds as well as sound at the extreme ends of the frequency range. Cutting the bass will remove rumbles coming through the wall, or from underground trains. Mains hum can be a problem and special 'notch' filters are available to remove the worst of this. Different settings are needed for American (60 Hz) and European (50 Hz) mains system.
- *Equalization.* This is a sophisticated form of 'tone' control usually referred to as 'EQ'. This can be used to brighten up muffled recordings, or to reduce sibilance from some interviewees. Domestic tone controls normally only affect the top and bottom frequencies; equalizers can also do useful things to the middle frequencies.

18.12 Mastering

Mastering refers to the process of creating a recording that is usable away from the PC. It is possible to play, or broadcast, your mix from the computer directly to an audience. However, this does tie up your computer, with all its expensive resources.

You need to put your final mix onto a 'removable medium' in a conventional stereo form (or on DVD 5.1 channel surround sound). This also has the, not inconsiderable, advantage that no one else can 'muck about with' your item.

Mastering can be as simple as playing the mix out of the computer's sound card onto a normal domestic medium like compact cassette. This format has the major advantage of being playable almost universally, as the compact cassette is the most widely available sound recording medium yet devised.

Professionals may copy to DAT (preferably using a digital link) or to quarter inch tape. Both these operations have to be done in 'real time' –30 minutes programme takes half an hour to dub.

In a broadcast environment, even now, dubbing to quarter inch has the advantage that, if an edit has been missed, or a retake forgotten, then a quick, old fashioned razor blade cut can be made. This is quicker than correcting the edit on the computer and then having to start copying to DAT from the beginning.

18.13 CD recording software

CD has become so widespread that it becomes a logical medium for distributing your item. Commercial CDs are pressed, not recorded individually and, for a long time, this route was not practical.

The advent of the recordable CD (CDR) has therefore been a boon to all creators of audio material. However, it is still constrained by a number of factors.

Some early ordinary audio CD players may not recognize a CDR. This, in fact, is relatively unusual and is unlikely to be a problem in a professional environment.

There are many different types of CD: CD-ROM, multisession CD, etc. Domestic CD audio players (and many professional ones as well) only recognize one CD format, the original audio format. A consequence of this is that any CD that is to be used on a standard audio player needs to be recorded in a tightly defined way (as the process involves a laser making physical changes to the disc, it is usual to talk of 'burning' a CD).

There are also stand-alone CD recorders that take an audio or digital audio signal and record discs in real time. Many of these will only use so-called 'copyright paid' consumer blanks which cost up to six or eight times that of computer data blanks.

You can also buy 'rewritable' CDs that can be erased when the data they contain is finished with. They can be recorded as standard audio CDs but, while most computer CD-ROM drives and DVD players can play them, very many audio CD players cannot.

Most CD writers come bundled with basic software. This is usually a 'sawn-off' version of generally available software with some of its cleverer options removed, so that you will be encouraged to go out and buy the full edition.

Simple software often gives you two main options for burning a complete CD. These are Track At Once (TAO) and Disc At Once (DAO). TAO means literally this: each CD track is burnt as a separate action by the software, with the laser being turned off between tracks. This usually imposes 2-second pauses between tracks. DAO burns the entire disc in one go without switching off the laser. With simpler software this provides no gaps at all between tracks.

For more information see Chapter 12 on Burning CDs.

18.14 DVD

DVD is as much an audio format as its more common video usage. At the time of writing, DVD is only beginning to show its full potential although it promises to have most of the advantages of CD but with five to ten times the capacity. The standard also makes provision for more sophisticated audio formats, notably '5.1' channel surround sound. Audio CDs are restricted to two-channel stereo. (They do have a four-channel option but that never caught on.) Playback machines, which can also play audio CDs, have rapidly becoming as widespread as CD players. DVD recorders have become common, particularly useful are the machines that combine a hard disk with the DVD recorder so it is possible to 'top and tail' edit material and then to transfer to DVD.

18.15 MIDI

Many digital audio editors are also set up to handle MIDI. If you are only using audio, then these facilities can be ignored. MIDI is only useful if you are planning to include synthesized music sessions within your programme, or you are synchronizing the audio to other events, such as a *Son et Lumière*.

MIDI data records performance data NOT audio. It is the modern electronic equivalent to the punched sheets that control player pianos and fairground organs.

MIDI can be used to synchronize different PC programs together. For specialized purposes, like exhibitions and *Son et Lumière*, MIDI can be used for external control of the audio output.

For more information see Appendix 2 and 3 on MIDI and Time code.

18.16 Control surface

Audio software often has lots of graphical displays of knobs and sliders for the user to operate. With a mouse you only have one 'finger' to operate the controls. What is becoming increasingly available at less and less cost are external specialized control surfaces. These contain physical knobs and switches which operate the graphical ones on the screen. Some of these look like full-scale mixers although no actual audio goes the surface (just as no audio goes through the mouse when you are controlling sound with it). These connect via USB, Firewire or occasionally MIDI and send control signals to the program.

A second use of these is to get you physically away from the computer. If you are playing an electric guitar, its pickup can also pickup buzzes from the monitor. For this sort of use you don't actually need lots of faders but just a few knobs and switches to control the recording software.

18.17 Automation

Automation of multitrack mixes makes the task much easier, especially in the 'one-finger' environment. It tends to fall into two basic categories. The first provides you with a very pretty graphic of a mixer. You play back the mix and move the knobs. The next time you play back the mix the knobs will move of their own accord (very impressive to the newcomer). You can update the mix by grabbing the faders whenever you wish. This has the disadvantage that in order to 'read' a mix you have to play back the entire item.

The other method is to use graphical envelopes where lines representing level, pan, etc. are drawn over the track. This system allows you to open the mix on to a screen and 'read' it and see what you (or someone else) did previously.

Adobe Audition 2.0 has both systems available giving an extraordinary degree of flexibility. As icing on the cake, the data produced by moving fader automation can also be displayed in envelope form and updated using the envelopes rather than the faders.

Appendix 1

Clicks and clocks

When you copy digitally to your computer, are you plagued with clicks and plops? Is there a regular beating sound where there should be digital silence? If so, you are probably suffering from unsynchronized word clocks. Arcane though it sounds, the problem is simply solved.

Copying digitally from one device to another is, in many ways, easier than using an analog link. There is no level adjustment to be made and quality is assured, provided the link is made by good quality cable.

Yet, there is an extra factor that it is vital to get right; this is what is known as word clock synchronization. You will be using 32 000, 44 100 or 48 000 samples per second sampling. These figures are also known as the word clock rate. 48 000 samples per second also means 48 000 digital words per second. For a 16-bit stereo signal each word will consist of two 16-bit samples (left and right) packaged together with additional 'housekeeping' data.

The word clock synchronizes the whole system; it can be thought of as the conductor of an orchestra. When you are copying from digital recorder to digital recorder, there are no complications. It is only when you have a device handling several digital signals at the same time – like a digital mixer or a computer sound card – that problems arise.

Returning to the conductor of an orchestra analogy, there are some pieces of music – Mahler symphonies, etc. – that feature offstage bands that cannot see the conductor in the hall. Unless something is organized, they will not be able to play in time with the hall orchestra. A common modern solution is to have a closed circuit TV connection, so that a second conductor can synchronize his beat to the image of the conductor in the hall.

So it is with digital systems. If your card can handle other inputs at the same time as the digital one then you need to be concerned with word clocks. The 'conductor's beat', *aka* the word clock, is sent as part of the digital signal; so we have our 'closed circuit TV'. Normally the sound card will generate its own 'beat'. This you set when you choose the sampling rate. While this is set to match the rate coming in from the external playback, there is no synchronization between the two. Even with crystal control, the chances of both clocks being precisely the same frequency are remote, so every few seconds they drift one word apart. The error resulting from this is what causes the clicks or plops. The answer is to tell the sound card to 'look' at the word clock incoming from the DAT, or Minidisc machine, feeding it. The sound card management software should have this as an option, alongside the sampling rate selection.

Figures A1.1 and A1.2 show drop down menus for a multitrack card which can handle either optical or electrical digital inputs (the selection between the two is made in a separate menu).

If you select the S/PDIF or TOSLINK option, then the sound card is controlled entirely from the external bit stream; it will follow the sampling rate as part of the synchronization.

Beware, that you have to switch synchronization back to internal once you stop using the external machine. As soon as it is disconnected, you will have no clock and the sound card may either not produce any sound, or sound at the wrong rate and pitch.

Another trap is that an external machine may default to a different clock, when it is not playing back. This means that, having very successfully copied a 44.1 kHz recording, the clock may go to 48 kHz when you take the DAT recording out. If you have not reset your sound card back to an internal clock setting, then the files played back through your editor may well play at the wrong speed and pitch – in this case faster and higher.

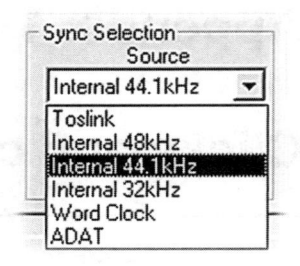

Figure A1.1 Drop down menus for a multitrack card: TOSLINK

Figure A1.2 Drop down menus for a multitrack card: S/PDIF

A1.1 Word clock in/out

The illustrated sound card can also synchronize to an incoming ADAT 8-track feed. It has another option which is specifically labelled word clock. This is for a fully professional installation where a separate high quality word clock generator is used. This separately synchronizes a number of separate digital devices, essential when dealing with a digital mixer.

Equally, when you have more than one digital sound card, you have to synchronize them by connecting the word clock output of the card you have designated as the master to the word clock in of the second card. The second card will always synchronize to the master provided it is set to use the word clock input, including following the sampling rate settings. If a third or fourth, card is added, then the word clock can be 'daisy-chained' from output to input of the next. Word clock connections are usually made with professional video-style BNC connectors. They keep the cards in synch even when there are no audio data present.

Appendix 2

MIDI

Audio and MIDI DIN plugs

Stereo Audio | Mono Audio | MIDI

MIDI Sockets

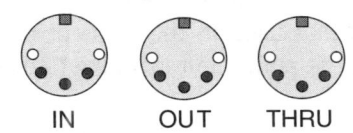

IN | OUT | THRU

Figure A2.1 Top: Showing the difference between audio and MIDI connections (white pin indicates not used). Bottom: Standard MIDI sockets

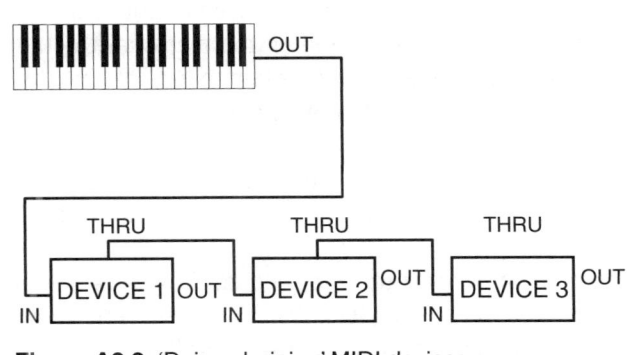

Figure A2.2 'Daisy-chaining' MIDI devices

The term MIDI stands for Musical Instrument Digital Interface. Do not confuse it with the term Midi as applied to Hi-Fi units, which is a description of their size (Midi as opposed to Biggi). MIDI is what is known as a serial connection. The data are sent as a series of long and short pulses in groups of eight. Additionally there is a pulse on either side known as 'start' and 'stop' bits. Just over 30 000 of these pulses can be sent every second – 3000 'groups' known as 'bytes' (31.25 kbaud in tech-speak).

The connections are via Deutsche Industrie-Norm (DIN) plugs which, although physically identical to audio DIN plugs, are differently wired (Figure A2.1, top).

A typical synthesizer will have three sockets as shown. The 'IN' will drive the synthesizer (sound making) circuits. The 'OUT' will carry the output of the keyboard. The MIDI system allows up to 16 instruments to be connected together in a 'daisy-chain' (see Figure A2.1, bottom). The 'THRU' socket (American spelling of 'through') relays the data arriving at the IN socket, unchanged, onto the next instrument. In the process the electrical signal is 'cleaned up' and isolated so that a fault on one machine will not necessarily adversely affect the rest of the instruments in the chain (Figure A2.2).

In general, each command to an instrument (Figure A2.3) takes three bytes of data to send. There is a possibility that an individual instrument may get sent too much data for it to cope with. Because of the 'live' nature of performance, it is vital that it can recover from this very quickly and does not 'go out of synchronism' with the data. This is achieved by uniquely marking the bytes of data. The first

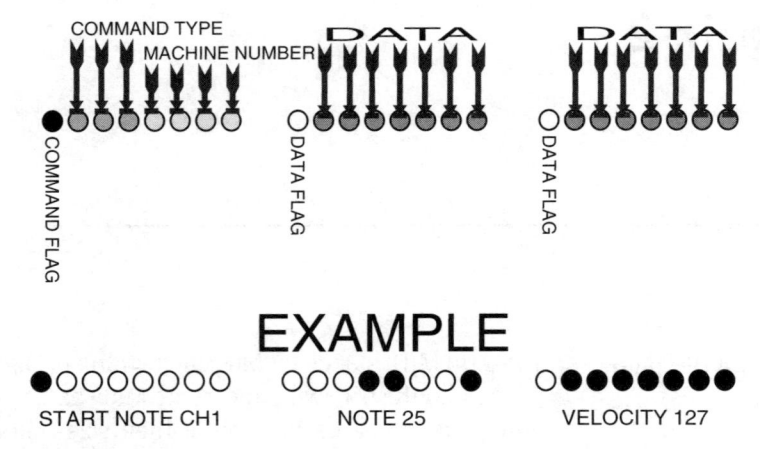

```
                COMMAND TYPE
               ↓↓↓  MACHINE NUMBER   ↓↓↓↓ DATA ↓↓↓↓     ↓↓↓↓ DATA ↓↓↓↓
              ●○○○○○○○              ○○○○○○○○            ○○○○○○○○
              │                      │                   │
              COMMAND FLAG           DATA FLAG           DATA FLAG
```

EXAMPLE

●○○○○○○○   ○○○●●○○●   ○●●●●●●●
START NOTE CH1   NOTE 25   VELOCITY 127

Figure A2.3 Commands to an instrument

byte of a group of three is the 'command' byte and has three sections. The bit corresponding to the highest value represented is always ON, and this means that only numbers from 128 to 255 are sent. The 4 bits that correspond to the lowest values (0–15) are allocated to indicate which machine the command is for. Because people are not used to the concept of 'machine 0' the channels numbers are named 1–16 although the actual numbers sent are 0–15. The remaining 3 bits are used to define the command.

Engineers notate the numbers represented by these 3 bits combined with the 8th bit which is always set as 8, 9, A, B, C, D, E, F (hexadecimal). The most important commands are 'note on' = 8, and 'note off' = 9.

The other two bytes are number values from 0 to 127. The 8th bit of a data byte is always 0 to indicate that the byte is data, not a command. What these data represent will depend on the command but a typical example is 'note on'.

Here the first data byte is a number that represents the note to be played on the standard western music scale where there are 12 notes to an octave (including the black keys). Thus two note numbers that are different by 12 are an octave apart. Other commands can modify the pitch for special effects. The second data byte represents the velocity – how 'hard' the key is pressed.

In practice most synthesizers are polyphonic; they can sound more than one note at a time and can respond to more than one MIDI channel at a time. They may also contain drum machines, etc. with their own MIDI channel number.

Synthesizers usually send out a data 'heartbeat' several times a second. This can distract computers and sound cards when being switched on causing lockups or crashes. Unless you are sure that your machine is free from this problem, it makes sense to switch off or disconnect your synthesizer when you switch on your computer at the beginning of a session.

Appendix 3

Time code

All modern audio systems have a time code option. With digital systems, it is effectively built in. At its simplest level, it is easy to understand; it stores time in hours, minutes and seconds. As so often, there are several standards.

The most common audio time display that people meet is on the CD; this gives minutes and seconds. For professional players, this can be resolved down to fraction of second by counting the data blocks. These are conventionally called *frames* and there are 75 every second. This is potentially confusing as CDs were originally mastered from 3/4 inch U-Matic videotapes (VT) where the data were configured to look like an American television picture running at 30 video frames per second (fps).

The need accurately to edit VT drove the requirement for a standard. VT editing is done by copying from source tapes to the final edited version. With modern microprocessors, tape synchronizers and control gear, this can now be done to frame accuracy (subject to some technical constraints outside the scope of this book).

In 1967, the *Society of Motion Picture and Television Engineers* (SMPTE) created a standard defining the nature of the recorded signal and the format of the data recorded. Data are separated into 80-bit blocks, each corresponding to a single video frame. The way that the data are recorded (*Biphase modulation*) allow the data also to be read from analog machines when the machines are spooling, at medium speed, with the tape against the head, in either direction. With digital systems, the recording method is different but the code produced stays at the original standards.

The *European Broadcasting Union* (EBU) adopted the same standard using the European TV 25 fps frame rate. The core of the format is the actual time code expressed in 24-hour clock mode of HH:MM:SS:FF. It also has eight groups of four *user bits* that are up to the user to decide how to use. The two have been combined in BS 6865:1987/93, IEC 421:1986/90.

SMPTE/EBU time code can be recorded as audio on a track of a multitrack tape machine. By convention the highest numbered track is used; track 4 on a 4-track; track 16 on a 16-track, etc. It is a nasty screeching noise best kept as far away from other audio as possible.

A3.1 MIDI

It can also be sent to a sequencer, via a converter, as MIDI data allowing the sequencer to track the audio tape. To save too much data overhead MIDI time code (MTC) is sent only every two frames and even then uses up just over 7% of the available MIDI data capacity.

The basic time code is sent as eight separate 2-byte MIDI messages. This also includes a code indicating what frame rate is being used. There are four options: 24 fps (Cinema film), 25 fps (video and film for European TV), 30 fps drop-frame (USA/Japan video), 30 fps non-drop-frame (used rarely for non-video applications). Sending other information, such as user bits, is optional. It is important to choose the format that is appropriate for the medium for which you are creating the audio. In the UK, the most likely standard is 25 fps.

Film has 24 separate pictures per second and these individual pictures are known as frames. Television has adopted a different practice. A European 625 line picture has 25 frames per second but these frames are divided into two 312½ line fields, which are interlaced together to make the final picture. This doubles the frequency of flicker from 25 Hz to 50 Hz and makes it less visible. Originally, with early technology, TV pictures had to be the same frequency as the local mains to avoid moving 'hum bars'. This was why 25 fps rather than film's 24 fps was chosen. Because USA mains is 60 Hz, their picture standard is 30 frames per second (60 fields per second).

The simple relationship between bars, tempo and SMPTE time as shown by sequencers like *Cubase* is only valid for 120 beats per minute 4/4 time. MIDI time code generators need to be programmed with the music tempo and time signature used by the sequencer, so that they can operate, in a gearbox fashion, so that the sequencer runs at the proper tempo.

A3.2 Drop-frame time code

This used in the USA to deal with a fudge that was made when they introduced color TV. This involves dropping frames so that time code agrees with real time. While American B&W TV ran at 30 fps, the color frame rate is nearer 29.97 fps. This means that a programme timed at an hour, using 30 fps time code, will actually run 3.6 seconds longer. This may not seem much, but its worries broadcasters. The SMPTE decided to standardize a way that the 108 'extra' frames could be 'dropped' every hour. It's similar to the way that the calendar has leap years to keep it in synch with the orbit of the Earth.

The basic rule is that whenever the time code ends a minute, it drops the first two frames on the next minute, for example. The 12:25:59:29 is followed by 12:26:00:02; frames 00 and 01 are dropped. Because this would drop 120 frames rather than 108 frames per hour, there is a further tweak to the rule, so that every 10th minute does not drop any frames, thus 12:29:59:29 is followed by 12:30:00:00. This makes drop-frame time code accurate to 75/1000th of a second in 24 hours, which is good enough for most people.

A3.3 Other systems

Modern digital systems like CD, Digital Audio Tape (DAT) and Minidisc all come with their own internal time codes. Their internal digital audio data is divided into blocks also called frames. Their connection to 'outside world' time code systems will translate to the SMPTE standard. DAT, for example store any incoming. SMPTE time code in an internal format that it can then output in any frame rate regardless of the original.

Film has traditionally used a totally different system for dubbing known as 35 mm feet. This increments faster than every second. As you would expect this corresponds the number of feet of 35 mm film but the measure is also used when the film is actually 16 mm!

A3.4 What time?

Time of day recording can be useful when recording of an event. Multiple cameras can stop and start and everything pieced together later. Production assistants can log events using an ordinary accurate clock or watch.

In practice a variation on duration coding is used but it is always unwise to start your recording at zero. There is no negative time so this would give no space for a *preroll* to allow times for separate equipment to synchronize. Very often the 'hours' are set to indicate which film roll or VT is being used. Roll 1 will be given a 01:00:00:00 start time, roll 2 gets 02:00:00:00, etc.

The 80 bits of each SMPTE/EBU frame block are allocated as follows:

Bits 0–3	Frame units
Bits 4–7	User bits
Bits 8–9	Frame tens
Bit 10	Drop frame flag/N.U
Bit 11	Color frame flag
Bits 12–16	User bits
Bits 16–19	Second units
Bits 20–23	User bits
Bits 24–26	Seconds tens
Bit 27	Group flag 2/parity
Bits 32–35	Minutes units
Bits 36–39	User bits
Bits 40–42	Minutes tens
Bit 43	Group flag 2/0
Bits 44–47	User bits
Bits 48–51	Hours units
Bits 52–55	User bits
Bits 56–57	Hours tens
Bit 58	Group flag 1
Bit 59	Parity/group flag 2
Bits 60–63	User bits
Bits 64–79	Sync word (0011111111111101)

Appendix 4

Adding RIAA to FFT filters

The FX presets are contained in the file

`Effects_settings.xml`

in the folder

`C:\Documents and settings\[username]\Application Data\Audition\2.0\`

where [username] is your Windows log-in user name.

You can obtain XML editors but they are rather specialist in application. Fortunately, this file can be edited in a text editor like Textpad or Notepad. The file has only two lines with no line breaks. The FFT presets are at the end of a very very long line! In practice, you can add line breaks which are ignored. To add the RIAA presets becomes easy.

The following shows the end of the file with the additional data separated by line breaks:

</KeyVal><KeyVal Key="Item8" Type="string">Kill The Subharmonics,3,4,0,0,0,0,351,50,4096,50,2,0,50,4096,50,1,1,2048,1,4,0,0,648,31, 831,57,1000,100,3,0,200,-15,15,8192,2,1,1,0,44100 </KeyVal><KeyVal Key="Item9" Type="string"> Mastering - Gentle & Narrow,3,6,0,35,1,35,365,50,3686,50,3925,70,4096,70,2,0,50,4096,50,1,1,2048,1, 4,0,0,648,31,831,57,1000,100,3,0,200,-15,15,8192,2,1,1,0,44100

In a text editor like Textpad (not Wordpad) search for "Mastering - Gentle & Narrow"

To find your place and insert the RIAA data block at the point shown. (This data is on the CD-ROM that came with this book.)

</KeyVal><KeyVal Key="Item22" Type="string"> RIAA,3,30,0,97,32,97,188,96, 330,95,541,91,696,87,855,83,1007,80,1128,76,1246,72,1375,68,1515,65,1656,61 ,1819,57,2024,53,2281,50,2535,46,2738,42,2901,38,3041,35,3168,31,3286,27,34 00,23,3509,20,3617,16,3723,12,3827,8,3931,5,4035,1,4096,0,30,0,0,1,0,2,98,330 ,95,541,91,696,87,855,83,1007,80,1128,76,1246,72,1375,68,1515,65,1656,61,18 19,57,2024,53,2281,50,2535,46,2738,42,2901,38,3041,35,3168,31,3286,27,3400, 23,3509,20,3617,16,3723,12,3827,8,3931,5,4035,1,4096,0,2,1,4096,1,4,0,0,648,3 1,831,57,1000,100,3,0,100,-20,20,16384,2,1,1,0,44100 </KeyVal><KeyVal Key="Item23" Type="string"> RIAA REVERSE,3,29,0,0,88,1,344,5,567,8,735,12, 878,16,1009,20,1135,23,1259,27,1385,31,1517,35,1660,38,1826,42,2029,46,228 4,50,2537,53,2739,57,2902,61,3042,65,3169,68,3287,72,3401,76,3510,80,3618,83,

3724,87,3828,91,3932,95,4036,98,4096,100,30,0,0,87,0,88,1,344,5,567,8,735,12,878,16,1009,20,1135,23,1259,27,1385,31,1517,35,1660,38,1826,42,2029,46,2284,50,2537,53,2739,57,2902,61,3042,65,3169,68,3287,72,3401,76,3510,80,3618,83,3724,87,3828,91,3932,95,4036,98,4096,100,2,1,4096,1,4,0,0,648,31,831,57,1000,100,3,0,100,-20,20,16384,2,1,1,0,44100
</KeyVal></Section><Section label="FFT Filter"><KeyVal Key="LocalizedName" Type="string">FFT Filter</KeyVal><KeyVal Key="Version" Type="int32">2</KeyVal></Section></Section>

Once Audition is run it will resave the file and remove line breaks. Note that the <KeyVal Key="ItemNN"> of the added data, where NN is a number, must be increased to the next highest value in that part of the table.

Glossary

AB Stereo Often used to distinguish MS stereo from the convention signal using left and right signals. In the context of microphones it often implies the use of spaced omnidirectional mics rather than a coincident pair.

ADAT A digital multitrack recording system that gives eight tracks on an S-VHS videocassette.

ADPCM (Adaptive Differential Pulse Code Modulation) Conventional Pulse Code Modulation stores the values of a waveform as a series of absolute values. Differential PCM does not do this but, instead, sends the data as a series of numbers indicating the difference between successive samples.

ADSL (Asymmetric Digital Subscriber Line) Sometimes known as a 'cable modem' which is telephone data service where higher data speeds are available. The connection is permanently open. The 'asymmetric' refers to the fact that the data rate is slower for uploads compared with downloads. The actual data bandwidth that is available is shared between a number of subscribers (contention ratio) and will varying depending on how many are sending or receiving data at any moment.

AES (Audio Engineering Society) Among other things, the AES lays down technical standards. In the context of this book they are best known for a professional standard for conveying digital audio from machine to machine which has also been adopted by the EBU.

AES/EBU A professional digital audio standard for transferring digital audio between machines. This is balanced and uses XLR connectors. The data format is similar but not identical to S/PDIF which can see an incoming AES/EBU format signal as copy prohibited.

AIFF Apple AIFF (.AIF, .SND) is Apple's standard wave file format and is a good choice for PC/Mac cross-platform compatibility.

ALC (Automatic Level Control) See AVC.

Aliasing Spurious extra frequencies generated as a result of the original audio beating with a frequency generated within the audio processing system (usually the sampling frequency in a digital system). Filters are used on the input to prevent this but these filters themselves can produce degradation of the signal unless very well designed.

Analog Audio Until digital techniques came along audio was conveyed and recorded by using a property that changes 'analogously' to the sound pressure. This property might be electrical voltage, magnetization or how a groove wiggled. Digital audio replaces this by a series of numbers.

AV (Audio Visual) AV standard hard drives are able to cope with long runs of data, such as a long continuous audio recording, without stopping to recalibrate themselves for temperature variations.

AVC (Automatic Volume Control) Often found as an option on portable recorders. This automatically adjusts the recording level from second to second. This can cause trouble when editing as the background noise will be going up and down. However, modern AVCs work surprisingly well. The background matching becomes a trivial problem when editing with a PC digital audio editor.

Balanced Normal domestic audio connections are unbalanced; a single wire carries the audio which is surrounded by a screening braid connected to earth as the return circuit. These circuits are prone to pickup of unwanted signals as well as high frequency loss when used beyond about 5 metres. Balanced circuits use two wires to carry the audio (still within a screening braid). The audio in the wires is going in opposite directions, as one wire goes positive the other goes negative. Interfering signals induce in the same direction on both wires. The input circuit is designed only to be sensitive to the difference in voltage between the two wires and therefore ignore the induced interference.

Barrier mics Barrier mics are designed to be placed on large flat surfaces rather than suspended in free air. These are often referred to as PZMs (Pressure Zone Microphones) after a commercial version.

bel See Decibel.

Binaural Current practice is to use this term to mean two channel audio balances intended to be heard on headphones.

BIT (BInary digiT) Digital/PCM systems use pulses which indicate either an 'ON' or 'OFF' state. Each individual piece of data is known as a bit.

BNC A professional video/digital audio connector with a locking collar.

Byte A group of 8 bits (allegedly a contraction of 'By Eight'). This is the standard measure of the capacity of digital systems.

Capacitor An electrical component that can store electrical charge, formerly known as a condenser. They consist of two parallel 'plates' separated by an insulator. The plates are so close together that when they are charged the positive charge on one plate is attracted to the negative charge on the other. The closer they are together, the greater the attraction. This increases the amount of charge that the device can store. A practical capacitor's plates are in fact metal foil sheets separated by a sheet of thin insulator rolled into a cylinder, rather like a Swiss roll. This reduces their size and gives them a cylindrical appearance. A specially constructed capacitor forms the basis of electrostatic microphones.

Cardioid (heart shaped) The most common of microphone directivity shapes (see Page 189).

CD (Compact Disc) When used without qualification this is taken to mean a standard audio CD. Subsequently adopted for data use in computers, this has led to many variants (see below).

CDR (Compact Disc Recordable) A recordable CD which cannot be erased.

CD-ROM (Compact Disc Read Only Memory) A CD containing data rather than audio.

CD-RW (Compact Disc Read-Write) A recordable CD that can be erased. While they can be recorded in audio format, most domestic CD players cannot play them.

Chequer boarding A technique for mixing sections of audio or video (see Page 71).

Cinch plug Another name for phono plug.

Clock All digital systems have a reference clock which acts rather like the conductor of an orchestra to keep everything in sync (see Appendix 1).

Clone In the digital audio context this is used to mean making an exact sample for sample copy. This is not possible with systems like Minidisc, which use lossy compression systems.

Coax, Coaxial plug A generic term for a connection that uses a cable where one or more conductors are surrounded by a wire braid that helps screen out interference. In the UK, this term is most often used for the plug used for television aerials. In the audio context, some companies use this term for a phono plug.

Coincident pair A stereo microphone technique using directional mics placed as close together as possible.

Compact Cassette The proper name for the ordinary analog audio cassette; undoubtedly the most commercially successful audio recording medium ever invented.

Compression, audio Compressor limiters are the most used effects devices in the studio. They can be thought of as 'electronic faders' which are controlled by the level of the audio at their input (see Page 90).

Compression, data There are two types of data compression. *Non-Lossy*: The first, and traditional, form of compression reduces the data to be stored on a disk or sent via a modem. The best known format is the Zip format. A Zip file can be uncompressed to recreate the original data without any change or error. *Lossy*: Graphics and audio are often compressed using formats that approximate the data using assumptions about how we see and hear (see Page 154).

Condenser Microphone Condenser is an old term for capacitor.

Cool Edit Pro A commercial audio editing package for Windows, combining both linear and non-linear editing. This was bought by Adobe and transformed into Audition.

DAC see Digital-to-analog converter.

DAO (Disc At Once) A technique of burning a CD in one go. This has a number of technical advantages, notably giving the ability to control inter-track gaps or even recording audio into those gaps.

DAT (Digital Audio Tape) Originally a generic term for 'Digital Audio Tape Recorder'. This is now used specifically for a format developed by Sony, supported by scores of other manufacturers, that has become popular amongst professionals and semi-professionals alike for mastering digital audio. This is now becoming obsolete.

Data Plural of 'datum'. From Latin 'things given'. The word is often erroneously used as a singular, that is 'The data are' not 'The data is'.

DAW (Digital Audio Workstation) A dedicated computer audio editor with specialized controls and software.

dB Decibel.

dBA Decibels are used widely in the field of acoustics, audio and video. Subtly different scales are used and the different types of decibel are indicated by a suffix. The dBA is used in the field of acoustic measurement.

dbx A commercial company that is best known for developing a popular analog noise reduction system.

DC Offset Poor analog to digital converters can have a DC offset that 'pushes' the audio away from being centred around zero volts towards either the positive or negative. This is a major cause of clicks on edits (see Chapter 4.2, Figure 4.9).

Decibel   One-tenth of a bel. The normal way of measuring audio. It is a logarithmic system using a standard reference level. Values are expressed as a ratio of a standard level. The bel itself is too large a unit to be convenient for audio. The Richter units used for measuring earthquakes are identical to bels but with a rather louder reference level! (see Page 6).

Delta modulation   A technique where the difference between samples is sent instead of the absolute values usually sent by PCM.

Digital audio   Sound pressure level variations are represented by a stream of numbers represented by pulses.

Digital audio workstation   A dedicated computer audio editor with specialized controls and software.

Digital Versatile Disc (DVD)   The successor to the compact disc. It uses similar technology but takes advantage of technical developments since the CD was introduced. Recordings are made at a much higher density, eight times greater than CD. Additionally the DVD can not only be double sided but also each side can be made up of two layers. This gives massive data capacity, enough for full-length films with 5.1 channel audio. The '0.1' is a, not very good, engineering joke. There are, in fact, six channels of audio but the sixth channel is a low bandwidth one used for low-frequency effects.

DIN (Deutsche Industrie-Norm)   German industrial standard. This includes the audio/MIDI/computer DIN plug. A standard sized case and connector containing a number of pins. The most common type met in the audio and MIDI context is the 5-pin 180°.

Disc, Disk   A convention has grown up where disc-based media using a magnetic medium are spelt as 'disk' (with a 'k'). Optical recordings like CD and Minidisc, as well as gramophone records, remain spelt as 'disc' (with a 'c').

Dither   A low-level signal, usually random noise, which is added to the analog signal before conversion to digital. Its effect is to reduce the distortion caused by quantization.

Dolby   Dr Ray Milton Dolby is possibly the most influential individual in audio. His company, Dolby Laboratories, began by making audio noise reduction systems. The first, a professional system, became known as Dolby A-type. A simplified system, called Dolby B-type, revolutionized the compact cassette medium for consumers. Dolby C-type gives about 20 dB noise reduction compared with the B-type system's 10 dB, albeit at the cost of poorer compatibility when played on machines without decoding. Dolby SR (Spectral Recording) is an enhanced professional system that can give better than 16-bit digital performance from analog tape. Dolby S is a powerful compact cassette system based on a simplified version of SR. As other companies have to obtain a Dolby trademark, the company has become the effective setter of standards for cassette machines, as minimum audio performance is set by Dolby Laboratories to allow their systems to be used.

Dolby Digital   Started in the cinema industry, this is the digital multi-channel sound format most widely used on DVD and with digital television.

Dolby Stereo   Dolby Laboratories devised a system for putting stereo onto optical release prints in the cinema. This was combined with a phase coding system that allows surround information to be heard. The Left total and Right total channels can be decoded to give Left, Centre, Right and Rear loudspeaker information. In the early 1990s the term 'Stereo' was dropped, and the analog film format is now simply referred to as 'Dolby'.

Dolby Surround The domestic version of Dolby Stereo as found on video cassettes, CDs, TV broadcasts and video games.

Drop in Switching from play to record while running to make an electronic edit.

Drop out (1) A momentary loss of sensitivity in an analog recording medium. Digital systems can correct or conceal errors resulting from drop outs in the medium. However, if they fail you will hear a mute instead. (2) Switching from record to playback to end a drop-in.

DTMF (Dual Tone Multi-Frequency) These are the tones that modern phones make when dialling. They are used to code the digits 0–9 as well as the special system codes of '*' and '#'. Four extra codes are also available known as 'A', 'B', 'C' and 'D'. The sixteen possible combinations are achieved by sending two frequencies (hence 'dual tone') out of a possible selection of a total of eight.

DVD See Digital Versatile Disc.

Dynamic microphone An alternative word for a moving coil microphone.

Dynamic range In audio systems the dynamic range available is determined by the number of bits used to measure each sample. In theory for every bit extra another 6 dB of signal-to-noise ratio is gained.

EBU (European Broadcasting Union) A trade association for European broadcasters, which also sets technical standards.

Echo Although this is often used interchangeably with the term reverberation (or 'reverb'), they are technically different. Echo is where you can distinguish individual reflections (echo … echo … echo) while reverberation is where there are so many reflections that they merge into one continuous sound (see Page 99).

Edit Decision List (EDL) Used in a non-linear editing process where the audio files are not altered. Instead a list of instructions (the Edit Decision List) as to what section to play when, at what level, etc. is created which causes the hard disk to skip around and produce apparently continuous audio.

EDL See Edit Decision List.

EIDE (Enhanced Integrated Drive Electronics) The most widely used (and hence cheapest) form of hard drive. Most PC motherboards are equipped to handle four drives. The usual alternative is SCSI but see also SATA.

Electret A form of capacitor which remains charged permanently. This means that when constructed as a microphone, it does not need a polarizing voltage. Typically a simple 1.5 V AA battery is used to power the built-in amplifier.

Electromagnetic Devices where magnetism is used to create electricity or electricity is used to create magnetism. Within audio, a moving coil (dynamic) mic is used to create electricity (the audio signal) by the diaphragm pushing and pulling a coil of wire between the poles of a magnet.

Electrostatic Electrostatic microphones use a diaphragm that is one of the plates of a capacitor held charged by a polarizing voltage (see Page 192).

EQ Pronounced 'Eee' 'Cue'. A widely used abbreviation for 'equalizer', a term for devices like tone controls which modify the frequency response of an audio system. Such devices were originally used by engineers actually to equalize, or correct audio deficiencies in land lines and recording systems. They were then borrowed by studio operators to improve their recording mix. Nowadays equalizers are designed specifically for studio use.

Error concealment   A technique where errors can be detected but not corrected. Instead they are concealed often by replacing the sample with an average of the samples either side. This is called interpolation (see below).

Error correction   As digital audio is a series of numbers, extra numbers can be added having been generated by various mathematical means. At the receiving end these numbers can be generated again from the incoming data. If they are different from the extra numbers sent, then error has been detected. With suitable maths, the errors can often be corrected. As the binary nature of the signal represents only '0' or '1', it is clear that, if the incoming value is established as being wrong, then the correct value must be the only other value. *Interpolation*: is a technique where when an error is detected a value between the preceding and following value is substituted, whereas full error correction actually reconstitutes the data. *Interleaving*: Errors can be made easier to correct by interleaving the data so that they are physically spread out on the medium so that a single drop out in the medium does not produce a single burst of errors.

fff   'f' is a music term for loud (Italian: *forte*); the degree of loudness is indicated by the number of 'f's, thus fff is very loud.

FFT (Fast Fourier Transform)   A mathematical way of defining a filter.

Figure of eight   Term used to describe a microphone that is sensitive front and back but dead at the sides.

Firewire   A standard way of connecting apparatus to a computer that is supported by both PCs and Apple Mac computers. The devices are 'daisy-chained' together and can be connected or disconnected without having to reboot the computer. This is faster, although presently less common, than USB.

Flanging   Phasing with continuously varying delay.

Flutter   Rapid variation of pitch, often caused by a dirty or damaged capstan pulley on an analog tape machine. A digital recording will be free from this fault.

Flutter echo   Term used by those of us who tend to use the terms 'echo' and 'reverb' interchangeably to show that we are really talking about echo.

FM (Frequency Modulation)   A way of sending data, or audio, using a carrier frequency which is varied in pitch. This is used by FM Radio, VHS video cassette, Hi-Fi sound, many hard disks, etc.

FX   A widely used abbreviation for 'effects'.

Gain   Another word for amplification.

Giga   1000 million, hence 1 gigahertz = 1 000 000 000 hertz = 1 GHz.

Glitch   A discontinuity in sound, due to data errors.

Gun mic   A generic term for very directional mics using a long tube and phase-cancellation techniques. Also known as shotgun or rifle mics. As they are very sensitive to wind noise they are usually concealed in long furry sausage-like windshields.

Hexadecimal   A numbering system based on the number 16 instead of 10. The characters 0–9 and A–F are used:

        0, 1, 2, 3, 4, 5, 6, 7, 8, 9, A, B, C, D, E, F, 10, 11, 12 etc.

It is a convenient notation for binary numbers as used by computers and for MIDI. FD is easier for a person to distinguish from FB than 11111101 from 11111011.

Hypercardioid   A microphone which has a slightly narrower front pick up compared with a cardioid. The penalty is that there is a reduced sensitivity lobe at the back, giving a dead angle a few degrees to the side of this.

IDE (Integrated Drive Electronics)   A way of connecting hard drives to personal computers. It is very cheap and has become very popular and hence become even cheaper. Although the technology has much improved it is regarded, by many, as inferior to SCSI as it uses the computer's processor and slows down programs that are being run at the time. It has been said that IDE works in a way analogous to arriving at a shop counter and asking for an item and saying 'I'll wait' whereas SCSI allows you to go away and do something else because it will 'deliver'. See also SATA.

Image   The perceived location of a single source within the sound stage. The image may be narrow (panned mono) or wide (string section). It may be precise or blurred.

Interleaving   Error correction technique (see under Error correction above).

Interpolation   Error concealment technique (see under Error correction above).

ips   Inches per second.

ISDN (Integrated Services Digital Network)   Sometimes ironically referred to as the 'It Sometimes Doesn't Network', which gives you a direct digital connection to the phone network. There are various audio coders that allow good quality down, what is effectively, a telephone line. Different organizations have standardized on different systems and some manufacturer's gear is not fully compatible.

Kilo (K)   The word kilo when used in the context of digital systems usually means 1024 *not* 1000. This represents the maximum value of a 10-bit word (i.e. 2 to the power of 10). This is not cussedness, but is used because it represents a very convenient unit. So the term 64K will mean 65 536. Some publications use the abbreviation 'K' to mean 1024 and 'k' to mean 1000.

Limiting   Compression of greater than 10 : 1.

Line in   An input to be fed by amplified audio rather than a microphone.

Line level   Domestic outputs tend to be $-8\,dB$ with professional outputs being $+4\,dB$; $0\,dB$ is 0.775 volts. This strange value dates from telephone technology, where 0.775 volts gave 1 milliwatt into a 600 ohm circuit.

Line out   Amplified output of a device.

Linear editor   An editor where the audio files themselves are altered by the editing process.

Lossy/non-Lossy   See Compression, data.

LP (Long Play)   Usually used to refer to 12 inch gramophone records. However video cassettes, DAT and Minidiscs have a long play mode.

Mastering   The general term for transferring studio material to a final stereo 'master' which will be used to generate the copies sold to the public.

Mega   One million, hence 1 megahertz = 1 000 000 hertz = 1 MHz (note upper case 'M'; lower case 'm' means milli = 1/1000th when used as a prefix).

Mic   Preferred British abbreviation of the word microphone.

Micro   One-millionth, hence one microsecond (1 μs). Used colloquially to mean small computer.

MIDI (Musical Instrument Digital Interface) A standard system for communicating performance information between synthesizers and computers.

Milli One thousandth, hence 1 millimetre = 1 thousandth of a metre (1 mm).

Monaural Sometimes used to mean monophonic but really means listening with one ear!

Monitoring speaker Monitoring speakers are designed to be analytical and reveal blemishes so that they can be corrected. As a generalization, Hi-Fi loudspeakers make the best of what is available.

Mono Contraction of monophonic; also contraction of monochrome (i.e. black and white pictures).

Monophonic Usually contracted to 'mono', this is conventionally derived from the stereo signal by a simple mix of left and right channels. Beware of some portable tape machines which have a switch labelled mono. This often means that only one input is fed to both legs of the recording *not* that the two inputs are mixed and fed to both legs. It is very easy to end up with an interview tape that only has questions on it if a reporter is not aware of this.

Moving coil The most common type of microphone and loudspeaker. A coil of wire attached to a diaphragm is suspended in a magnetic field. If the coil is moved by pressure on the diaphragm a small voltage is generated. If a current is passed through the coil, the diaphragm will be moved.

MPEG (Motion Picture Experts Group) MPEG is a series of Lossy compression systems for audio and video.

MS (Middle Side) A stereo technique where instead of the left and right information being used for the two audio channels, the middle and side are used. The middle signal corresponds to the mono signal. The side signal corresponds to the amount of 'stereoness'. It is zero for centre signals and at a maximum for sounds coming from the extreme left or right.

Multiplex Generally any method of carrying several signals (e.g. stereo left and right) on a common circuit. Digital radio and television are broadcast using multiplexes that carry a number of channels; how many depends on the quality required.

Nano One-thousand-millionth, hence one nanosecond (1 ns).

NICAM (Near Instantaneously Companded Audio Multiplex) Digital system used in the UK to add stereo sound to television broadcasts.

Noise reduction Analog recordings are often made using noise reductions systems like Dolby and dbx. They boost the signal on record and apply a correction to this on playback. In the process noise and hiss are correspondingly reduced. In the digital editor context noise reduction usually refers to various software solutions that remove clicks, hiss or noise from an existing recording, usually a transfer from an analog original.

Non-linear editor An editor which does not alter the original audio files. Instead it uses some form of edit decision list to instruct the computer to jump around the hard disk reproducing and mixing audio as required.

NTSC (National Television System Committee) American color television system.

Nyquist limit This is named after Harry Nyquist the Bell Telephone Laboratories' theoretician who first enunciated the principle that you have to sample at a frequency at least twice that of the highest you intend to transmit.

Omnidirectional Omnidirectional microphones are equally sensitive to audio from any direction.

Out-of-phase If loudspeakers are out-of-phase, this means that their diaphragms, instead of moving in and out together, move in opposite directions. This has the effect of blurring the sound image and reducing the bass response of a system.

PAL (Phase Alternate Line) Color television system used by most countries in Europe. This is often used to imply a 25 fps frame rate, although 30 fps versions do exist. Many video-cassette machines will output a PAL signal at 30 fps when playing an NTSC tape.

PATA (Parallel ATA) Used to distinguish the original IDE (Integrated Drive Electronics) bus; also known as the *ATA* (Advanced Technology Attachment) specification (ATA Bus). The IDE bus is a Parallel bus. A new form is being introduced called Serial ATA (SATA), as a result the parallel form is now commonly called *parallel ATA* (*PATA*).

PC Originally a generic term for 'personal computer' this has been hijacked to mean a computer based on the original IBM design. While sometimes used to imply using a Microsoft operating system such as Windows, as opposed to Apple Mac, they can also be used for other operating systems such as BeOS, Unix, Linux, etc.

PCM (Pulse Code Modulation) The technique of sending numbers as a series of pulses.

Phantom volts Electrostatic (capacitor or condenser) mics need to be powered to work. Balanced studio mics are often powered by adding the DC voltage (usually 48 volts) to the audio wires. Circuitry at each end separates the audio. Although the circuit behaves as if there is an extra wire for the power, it has no physical existence, hence the term phantom.

Phase A measure of the relative delay between two waveforms at the one cycle level. The positive-going zero crossing is described as 0° and the negative-going zero crossing as 180°. Where the difference between the two waveforms is exactly reverse – positive in one is matched by negative in the other – they are said to be 180° out-of-phase, or just out-of-phase.

Phasing This is caused by selective cancellation of some frequencies either using a comb filter or, more often, by mixing two nominally identical signals with a short delay between them. In the late 1960s this was a much loved 'psychedelic' sound. If the delay is continuously varied this is called flanging.

Phono plug The RCA phono plug was originally designed to connect phonograph (gramophone) turntables to amplifiers. This has become a universal way of connecting unbalanced audio (and, often, video) largely because of the cheapness of the connector.

POTS (Plain Old Telephone System) A general term for ordinary phone lines which have limited capabilities compared with specialized systems using newer technology such as ISDN.

ppp 'p' is a music term for quiet (Italian: *piano*); the degree of quietness is indicated by the number of 'p's, thus ppp is very quiet.

Pre-emphasis A technique of boosting high frequencies on record or transmission; they are then restored on playback or reception. In the process any added hiss has top cut applied to it.

Program, programme In this book 'program' is used in the context of computers. The British spelling is used in the context of radio and TV programmes.

PZM (Pressure Zone Mic) See Barrier mics.

Quantizing   (1) The process of turning an analog signal, like audio, into numbers. (2) In MIDI sequencers, the automatic moving of notes onto the beat.

Quantizing interval   The difference in voltage between quantizing levels.

Quantizing levels   The number of possible values into which an analog signal may be divided or quantized.

RCA plug   The RCA company originally devised a plug for connecting phonographs (gramophone turntables) to amplifiers. These plugs became known as phonograph plugs, abbreviated to phono plugs.

Reverb   Contraction of 'reverberation'.

Reverberation   The sound made by reflections from a room, or an electrical simulation of this. It differs from echo in that there are so many reflections that individual reflections are not discernible. Reverberation time is measured as the time taken for the reflection level to decrease by 60 dB (RT60).

RIAA (Radio Industries Association of America)   A trade association that also sets standards. Most often met is the RIAA standard for equalization of long playing gramophone records. When the disc is cut the high frequencies are boosted and the bass frequencies cut. This is corrected by reversing this on playback.

Ribbon mic   A microphone using a corrugated aluminium ribbon as a diaphragm. This is placed within the field of a powerful magnet. Once the standard microphone used by the BBC and other broadcasters, it is capable of extremely good quality especially on strings. It is advisable to take off your wristwatch before handling one as the strong magnetic field may stop it.

S/PDIF (Sony/Philips Digital Interface)   Standardized digital audio connection format using an unbalanced audio connection and phono plugs.

SATA drive (Serial Advanced Technology Attachment)   Nominally called the IDE (Integrated Drive Electronics) bus; it's also known as the *ATA* (Advanced Technology Attachment) specification (ATA Bus). The IDE bus is a Parallel bus. A new form is being introduced called Serial ATA (SATA), as a result the parallel form is now commonly called *parallel ATA (PATA)*.

SCMS (Serial Copy Management System)   A specification for copy protection flags contained within domestic digital audio connections. While the system allows for no copy inhibition or full copy protection, its default is usually to allow only one digital copy; a SCMS equipped digital recorder will copy a digital recording set to allow one copy but the copy it makes will be set to no copying allowed. In practice the system is merely an annoyance for serious recordists working with original material. It can easily be defeated by analog copying, copying via a computer, or via an interface box that allows the resetting of the SCMS flags.

Scrub editing   A term now used to describe the traditional way of finding edits on reel to reel tape which involved 'scrubbing' the tape back and forth. Many users familiar with tape editing like to have this facility on a digital editor, although this usually soon turns out to be a security blanket.

SCSI (Small Computer Systems Interface)   Usually pronounced 'Scuzzy'. A way of connecting personal computers. Originally used on Apple Macintosh computers, it is regarded as having many technical advantages over its rival, IDE, universally used by PCs. However, there is no reason why a PC cannot use SCSI (some PC mother boards have it built-in, others can have a SCSI card fitted). Many people regard the extra expense worth the extra reliability it gives to audio recording. Both systems can be used within the same machine. The SCSI interface is also used for other devices like scanners. Seven devices plus the controller can be accommodated by a simple SCSI system, with fifteen devices available on more sophisticated systems.

SECAM (System En Couleur Avec Memoire) French color television system. SECAM VHS tapes will play in monochrome on a PAL video cassette machine.

Serial A term used when data is sent down one wire, one bit at a time.

SMPTE (Society of Motion Picture and Television Engineers) In conversation, often referred to as 'Simty'; this body sets standards.

SMPTE/EBU The combined American and European time code standard.

Stereo, stereophony Literally 'Solid Sound' from the Greek. While there is argument over a precise definition it seems generally to mean using two channels to give a directional effect. Good stereo will do more than this, giving a sense of depth as well as direction.

Table of contents Compact Disc and Minidisc use a table of contents to tell them where each track and index begins and ends. If this is lost then the whole disc becomes unplayable.

TAO (Track At Once) Where a CD is burned one track at a time with one wave file per track. The laser is turned off between tracks.

Time code A code contained within a recording that identifies each part of the tape uniquely in terms of time.

Tinnitus A distressing condition caused by hearing damage where noises are generated within the ear, often perceived by the victim to be at a very loud level.

TOSLINK (TOShiba LINK) Optical connector used by many domestic digital audio devices.

Transient A short-lived sound, typically the leading and trailing edges of a note.

USB (Universal Serial Bus) A standard way of connecting apparatus to a computer that is supported by both PCs and Apple Mac computers. The devices are 'daisy-chained' together and can be connected or disconnected without having to reboot the computer. The presently less common Firewire system is operationally similar but faster.

Variable pattern mic This contains two mic capsules that can be combined in different ways to give a range of directivities (see Page 191).

Varispeed A control for varying the speed of an analog recorder, usually on playback. This not only changes the rate but also the pitch. Digital editors usually have software that will allow change to rate or pitch independently of each other. Small changes of less than 10 per cent are usually very successful. Larger changes can suffer from glitches and artefacts. As a result you are often offered a choice of software methods (algorithms) which have different strengths and weaknesses.

VHS (Victor Home System) Presently the most popular home video tape format. In its NTSC version this has long play and extended play modes in addition to standard play. PAL systems only have a long play mode. The audio is recorded in two forms. There is a low quality recording on a linear track on the edge of the tape; this is usually mono but some machines provide stereo sometimes with Dolby noise reduction. Additionally many machines record in stereo using rotating heads within the video signal. The actual sound uses an FM carrier with a noise reduction using high-frequency pre-emphasis and 2 : 1 compression.

WAV The .WAV format originated from Microsoft as a simple format for storing audio for games, etc. The format has been expanded for professional use adding text fields, etc. If an exported .WAV file begins with a click on another piece of software then it is possible that this software is not reading the file correctly. There is usually an option to export the files without the text information which can solve the problem. There is also an ADPCM version of this format giving 4 : 1 lossy compression. This is best regarded as an end-user format.

WMA Windows Media Audio. A Microsoft rival to MP3 files for streaming over the Internet.

Wow Slow variation of pitch, traditionally caused by an off-centre gramophone record or a slipping capstan pulley on a tape machine. Digital audio systems are entirely free of this unless it is deliberately introduced as a special effect.

XLR Connector widely used in the professional audio field.

XY stereo In the context of microphones this is sometimes used to indicate the use of a coincident pair of microphones as opposed to spaced (AB).

Zip drive Proprietary form of removable hard disk cartridge.

Zip format Lossless data compression format much used to reduce the size of program and data files on disk and sent via modems. Unfortunately the assumptions that it makes do not apply to audio and files can end up larger after Zip compression than before.

Index

32-bit files, 54, 118

Acoustics:
   booths/separate studios, 188
   directional microphones, 188
   microphone distance, 188
   multimiking, 187–8
   multitrack recording, 188
   orchestral music, 188–9
   separation, 187
   see also Microphone; Transducer
Actuality, 185–6
Adaptive Differential Pulse Code Modulation (ADPCM), 155
ADSL (Asymmetric Digital Subscriber Line), 154
AES/EBU, 17, 18
Alias frequencies, 10
Ambisonics, 85
Amplitude statistics, 168–9
   DC offset, 169
   histogram tab, 169
   peak amplitude, 169
   possibly clipped samples, 169
   RMS power, 169
   sample value, 168
Analog audio, 8
Analog materials, archiving, 201–2
Analog recording, 16, 90
Analog transfer, 22
Anti-aliasing filters, 10
Archiving, 200–3
   analog media, 201–2
   and backup, 202–3
   CD (compact disc), 200
   CD-ROM, 200–1
   DAT (digital audio tape), 201
   selection, 200

ASIO *see* Audio Stream In/Out
Asymmetric Digital Subscriber Line (ADSL), 154
ATA bus *see* IDE bus
Attack and release times, 95
Audio CDs, 45, 170, 201, 223
   burning, 171–2, 173
Audio design:
   basic effects, 138–9
   basic level adjustment, 88–90
   compression/limiting, 90–3
   dangers, 86
   expanders and gates, 93–5
   filters, 114–23
   multiband compressor, 95–8
   multitrack effects, 140–2
   normalization, 86–8
   real-time controls and effects, 142–6
   restoration, 124–31
   reverberation and echo, 99
   spatial effects, 136–8
   special effects, 131–3
   time/pitch, 133–6
Audio editors, 219, 220
Audio processing, 220–1
Audio Stream In/Out (ASIO) drivers, 209, 212
Audition-Theme.ses, 29
Automation:
   clips, 66–7
   lanes, 65, 98
   of multitrack, 64–7, 223–4
AVC (automatic volume control), 90, 91
Azimuth error, 166–7

Backtimed and prefaded music, 76
Bar codes, 178, 179–80
Barrier microphone, 190
Bass rumbles, 115

Index

Batch files, 43, 48–53, 123, 139
  batch processing dialog, 48, 49, 50–3
  scripts, 48–9
BBC practice, 149
Bel, 6
Binaural sound, 82–3
Biphase modulation, 12, 229
Blue bar blues, 42–5
Booths/separate studios, 188
Brown noise, 151
Buffer under run, 171
Burnproof® technology, 172
Busbars, 58, 59, 65

Capacitor microphones, 192–3, 115
Cardioid microphones, 189–90, 191
CD (compact disc), 9, 12, 20–2, 25, 28, 33, 43–5, 83, 91, 153, 154, 200, 229
  archiving, 200
  audio CD, 45, 170, 171–2, 173, 201, 223
  burning, 25, 47, 153, 170–80, 200, 201, 217, 222
  CD-ROM, 10, 149, 170, 200–1, 213–14
  CDR (Compact Disc Recordable), 10, 153, 154, 172, 173, 194, 222
  cleaning, 21
  labelling, 153–4
  multi-session CDs, 172, 173
  rewritable CDs (CD-RWs), 172–3
  writers, 153, 222
CD recording software, 173–9, 222
  Adobe Audition 2.0, 173–4
  CD burning dialog, 174
.CDA files, 45
CDR (Compact Disc Recordable), 153, 154, 172, 173, 194, 222
  error, 10
.CEL file, 79
Center channel extractor, 114–15
Chequer boarding, 71–3
Chorus effect, 103–4, 135
Click/Pop eliminator, 125–6
Clickfix, 126, 127
Clicks and clocks, 225–6
  word clock in/out, 226
Clip restoration, 126–8

Clipping, 126–7, 128, 169
Clips, 58, 66–7, 176
Clocks, 19
  and clicks, 225–6
Coaster, 171–2
Comb filter effect, 105, 106
Compact Disc *see* CD
Compression, 8
  audio, 26, 88, 90, 155, 156
  data, 14, 26, 91, 92, 93, 95, 96, 141, 186, 221
Compression/limiting, 90–3
  attack, 92
  hard limiting, 92
  in/out, bypass, 93
  input adjusting, 93
  link/stereo, 93
  output adjusting, 93
  release/recovery, 92
  threshold, 92
Compression ratio, 91–2, 156
Computer backup:
  password, 202
  storage systems, 202–3
Condenser microphones, 192–3
Control surface, 223
Convolution effect, 131–3
Cool Edit Pro, 155
Copy, 43, 45–6
Copy protection, 18, 178
Cross fade, 2, 46, 69–71, 77
  functions, 82
Cubase, 12, 199, 230
Cut, 3, 43, 45–6, 69

DAO (Disc At Once), 222
DAT (Digital Audio Tape), 11, 17, 18, 149, 181, 230
  and analog masters, 153
  archiving, 201
  counter modes, 14–15
DC offset, 32, 33, 169
Decibel, 6–7
Decibels format option, 32–3
Declicking, 4, 23, 125, 226
Delay effect, 104–5

Index

Digital audio, 8–12
  dither, 12
  errors, 10–11
  errors correction, 11
  sampling rate, 9–10
Digital Audio Tape see DAT
Digital delay circuit, 136
Digital reverberation, 100–3
Digital transfer, 16–22
Direct Sound drivers, 209
Directional microphones, 188
DirectX plug-ins, 112, 124
Distortion effect, 133
Dither, 12, 52
Documentaries, 185
Dolby laboratories, 83–4
Dolby stereo, 84, 136
Doppler Shifter, 133–4
Drama, production of, 198
Drop-frame time code, 230
DVD (Digital Versatile Disc), 82, 84, 136, 223
Dynamic delay effect, 105–7
Dynamic EQ, 115–17
Dynamic microphones, 192
Dynamics transform, 94–5

EBU see European Broadcasting Union
Echo see Reverberation
Edge-in, 80–1
Edges and fades, 74–6
Editing, 29, 38–9, 78, 147, 182, 186, 206, 214
  audio editing, 1, 7, 39, 215
  file loading, 30–5
  hazards, 7
  linear editors, 71, 219
  non-linear editors, 66, 198, 219–20
  scrub editing, 38
  spectral editing, 157, 221
  transport, 35–8
  visual editing, 1, 39–41
Electrical connector (S/PDIF), 17
Electromagnetic microphone, 192
Electrostatic microphone, 189, 192–3
Email and World Wide Web, 154–6
Envelopes, 64, 67–9, 75
  Envelope Follower, 140, 141

Equalization (EQ), 62–3, 115, 221
  see also Filters
Errors, 10–11, 153
  Azimuth error, 166–7
  correction, 11, 78, 170
European Broadcasting Union (EBU), 84–5, 149, 150, 229
Expanders and gates, 93–5
  dynamics transform, 94–5
Expansion ratio, 93

Fades and edges, 74–6
Fast Fourier Transform filter see FFT filter
Favorites menu, 207
Features, production of, 198
FFT filter, 117–18
  RIAA addition, 232–3
Figure-of-eight microphone, 85, 190–1
File loading, 30–5, 43
Filters, 221
  center channel extractor, 114–15
  dynamic EQ, 115–17
  FFT filter, 117–18, 232–3
  graphic equalizer, 118, 119
  graphic phase shifter, 118–20
  notch filter, 120–1, 221
  parametric equalizer, 121–2
  quick filter, 122
  scientific filters, 122–3
Firewire, 19, 218
First action level (noise), 7
Flanging, 106, 108–9
Flash drives, 20
Frames, 12, 229, 230
Frequency, 5
  audible range, 5–6
Frequency Analysis window, 163–4
Frequency Band Splitter, 140–1
FX, 61–4
  sends panel, 62

Gramophone records, 23–4, 125, 126
Graphic equalizer, 118, 119
Graphic phase shifter, 118–20
Graphical tab, 94–5
Grot speaker, 148

Index

Hamming code, 11
Hard disks, 217–18
Hard interview, 182
Hardware controller, 204–5
Hardware requirements, 215–24
Harmonics, 6
Head alignment, 16
Headphones, 216–17
   spill, 131, 132
Hearing, safety, 7
Hemispherical pick-up pattern microphone, 190, 191
Hi-Fi speakers, 148
Hi-MD, 25, 181
Histogram tab, 169
Human ear, 5–7, 80, 151
Hypercardioid microphone, 190

Icons, 37, 59, 212
IDE, 20, 28, 215
IDE bus, 217
Informational interview, 182
Insert, 45, 71–2
Integrated Services Digital Network (ISDN), 154, 195, 203
Internet *see* Email and World Wide Web
Interviews:
   on the move, 185–6
   oral history, 185
   preparation, 182
   pre-recorded, 184
   techniques, 183–4
   on telephone, 195
   types, 182–3
   vox pop, 184
Inverse square law, 102
Invert option, 89, 138–9
Invert tick boxes, 46
ISRC codes, 180

Jitter, 171

Keyboard shortcuts, 37, 206

Level matching, 80–1
Limiting settings, 92
Line-up tones, 149–50

Linear editors, 71, 219
Loop paste, 46
Loops, 78–9
Loudness, 5, 6, 87
Loudspeakers, 216–17

Magazine item, 197
Magazine programmes, 197
Masterbus, 58
Mastering, 220
   analytical functions, 156–7
   CD labelling, 153–4
   changeovers, 152–3
   Email and World Wide Web, 154–6
   line-up and transmission formats, 149–50
   noise, generating, 151–2
   process of, 153
   silence, generating, 150–1
   spectral editing, 157
   tones, generating, 152
Memory cards, 19
Microphone, 196
   directivity, 188, 189–91
   quality, 189
   variable pattern microphones, 190, 191
   see also Acoustics; Transducer
MIDI (Musical Instrument Digital Interface), 208, 223, 227–8, 229–30
   triggering, 206
   and video files, 78
Minidisc, 15, 46, 83, 155, 186, 230
Missed edits, 147
Mix paste, 3, 71, 45–6
Mobile phones (beware), 28, 184
Modulate, 46
Monitoring speakers, 148
Mono audio, 2
Morse code, 8
Mouse options, 204
MP3, 155
MP3Pro decoder, 155
MPEG (Motion Picture Experts Group), 155
Multiband compressor, 95–8
   tube-modelled compressor, 97–8
Multi-layer mixing, 85
Multimiking, 187–8
Multi-session CDs, 172

Multitap delay effect, 110
Multitrack, 140–2, 194, 220
    automation, 64–7
    chequer boarding, 71–3
    envelope follower, 140, 141
    envelopes, 67–9
    fades and edges, 74–6
    frequency band splitter, 140–1
    FX, 61–4
    loading, 55–6
    loops, 78–9
    MIDI and video, 78
    music, 77–8
    prefaded and backtimed music, 76
    recording, 60, 61, 188
    for simple mix, 69–71
    tour, 56–61
    track bouncing, 73–4
    transitions, 76–7
    vocoder, 142
Music:
    balance, 56, 57
    on multitrack recorder, 77–8
    orchestral, 188–9
    prefaded and backtimed, 76
    production of, 198–9
Mute, 139

NICAM system, 6, 9, 124
Noise, generating, 151–2
Noise levels, 7
Noise reduction, 124, 221
    options, 128–31
Noise-shaping, 12
Non-linear editor, 66, 198, 219–20
Normal waveform display, 160–2
Normalization, 86–8, 220
Normalize option, 32, 33
Notch filter, 120–1, 221
Nyquist limit, 9

Omnidirectional microphone, 190
Optical connector, 17–18, 216
Optical leads, 18
Oral history interviews, 185
Orchestra, 188–9, 225
Overlap, 15, 45

Pan and volume *see* Volume and pan
Pan envelope, 67, 68, 138
Panning, 81
Parallel ATA (PATA), 217
Parametric equalizer, 121–2
Parity, 11
Password, 202
Paste, 43, 45–6, 208
PC, 1, 2, 195, 215
Peak amplitude, 169
Personality interview, 183
Phase analysis, 164–8
    Adobe Audition 2.0, 166
    display interpretation, 166–7
    phase check test file, 165–6
Phaser, 114
Phasing, 105
Pink noise, 151, 152
Pitch transform *see* Time/pitch transform
Plain Old Telephone System (POTS), 154
Plug-in reverb, 104, 112–13
Post-production, 80–5
    cross fading, 82
    level matching, 80–1
    multi-layer mixing, 85
    panning, 81
    stereo/binaural, 82–3
    surround sound, 83–5
    timing, 80
Prefaded and backtimed music, 76
Preferences dialog, 208–12
Pre-recorded interviews, 184
Production, 195–9
    drama, 198
    magazine item, 197
    magazine programmes, 197
    multi-layer feature, 198
    music, 198–9
    programme types, 195–6
    simple feature, 198
    talks, 196–7
Pulses, 8–9

'Q', 63
Quadraphonic audio, 83
Quarrying material, 42–6
Quick filter, 122

Index

RCA phono plugs, 17
Real-time controls and effects, 142–6
Recorders, 186–7
Recording, 24–8, 49
   codes, 179–80
   file, 27
   level setting, 26–7
   space, 27–8
   topping and tailing, 28
Red Rover, 205
Replace, 46
Restoration, 124–31
   Click/Pop Eliminator, 125–6
   clickfix, 126, 127
   clip restoration, 126–8
   noise reduction options, 128–31, 132
Reverberation, 99, 221
   chorus effect, 103–4
   delay effect, 104–5
   devices for, 99–100
   diffusion, 99
   digital reverberation, 100–3
   dynamic delay effect, 105–7
   echo, 99, 107, 108
   echo chamber, 107, 108
   flanging, 108–9
   full reverb, 109–10
   multitap delay, 110, 111
   perception, 99
   phaser, 114
   plug-in reverb, 112–13
   reverb, 110–11
   reverberation time, 99
   studio reverb, 111–12
Reverse option, 139
Reviewing material, 14–16, 147–8
   assessing levels, 147
   full quality speakers, 148
   missed edits, 147
Rewritable CDs, 172–3, 222
RIAA equalization, 23–4
   addition to FFT filters, 232–3
Ribbon microphone, 191, 192
RMS power, 31, 169
RNID (The Royal National Institute for Deaf People), 7
Rubber band method, 43, 44

Safety hearing, 7
Sample value, 168
Sampling rate, 9–10, 25, 65, 155, 215
Scientific filters, 122–3
Script Starts from Scratch, 48, 49
Script Works on Current Wave, 48–9
Script Works on Highlighted Selection, 49
Scrub editing, 38
SCSI, 20, 28, 217
Second action level, 7
Sends panel, 62
Serial ATA (SATA), 217, 218
Serial Copy Management System (SCMS), 18
Short peaks, 33
Side fire microphones, 191, 192
Silence, generating, 150–1
Simple mix, 58, 59, 137
   multitrack for, 69–71
Society of Motion Picture and Television Engineers (SMPTE), 12, 84–5, 208, 229, 230, 231
   SMPTE synchronization, 207
Software requirements, 215–24
Sony/Philips Digital Interface (S/PDIF), 16–22, 226
Sound, 5, 6
   binaural sound, 82–3
   stereo sound, 82
   surround sound, 83–5
Sound card, 215–16, 225–6
Spatial effects, 136
   faking stereo, 137–8
   MS stereo, 137
Speakers, 148, 155, 164, 197, 216
Special effects:
   convolution, 131–3
   distortion, 133
Spectral editing, 157, 221
Spectral view, 157
   analytical function, 156
Standardize format, 47–8
Stereo sound, 82
   Faking, 137
Storage systems, 202–3
Studio reverb, 111–12
Surround sound, 83–5
Synthesizer, 227, 228

Table mat, 172
Table of contents, 21, 170, 172
Talks, production of, 196–7
TAO (Track At Once), 222
Third-order filter, 122
Thunder peak, 33–4
Timbre, 6
Time code, 12–13, 229–31
   Adobe Audition 2.0, 13
   drop-frame time code, 230
   MIDI, 229–30
Time/pitch transform:
   digital delay circuit, 136
   Doppler shifter, 133–4
   Pitch Bender, 134
   pitch changer, 136
   Pitch Correction, 135
   pitch manipulation, 135–6
   Pitch Shifter, 135
   uses, 136
Timing, 80
Tinnitus, 7, 214
Tip, ring and sleeve jack, 22
Tone blips, 157–9
Tones, generating, 152
TOSLINK, 17–18
Track bouncing, 73–4
Track equalizer, 143–6
Traditional tab, 95
Transducer, 191–4
   electromagnetic microphone, 192, 193
   electrostatic microphone, 192–3
   environment, 193
   power requirement, 193
   reliability, 194
   robustness, 193
   size and weight, 193
   visibility, 193
   see also Acoustics; Microphone
Transfer, 14
   analog, 22–4
   digital, 16–22
   head alignment, 16
   recording, 24–8
Transitions, 76–7
Transmission formats, 149
Transport, 35–8
Tube-modelled compressor, 97–8

UPC/EAN codes, 178, 179–80
USB, 19, 218

Variable pattern microphones, 190, 191
Visual editing, 1, 39–41
Vocoder, 142
Volume and pan, 59, 62, 64
Volume envelope, 67, 69, 77
Vox pop interview, 184

.WAV files, 14, 45, 153, 155
White noise, 151, 152
Window/Amplitude Statistics, 31–2, 168
Windows Media Audio (WMA), 156
Word clock in/out, 226
Word processing, 1, 3

XLR connector, 22–3, 187

Zip files, 155
Zooming, 33, 37–8

"This guide has enabled me to learn the function of the program with extreme ease"
Michael Feinberg, Dialogue Editor

"Antony Brown's book is an easy to follow guide to mastering Audition"
Peter Baldock, Art4noise Ltd – Sound Design & Post Production

If you want to tap into the full power of Adobe® Audition™ 2.0 when creating digital audio files then this is the book for you. This quick reference guide tells you how to get up and running fast with the program covering the important features clearly and concisely, making it ideal if you are new the program or want a handy desk reference.

Highly illustrated throughout with full color diagrams and screenshots, Antony Brown takes you step by step through the essential areas: set up, editing, looping content, mastering, finalizing, working with video and making a CD, ensuring you understand the key tools and features of Adobe® Audition™ 2.0 to get the most out of the program.

Antony Brown is an experienced Adobe software trainer and demonstrator, as well as a specialist in sound design/editing and music composition and audio restoration.

June 2006: 148 × 210 mm: 200 color illustrations:
Paperback: 0240520181

To order your copy call +44 (0)1865 474010 (UK) or +1 800 545 2522 (USA) or visit the Focal Press website: www.focalpress.com

Also available from Focal Press

Sound for Digital Video

Tomlinson Holman

"This is an excellent technical manual, packed with hard practical information . . . the chapters on sound design, editing, mixing, and monitoring are first class." – BKSTS magazine

- Learn how to get great sound on a limited budget, from the recognized expert in the field
- Provides tips and instructions for improving sound, from planning through mixing
- Includes CD with recording and editing exercises

The distinguishing feature of many low-budget films and TV shows is often the poor sound quality. Now, filmmakers shooting DV on a limited budget can learn from Tomlinson Holman, a film sound production pioneer, how to make their films sound like fully professional productions. Holman offers suggestions that you can apply to your own project from pre-production through postproduction and provides tips and solutions on production, editing, and mixing.

Tomlinson Holman is President of TMH Corporation and one of the prominent figures in audio today. He is a Professor at the University of Southern California School of Cinema Television, Principal Investigator in the Integrated Media Systems Center of the University, and an honorary member of the Cinema Audio Society and the Motion Picture Sound Editors.

Sep 2005: 191 × 235 mm: 100 illustrations:
Paperback: 0240807200

To order your copy call +44 (0)1865 474010 (UK) or +1 800 545 2522 (USA) or visit the Focal Press website: www.focalpress.com